TRIBOLOGY IN ELECTRICAL ENVIRONMENTS

Vol. 27 Dissipative Processes in Tribology (Dowson et al., Editors)
Vol. 28 Coatings Tribology – Properties, Techniques and Applications in Surface Engineering (Holmberg and Matthews)
Vol. 29 Friction Surface Phenomena (Shpenkov)
Vol. 30 Lubricants and Lubrication (Dowson et al., Editors)
Vol. 31 The Third Body Concept: Interpretation of Tribological Phenomena (Dowson et al., Editors)
Vol. 32 Elastohydrodynamics – '96: Fundamentals and Applications in Lubrication and Traction (Dowson et al., Editors)
Vol. 33 Hydrodynamic Lubrication – Bearings and Thrust Bearings (Frêne et al.)
Vol. 34 Tribology for Energy Conservation (Dowson et al., Editors)
Vol. 35 Molybdenum Disulphide Lubrication (Lansdown)
Vol. 36 Lubrication at the Frontier – The Role of the Interface and Surface Layers in the Thin Film and Boundary Regime (Dowson et al., Editors)
Vol. 37 Multilevel Methods in Lubrication (Venner and Lubrecht)
Vol. 38 Thinning Films and Tribological Interfaces (Dowson et al., Editors)
Vol. 39 Tribological Research: From Model Experiment to Industrial Problem (Dalmaz et al., Editors)
Vol. 40 Boundary and Mixed Lubrication: Science and Applications (Dowson et al., Editors)
Vol. 41 Tribological Research and Design for Engineering Systems (Dowson et al., Editors)
Vol. 42 Lubricated Wear – Science and Technology (Sethuramiah)
Vol. 43 Transient Processes in Tribology (Lubrecht, Editor)
Vol. 44 Experimental Methods in Tribology (Stachowiak and Batchelor)
Vol. 45 Tribochemistry of Lubricating Oils (Pawlak)
Vol. 46 An Intelligent System For Tribological Design In Engines (Zhang and Gui)
Vol. 47 Tribology of Elastomers (Si-Wei Zhang)
Vol. 48 Life Cycle Tribology (Dowson et al., Editors)

Aims & Scope
The Tribology Book Series is well established as a major and seminal archival source for definitive books on the subject of classical tribology. The scope of the Series has been widened to include other facets of the now-recognised and expanding topic of Interface Engineering.

The expanded content will now include:
• colloid and multiphase systems; • rheology; • colloids; • tribology and erosion; • processing systems; • machining; • interfaces and adhesion; as well as the classical tribology content which will continue to include • friction; contact damage; • lubrication; and • wear at all length scales.

TRIBOLOGY AND INTERFACE ENGINEERING SERIES, 49
EDITOR: B.J. BRISCOE

TRIBOLOGY IN ELECTRICAL ENVIRONMENTS

HAR PRASHAD
Hyderabad, India

ELSEVIER
Amsterdam – Boston – London – New York – Oxford – Paris
San Diego – San Francisco – Singapore – Sydney – Tokyo

ELSEVIER B.V.
Radarweg 29
P.O. Box 211, 1000 AE
Amsterdam
The Netherlands

ELSEVIER Inc.
525 B Street, Suite 1900
San Diego, CA 92101-4495
USA

ELSEVIER Ltd.
The Boulevard,
Langford Lane
Kidlington, Oxford OX5 1GB UK

ELSEVIER Ltd.
84 Theobalds Road
London WC1X 8RR
UK

First edition 2006

ISBN-13: 978-0-444-51880-4
ISBN-10: 0-444-51880-0
ISSN: 1572-3364 (series)

06 07 08 09 10   10 9 8 7 6 5 4 3 2 1

Printed in Great Britain.

CONTENTS

PREFACE

Studies presented in this book deal with the behavior and response characteristics of lubricants, rolling-element, hydrodynamic journal and thrust bearings working under the influence of an electric current. The analysis reported in different chapters of this book fulfills the existing know-how gaps in the area of 'Tribology in Electrical Environments'. This is because genesis of intermolecular forces during tribological interaction involves electrostatic attraction or repulsion that creates electrodynamic, magnetic and exchange forces between atoms. Thus, hereby, all the tribological phenomenons, occurring in any interacting system is electrical in nature.

This book presents the summary of work done and reflects the experience gained by the author in the last three decades in the area of 'Tribology in Electrical Environments'. Basically, the initial investigations of a few case studies of the failure of bearings in heavy industry, diagnosed as caused by the passage of an electric current, have given direction and motivation to the author to go in depth to study the cause and effect of shaft voltages on the bearings. Gradually, joining one linkage of the bearing failure to another, since the available literature was very scare in this field, has lead the author to make deeper diagnosis in the area of tribology that opened a new era of 'Tribology in Electrical Environments'.

This book basically covers four main categories, i.e. rolling-element bearings, lubricants, hydrodynamic journal and thrust bearings. The various aspects of the effect of an electric current on these bearings and lubricants are covered in 21 chapters.

This book covers the various causes of shaft voltages, the flow of a current through bearings, the generation and flow of a localized current, the analysis of electrical parameters of different types of bearings, the deterioration of lubricants and their diagnosis, the threshold voltage phenomenon, the contact temperature and residual stresses leading to the initiation of slip bands, and the formation of craters/corrugations on the track surfaces of bearing elements besides the life estimation and the effect of starts/stops of a machine on the performance of a bearing.

Theoretical procedure to establish, capacitance, resistance, capacitive reactance, impedance and charge accumulation on surfaces of different types of hydrodynamic journal and thrust bearings, and rolling-element bearings is given in the book that originates the new methodology for the diagnosis. Also, the decrease in life span of bearings under the influence of different levels of shaft voltages has been established theoretically and the safe limit of the shaft voltage for a bearing has been identified. The dynamic coefficients of the bearings have been determined using an electrical analogy approach. Analysis of inductive effect of bearings under the influence of shaft voltages is also covered.

In short, this book covers original contributions of the author establishing the behavior of different bearings and lubricants under the influence of an electric current. Assessment of the flow of current through rolling-element bearings by the study of magnetic flux density distribution on the surfaces, theoretical evaluation of corrugation pattern, resistivity and recouping/recovery of resistivity phenomenon in the lubricants, theory that explains causation, morphology and rate of the formation of electric current damage and damage by unforeseen causes are some of the significant contributions that covers the major part of the book.

The research in the area of tribology in electrical environments is in the very initial stage of development concerning analysis of bearings, diagnosis and frictional processes. This book is a first step to compile the available published literature for the benefit of tribologists, researchers, students, practicing engineers, professionals, mechanical and electrical engineers.

Thanks are due to BHEL, Corporate R&D Management, for allowing me to continue research in this area for the benefits of engineers and technicians of heavy industry, designers, academicians and others who grapple with the problems of bearings diagnosis, and furthermore for the establishing the new field 'Tribology in Electrical Environments' in science, engineering and technology.

This book would not have seen the light of the day, however, without close cooperation of my colleagues and staff members of Tribology Laboratory of BHEL, Corporate R&D Division, during professional career. Furthermore, without silent sacrifice of my wife, Darshan and my beloved children Shwetlana and Poojan, who have allowed me to work without any interference at the time when they need more of my involvements for their needs, this task would not have been completed in the time frame.

Finally, my gratitude is due to Almighty, whose blessings, radiance, continued inspiration, pouring of booming energy and silent guidance that has nurtured me from time to time to follow the path of untiring research for the ultimate progress.

ABOUT THE AUTHOR

Dr. Har Prashad is a Senior Deputy General manager at the Bharat Heavy Electricals Limited, Corporate Research and Development Division, Hyderabad. He obtained an M.E. (Hons) degree in Mechanical Engineering in 1970, and subsequently, Ph.D. in Tribology. He worked with the Indian Institute of Petroleum, Dehra Dun, and with the Design Bureau at Bokaro Steel Ltd., Dahnbad, Bihar, before joining BHEL in 1974.

At BHEL, he was associated with setting up of the Tribology Laboratory at the Corporate R&D Division, Hyderabad, particularly design and development of various bearing test rigs for hydrodynamic and rolling-element bearings. He has done substantial design/development work on magnetic, dry and other special types of bearings. He has developed energy saving double decker high-precision rolling-element bearings, and established the performance of these bearings both theoretically and experimentally.

His areas of interest include diagnostic monitoring, failure analysis and bearing performance evaluation. He has done significant original work to establish the behaviour of different bearings and lubricants under the influence of electrical current. He has established electrical analogy for dynamic analysis of bearings. Assessment of flow of current through rolling-element bearings by study of magnetic flux density distribution on the bearing surfaces, theoretical evaluation of corrugation pattern, bearing life estimation, resistivity and recouping of resistivity phenomenon in lubricants, theory that explains the causation, morphology and rate of formation of electrical current damage are some of the outstanding original contributions of Dr. Prashad especially for the engineers engaged in the design or operation of heavy rotating electrical machinery.

Dr. Prashad has published more than 115 papers in both national and international journals, particularly in ASME-Journal of Tribology, STLE-Tribology Transactions, WEAR, Lubrication Science, STLE-Lubrication Engineering, Tribo-test, Tribology International, IE (I) and BHEL journals and delivered invited talks. He has patents to his credit. He is recipient of the Corps of Electrical and Mechanical Engineering Award and Corps of Engineers Medal Award – 1998 along with other various awards for his contributions and publications. He was convener and organising secretary of Second International Conference on Industrial Tribology, held at Hyderabad during December 1999.

He has served as Head, Tribology Department, at BHEL, Corporate R and D, Hyderabad. He is on the review panel for technical publications in the journals of Wear, Tribology International, Lubrication Basics, Tribo-test, International Journal of Comadem, Journal of Tribology Society of India, Journal of Institution of Engineers (India), and for assessment of the awards for National Research and Development Corporation of India. He is a Fellow of Institution of Engineers (India) and Member of Society of Tribologists and Lubrication Engineers, U.S.A. He is life member of Tribology Society of India and executive committee member of the Society. He was joint secretary of Tribology Society of India during 1997-99, and was the editor of Tribology Newsletter during this period. For his achievements, his name has been included in Marquis Who's Who in the world published in 2001 from USA. Also, his name is included in International Directory of Distinguished Leadership-2001, published by American Biographical Institute. His bio-data has been chosen for publication in the directory of "Contemporary Who's Who -2004" by the same institute.

Contact details:
Dr. H. Prashad
1-2-319/A, Gagan Mahal,
Domal Guda
302 Central View Apartments
Hyderabad-500029
India
e-. mail: har.prashad@gmail.com

(Ex.Senior Deputy General Manager (Tribology), Bharat Heavy Electricals Limited (BHEL), Corporate Research and Development Division, Hyderabad, India.)

Chapter 1

ANALYSIS OF BEARINGS AND LUBRICANTS IN ELECTRICAL ENVIRONMENTS: A STATE-OF-THE-ART REVIEW

1.1 Introduction

Investigations into the response and performance of bearings and lubricants under the influence of an electric current have given an insight into the behavior of bearings, the deterioration of lubricants and an approach to the evaluation of various electrical parameters of different types of bearings. Multifold bearing analysis and its correlation with the electrical analogy opens up a hitherto little explored area, tribology in electrical environments, which adds a different dimension to the tribological approach. This is because genesis of intermolecular forces involves electrostatic attraction or repulsion during tribological interactions, and thus it creates electrodynamic, magnetic and exchange forces between atoms. Thus, nearly all tribological phenomena, occurring in any interacting system, metal-lubricant-metal or metal-metal, are electrical in nature. This originates the science of tribology in electrical environment. If the precise nature of the microscopic contacts can be determined, then a better understanding of friction could lead to industrial innovation in the development of improved lubricants and wear resistant components. Furthermore, in any tribological interaction thermoelectrical, capacitance, acousto-electric, galvanic, inductive components of oil film thickness play a very significant role, in the understanding of friction and wear in the presence of lubricants having different properties, beyond conventional knowledge.

It is well known that a large number of components found in the degraded oils cannot be thermodynamically formed by the normal chemical transformations that are expected to take place in tribological systems. Owing to the interaction between metal surfaces and oil, the metal surface acquires electrostatic charge at random points, which emits low-charge particles. The frequency of emission depends on operating parameters that govern the tribological behavior of the interacting surfaces, and the need thus arises for fundamental microscale and nanoscale investigations.

Before dealing with the behavior and response of bearings and lubricants influenced by their electrical environments, it must first be understood how bearings and lubricants work under such conditions, because of the shaft voltage developed in a machine due to various causes. The causes of shaft voltages can be summarized as: asymmetry of faults, e.g. winding faults, unbalanced supplies, electrostatic effects, air gap fields, magnetized shafts or other machine parts and asymmetries of magnetized fields, etc. [1]. The causes of shaft voltages can be grouped into four categories:

- external causes,
- magnetic flux in the shaft,
- homo-polar magnetic flux, and
- ring magnetic flux.

For example, the friction between a belt and a pulley can set up an electrostatic voltage between shaft and bearings. Accidental grounding of a part of a rotor winding to the rotor core can lead to stray currents through the shaft and bearings, and results in permanent magnetization of the shaft. Shaft voltage and current can be generated when the machine is started. Furthermore, homo-polar flux from an air gap or rotor eccentricity can generate a voltage [2]. Another cause of shaft voltage is the linkage of alternating flux with the shaft. The flux flows perpendicular to the axis of the shaft, and pulsates in the stator and rotor cores. It is caused by the asymmetries in the magnetic circuit of the machine, such as [3]:

- uneven air gap and rotor eccentricity,
- spilt stator and rotor core,
- segmental punchings,
- axial holes through the cores for ventilation and clamping,
- key ways for maintaining the core stackings, and
- segments of different permeability.

All the causes listed above lead to a magnetic flux, which closes over a yoke and induces a voltage in the shaft as the machine rotates. This results in a localized current at each bearing rather than a potential difference between the shaft ends.

A current path, however, along the shaft, bearings and frame, results in a potential difference between the shaft ends [4]. This happens because of axial shaft flux caused by residual magnetization, rotor eccentricity and asymmetrical rotor winding.

To review the response and behavior of bearings in electrical environments, the subject is here categorized into four main topics:

- rolling-element bearings,
- lubricants,
- hydrodynamic journal bearings, and
- thrust bearings.

Understanding all the papers listed in the references cannot be summarized separately, furthermore, it would be tedious to read neutral summaries of the published papers, so the collective summary of each topic is given in a broader prospective.

1.2 Rolling-Element Bearings

In a rolling-element bearing, at each revolution of the shaft, part of the circumference of the inner race passes through a zone of maximum radial force, and Hertzian pressure between the rolling-elements and raceways at the contact points leads to a maximum shear stress. This occurs in the sub-surface at a depth approximately equal to half the radius of the contact surface [5]. It is generally at this point that any material failure will initiate. The process of deformation that leads to the formation of a corrugation pattern on the track surfaces is accelerated by the factors including the passage of electric current, corrosion and oxidation of surfaces, lubricant characteristics and quality of the bearing [6].

1.2.1 Corrugation Pattern

Mathematical formulations have been developed separately, because of the difference in contact profile, for roller bearing and ball bearing to evaluate the pitch of corrugations, which depends on the bearing kinematics, the frequency of rotation, position of the plane of action of radial loading, bearing quality, and lubricant characteristics. Formulations for the width of corrugations on the track surfaces have been worked out, and comparisons of experimental and theoretical data for the pitch and width of corrugations reported. The mathematical formulations show that the width of corrugations on the track surfaces is not affected by the frequency of rotation and depends only on load conditions and bearing kinematics [6, 7]. It has been determined that the passage of current causes local surface heating, which leads to low-temperature tempering, and accelerates the formation of corrugations with time for bearings using low-resistivity lubricant (10^5 ohm m). After operation for a long time, the softer tempered surfaces of the races become harder, and thus harder and rehardened particles due to localized high temperature and load are removed from the craters, which increases the depth of corrugations.

1.2.2 Flux Density Distribution

Under the influence of an electric current, a rolling-element bearing using low-resistivity lubricant develops a magnetic flux density distribution on the track surfaces due to the passage of high current density through the bearing [8]. A bearing using high-resistivity lubricant (10^9 ohm m) does not develop a significant flux density on its surface [9]. Theoretical and experimental investigations have been reported on the distribution of magnetic flux density on the track surfaces of races and rolling-elements of bearings [10]. By using the developed models and by experimental determination of the residual flux density on the track surfaces, the level of current flow through a bearing can be ascertained without the measurement of shaft voltage and bearing impedance [11]. The flux density distribution technique was used for a case study on the failure of the bearings of alternators [12].

1.2.3 Threshold Voltages, Impedance Response, Capacitance and Charge Accumulation

The phenomenon of threshold voltages and investigations pertaining to first/second threshold voltages for bearings operating under the influence of electric currents has been investigated, and it was shown that the threshold voltages depend on lubricant resistivity, oil film thickness and operating parameters [13]. The threshold voltages detected are primarily responsible for momentary flow of current and further increase in current density with a slight change in potential across the bearings. The threshold voltage decreases as the load on bearing increases at constant speed. For increase in current density reduces the bearing impedance significantly. For the reliable operation of a bearing, the safe limit of the potential drop across the bearing elements should be less than the first threshold voltage [13].

Under the influence of potential drop across a bearing, the minimum film thickness between races and rolling-elements offers a maximum capacitance and minimum capacitive reactance, depending on the permittivity of the lubricant. The electrical interaction between the races and the rolling-elements in the presence of an oil film resembles a resistor-capacitor (RC) circuit that offers impedance to the current flow. The capacitance and resistance for roller bearings and ball bearings, separately determined analytically, were found to depend on film thickness and width of deformation, governed by the permittivity and resistivity of the lubricant used [14, 15].

Charges are stored on the bearing surfaces, and the equivalent capacitance of a bearing decreases with increasing speed at constant load. A bearing lubricated with high-resistivity lubricant, as opposed to low-resistivity lubricant, with the same permittivity, behaves like a capacitor up to the first threshold voltage. Also, for a bearing to accumulate charges, the ratio of capacitive reactance to active resistance should be less than unity [14].

1.2.4 Contact Temperature, Developed Stresses, Slip Band Initiation and Time/Cycles for Appearance of Flutes and Craters on Track Surfaces

A current passing through a bearing at the line/point contact between the tracks and rolling-elements, and the corresponding impedance at the junction, generates heat and increases temperature instantaneously. This increases the contact stresses and allows determination of the number of cycles before the slip bands are initiated on the track surfaces of a bearing lubricated with low-resistivity lubricant [16, 17]. The contact duration and temperature rise between track surfaces of races and a rolling-element can be theoretically determined as a function of kinematics, the number of rolling-elements in the loaded zone, the material properties and the depth of slip bands. Based on the stress levels developed on the track surfaces, the time, before the slip bands are established, can be determined [16]. Furthermore, a theoretical approach has been developed using the continuum theory of Griffith to assess the time span/cycles for the development of flutes/corrugations on the track surfaces after the appearance of slip bands [18]. A similar approach using the effect of leakage of charge has been extended to the formation of craters on the track surfaces of a bearing lubricated with high-resistivity lubricant [19]. Also, crater formation criterion and mechanism of electrical pitting on the lubricated surfaces has been analyzed [20, 21].

1.2.5 Effect on Track Surfaces of Instantaneous Leakage of Stored Charge Energy

The effect of leakage of charge on the increase of contact temperature on the track surfaces of a bearing using high-resistivity lubricant can be established using mathematical formulations for the evaluation of contact stresses and minimum cycles for the appearance of craters [22]. Important here are the ratio of contact cycles required for charge accumulation and discharge of the accumulated charge on the bearing surfaces, and the ratio of bearing-to-shaft voltage, as the number of cycles and of starts and stops before initiation of craters on track surfaces versus the ratio of bearing-to-shaft voltage and other causes [23].

The diagnosis and cause analysis of failure of rolling-element bearings used in electric power equipments due to various unforeseen causes leading to current leakage have been examined [24]. Also, to highlight the experimental techniques and diagnosis of the data to assess the causes of various failures of bearings, the studies have been conducted [25].

1.2.6 Localized Electric Current

The cause of generation of localized current in presence of shaft voltage has been established. A theoretical model has been developed to determine the value of localized current density depending on dimensional parameters, shaft voltage, contact resistance, frequency of rotation of shaft and rolling-elements of a bearing. The time for appearance of flutes on the track surfaces can be estimated by bearing kinematics, existing potential difference between track surface of inner race and rolling-elements, value of localized current, properties of bearing material together with measured values of the pitch of corrugations on the track surfaces [26].

1.2.7 Bearing Stiffness

An alternative approach and model has been developed to determine the effective stiffness of ball and roller bearings through evaluation of electrical parameters under different operating conditions. These parameters have been theoretically evaluated using width of deformation on the track surface of races, minimum film thickness, lubricant's characteristics and bearing geometry. Besides this, by applying another developed alternative approach, the stiffness of ball and roller bearings has been assessed approximately. This alternative approach is based on the inverse of the change of width of bearing deformation on the track surfaces of races and applied load.

The variation of stiffness can be determined with the applied loads under different speeds of operation using electrical analogy. The values of effective stiffness determined by these alternative approaches have been compared with the values determined by the existing conventional procedure, and found to have matching trends [27, 28].

1.3 Lubricants

Investigations indicate that the resistivity of a lubricant depends on antiscuff properties, viscosity, torque characteristics and consistency. Resistivity is a function of applied voltage and time over which the voltage is applied to the lubricants. The difference in resistivities among lubricants can be as high as 10^5 times. The difference in resistivity depends on the nature of impurities or by-products, and the type of additives present in the lubricant as well as its density, compressibility and structure.

Detailed experiments into the behavior of the resistivity of commonly used industrial lubricants have shown that a low-resistivity lubricant tends to 'recoup/recover' its resistivity when the applied voltage is switched off [29, 30]. The amount of recovery from the original value varies, and depends on the stretching of the molecules [31]. Also, it has been found that the low-resistivity lithium base grease decomposes and lithium metal concentration in the aqueous solution increases. The decomposition of carboxylic acid leads to corrosion of the track surfaces before pitting process is initiated. However, the applied voltage does not affect the oil content of the grease, but the carboxylate anion stretching and carboxylic group, present in the soap residue of the grease, undergo changes [32]. Investigations using x-ray diffraction techniques have indicated that the original structure of the soap residue of the fresh grease, i.e. lithium stearate changed to lithium palmitate, and lithium iron oxide and lithium zinc silicate were formed [33]. However, such changes were not detected under rolling friction without the effect of electric current [34]. It has been further found that under the influence of electric current, the formation of lithium hydroxide and lithium carbonate make the dielectric alkaline and corrode the bearing surfaces, which finally lead to increased wear and bearing failure [35]. Rheological physical studies of lubricating greases before and after use in the bearings have also been reported [36].

1.4 Hydrodynamic Journal Bearings

In a hydrodynamic journal bearing, the zone of minimum film thickness, i.e. load-carrying oil film varies along the circumference of a bearing through its length. This has been found to form a capacitor of varying capacitance between the journal and the bearing depending on permittivity of the lubricant used, the circumferential length of load-carrying oil film, the bearing length, the eccentricity ratio and the clearance ratio. Besides this, load-carrying oil film offers resistance that depends on operating parameters and resistivity of the lubricant. Thus, the load-carrying oil film forms a resistor-capacitor (RC) circuit and offers an impedance to current flow. By analyzing the RC circuit, the behavior of a journal bearing can be predicted.

A mathematical model has been developed to determine capacitance, active resistance, capacitive reactance and impedance under different parameters of operation so as to predict bearing performance and safe load carrying capacity [37]. Variations occur in bearing life for different levels of shaft voltages before the initiation of craters on the bearing surfaces compared to that of the bearings operating without the influence of shaft voltages. This helps to explain the mechanism of formation of craters on the bearing surfaces. In general, for safe, reliable operation and adequate life of a bearing, a shaft voltage of 0.5 V must not be exceeded [38].

Repeated starts and stops of a machine operating under the influence of shaft voltage have a capacitive effect on bearing life [39]. The increase in charge accumulation on the bearing liner with time can be established theoretically when the machine is started, as can the gradual leakage of the accumulated charges from the liner, as the shaft voltage falls when the power supply to the machine is switched off. Under such conditions, the variation in shaft revolutions to accumulate charge and to discharge the accumulated charge from the liner surface at various bearing-to-shaft voltage ratios can be formulated mathematically. The limits of starts and stops as a function of the ratio of bearing-to-shaft voltage have been analyzed, showing that with an increase in ratio of bearing-to-shaft voltage, the ratio of shaft rotations to accumulate and discharge of the accumulated charge increases, but the number of starts and stops needed to initiate craters on the liner surface decreases [39]. There is an inductive effect of bearings under a start/stop regime of a machine, which affects the ratio of the time required for the growth of bearing current to that for its decay under this regime at various ratios of transient to steady-state values of bearing current, as well as self inductance on the bearing surfaces [40].

A theoretical model for assessing the minimum number of cycles for the appearance of craters of various sizes on the liner surface under different levels of shaft voltages has been developed that uses softening temperature of the liner at the 'high' points or 'prows' of its surface [41]. The number of cycles, established theoretically using this model, matched with that of the operating cycles of the hydrodynamic journal bearings of a synchronous condenser that failed within a few hours of operation [42, 43].

An approach such as that for a cylindrical bearing for the determination of electrical parameters can be extended to two-, three- and four-lobe bearings: the capacitance, resistance, capacitive reactance and impedance are evaluated for each lobe separately under different operating parameters, and equivalent values determined for two- and three-lobe bearings [44, 45]. Using electrical parameters, an electrical analogy can be applied to the evaluation of effective dynamic coefficients of two-, three- and four-lobe journal bearings [46–48], as well as to cylindrical and elliptical bearings [49]. Besides this, non-conventional electrical analogy approach has been further extended for determination of stiffness and damping coefficients in xx, yy, xy, yx directions and their comparison with the published conventional approach [50–53].

1.5 Hydrodynamic Thrust Bearings

Similar to journal bearings, in a thrust bearing, the variation in oil film between the pads and the thrust collar forms a capacitor of varying capacitance from leading to trailing edge. This capacitance depends on permittivity of the lubricant, the pad width, the angle of tilt and the ratio of oil film thickness at the leading to trailing edge. Besides this, a variable oil film thickness offers a variation in resistance along the pad profile depending on the resistivity of the lubricant. A theoretical approach has been developed which allows the variations of capacitance, capacitive reactance and other electrical parameters of a thrust bearing with the angle of tilt to be mathematically formulated [54]. The safe limit of shaft voltage can be assessed for reliable operation of the thrust bearings in which charge leakage between high 'points' of the thrust collar and a pad liner during momentary contact in the zone of load-carrying oil film is used to establish the heat generated and instantaneous temperature rise in each shaft rotation. The contact stresses caused by instantaneous temperature rise lead to crater formation: at a shaft voltage of 2 V, the reduction in bearing life is 15.6 times as much as that at a shaft voltage of 0.25 V [55]. The minimum number of cycles before the formation of craters due to leakage of charge energy on the liner surface of tilting pads of a thrust bearing can be calculated [56, 57]. A theoretical model developed for the capacitive effect and life estimation of pivoted pad thrust bearings on repeated starts and stops of a machine operating under the influence of shaft voltages, gives the time required for charge accumulation and increase of charge with time on the surface of liners. Gradual leakage of the accumulated charge with time as the shaft voltage falls when the power is switched off, allows determination of the number of repeated starts and stops before the initiation of craters on the liner surface of the pads of a thrust bearing [58].

1.6 Conclusions

Based on the state-of-the-art review given in this chapter, the following conclusions are drawn [59]:

It is evident that bearings working in electric fields undergo very specific processes that affect crater formation, lubricant deterioration and life. The research in tribology in the electrical environments is in an initial stage of development for analysis of bearings, diagnosis and frictional processes. This book is a first step to compile the available literature in this field for the benefit of mechanical engineers. As the need to conserve both energy and raw material is becoming very significant, the basic tribological electromagnetic phenomenon concerning friction/wear process and bearing behavior/performance need to be further explored and accelerated in engineering and scientific organizations.

REFERENCES

1. Anderson, S. (1968). Passage of electrical current through rolling bearings. *Ball Bearing J.*, **153**, 6–12.
2. Kaufman, H. N. and Boyd, J. (1959). The conduction of current in bearings. *ASLE Trans.*, **10**, 226–234.
3. Bradford, M. (1984). *Prediction of Bearing Wear due to Shaft Voltage in Electrical Machines*, ERA Technology Limited.
4. Chu, P. S. Y. and Cameron, A. (1967). Flow of electrical current through lubricated contacts. *ASLE Trans.*, **10**, 226–234.
5. Warnock, F. V. and Benham, P. P. (1965). *Mechanism of Solids and Strength of Materials*, Sir Isaac Pitman and Sons, London.
6. Prashad, H. (1988). Investigations of corrugated pattern on the surfaces of roller Bearings operated under the influence of electrical fields. *Lubric. Eng.*, **44**(8), 710–718.
7. Prashad, H. (1991). Theoretical and experimental investigations on the pitch and width of corrugations on the surfaces of ball bearings. *Wear*, **143**, 1–14.
8. Prashad, H. (1988). The effects of current leakage on electro-adhesion forces in rolling friction and magnetic flux density distribution on the surfaces of rolling-element bearings. Transactions of the ASME. *J. Tribol.*, **110**, 448–455.
9. Prashad, H. (2000). Determination of magnetic flux density on the surfaces of rolling-element bearings – An investigation. *BHEL J.*, **21**(2), 49–67.
10. Prashad, H. (1996). Magnetic flux density distribution on the track surface of rolling-element bearings – An experimental and theoretical investigations. *Tribol. Trans.*, **39**(2), 386–391.
11. Prashad, H. (1999). Determination of magnetic flux density on the surfaces of rolling-element bearings as an indication of the current that has passed through them. *Tribol. Int.*, **32**, 455–467.
12. Prashad, H. (1996). Diagnosis of failure of rolling-element bearings of alternators – A study. *Wear*, **198**, 46–51.
13. Prashad, H. (1987). Effect of operating parameters on the threshold voltages and impedance response of non-insulated rolling-element bearings under the action of electrical currents. *Wear*, **117**, 223–240.
14. Prashad, H. (1988). Theoretical evaluation of impedance, capacitance and charge accumulation on roller bearings operated under electrical fields. *Wear*, **125**, 223–239.
15. Prashad, H. (1993). Theoretical determination of impedance, resistance, capacitive reactance and capacitance of ball bearings. *BHEL J.*, **14**(2), 40–48.
16. Prashad, H. (1989). Analysis of the effects of electrical current on contact temperature, contact stresses and slip bands initiation on roller tracks of roller bearings. *Wear*, **131**, 1–14.
17. Prashad, H. (1990), Analysis of the effects of electrical current on contact temperature, residual stresses leading to slip bands and initiation and formation of corrugation pattern on ball tracks of ball bearings. *BHEL J.*, **11**(1), 39–47.

18. Prashad, H. (1998). Determination of time span for appearance of flutes on track surface of rolling-element bearings under the influence of electric current. Presented in World Tribology Congress, London, 8–12 September 1997 and published in *Tribol. Trans.*, **41**(1), 103–109.
19. Prashad, H. (2001). Appearance of craters on track surface of rolling-element bearings by spark erosion. *Tribol. Int.*, **34**(1), 39–47.
20. Chiou, Y. C., Lee, R. T. and Lin, C. M. (1999). Formation criterion and mechanism of electrical pitting on lubricated surface under ac electrical field. *Wear*, **235**, 62–72.
21. Prashad, H. and Venugopal, K. (2002). Formation of craters on track surfaces of rolling-element bearings due to spark erosion. *BHEL J.*, **23**(1), 34–37.
22. Prashad, H. (1990). Theoretical analysis of the effects of instantaneous charge leakage on roller tracks of roller bearings lubricated with high-resistivity lubricants. ASME Transaction, *J. Tribol.*, **112**, 37–43.
23. Prashad, H. (1992). Theoretical analysis of the capacitive effect of roller bearings on repeated starts and stops of a machine operating under the influence of shaft voltages. Transactions of ASME, *J. Tribol.*, **114**, 218–222.
24. Prashad, H. (1999). Diagnosis and cause analysis of rolling-element bearings failure in electrical power equipments due to current leakage. *Lubric. Eng.*, STLE, 30–35, May (1999). Also, presented in ASME/STLE Tribology Conference in Toronto, Ontario, Canada, 26–28 October 1998.
25. Prashad, H. (1999). Investigations and diagnosis of failure of rolling-element bearings due to unforeseen causes – A case study. *BHEL J.*, **20**(1), 59–67.
26. Prashad, H. (2002). Diagnosis of rolling-element bearings by localized electrical current between track surfaces of races and rolling-elements. *ASME J. Tribol.*, **124**, 468–473.
27. Prashad, H. (2003). Alternative approach to determination of stiffness of ball bearings. *BHEL J.*, **24**(3), 17–22.
28. Prashad, H. (2004). Determination of stiffness of roller bearings – An alternative approach. *IE (I) J.-MC*, **84**, 186–192.
29. Prashad, H. (1986). Experimental study on influence of electrical fields on behaviour of greases in statically bounded conditions and when used in non-insulated bearings. *BHEL J.*, **7**(3), 18–34.
30. Prashad, H. and Murthy, T. S. R. (1998). Behaviour of greases in statically bounded conditions and when used in non-insulated anti-friction bearings under the influence of electrical fields. *Lubric. Eng.*, **44**(3), 239–246.
31. Prashad, H. (1997). Variation and recouping of resistivity of industrial greases – An experimental investigation. Proceedings of International Symposium on Fuels and Lubricants, 8–10 December 1997, New Delhi.
32. Prashad, H. (1994). Investigations of damaged rolling-element bearings and deterioration of lubricants under the influence of electric fields. *Wear*, **176**, 151–161.
33. Prashad, H. (1989). Diagnosis of deterioration of lithium greases used in rolling-element bearings by X-ray diffractrometry. *STLE Trans.*, **32**(2), 205–214.

34. Prashad, H. and Murthy, T. S. R. (1998). The deterioration of lithium greases under the influence of electric current – An investigation. presented at tenth symposium on lubricants, additives, waxes and petroleum speciality products (LAWPSP), held on 30, 31 January and 1 February 1997, at IIT, Mumbai. Published in *J. Lubric. Sc.*, France, 10–4, 323–342.
35. Prashad, H. (1998). Variation and recovery of resistivity of industrial greases – An experimental investigation. *J. Lubric. Sc.*, France, 11–1, 73–103.
36. Mas Remy and Magnin Albert (1996). Rheological and physical studies of lubricating greases before and after use in bearings. Transactions of ASME, *J. Tribol.*, **118**, 681–686.
37. Prashad, H. (1991). Theoretical evaluation of capacitance, capacitive reactance and their effects on performance of hydrodynamic journal bearings. ASME Transaction, *J. Tribol.*, **113**, 762–767.
38. Prashad, H. (1991). Theoretical evaluation of reduction in life of hydrodynamical journal bearings operating under the influence of different levels of shaft voltages. *STLE Trans.*, **34**(4), 623–627.
39. Prashad, H. and Rao, K. N. (1994). Analysis of capacitive effect and life estimation of hydrodynamic journal bearings on repeated start and stop of machine operating under the influence of shaft voltages. *Tribol. Trans.*, **37**(3), 641–645.
40. Prashad, H. (1994). Analysis of inductive effect of bearings under the influence of shaft voltages. *BHEL J.*, **15**(1), 27–33.
41. Prashad, H. (1995). A theoretical model to determine the minimum number of shaft revolutions/cycles for appearance of craters of various sizes on the liner surface of a hydrodynamic journal bearing operating under the influence of shaft voltages of different levels. *BHEL J.*, **16**(2), 27–33.
42. Prashad, H. (1995). Diagnosis of bearing problem of synchronous condenser – An experimental and theoretical investigation. *Wear*, **188**, 97–101.
43. Prashad, H. (2000). A Study of electrical pitting of journal bearing with water contaminated lubricant. *Tribo-test J.*, Leaf Coppin (France), 115–124.
44. Prashad, H. (1996). Theoretical evaluation of electrical parameters of two-lobe journal bearings. *IE (I) J.-MC*, **77**, 47–51.
45. Prashad, H. (1997). Assessment of electrical parameters of three lobe journal bearings – An approach. *Mech. Eng. J. Inst. Eng. (I)*, **78**, 53–56.
46. Prashad, H. (1996). Evaluation of dynamic coefficients of a two-lobe journal bearing using an electrical analogy approach. *ASME J. Tribol.*, **118**, 657–662.
47. Prashad, H. (1997). Assessment of dynamic coefficients of three-lobe journal bearings through evaluation of electrical parameters – A new approach. *BHEL J.*, **17**, 40–48.
48. Prashad, H. (2000). Determination of capacitance, resistance and dynamic coefficients of four-lobe journal bearings through electrical-analogy. *IE (I) J.*, **81**, 30–36.
49. Prashad, H. (1997). Determination of effective stiffness of cylindrical and two-lobe journal bearings by an electrical analogy and its comparison with the conventional approach. IC-HBRSD 97, Second International Conference on Hydrodynamic Bearing Rotor System Dynamics, held on 20–22 March 1997 in Xi'an, P.R. China.

50. Prashad, H. (2000). Evaluation of stiffness coefficients of cylindrical journal bearings by electrical analogy – A non-conventional approach. *IE (I) J.-MC*, **81**, 55–61.
51. Prashad, H. (2001). Evaluation of damping coefficients of cylindrical journal bearings by electrical analogy – A non-conventional approach. *BHEL J.*, **22**(1), 51–60.
52. Prashad, H. (2003). Evaluation of stiffness coefficients of cylindrical journal bearings by a non-conventional approach. *BHEL J.*, **24**(1), 34–45.
53. Prashad, H. (2002). Evaluation of damping coefficients of cylindrical journal bearings by electrical analogy – A non-conventional approach. *IE (I) J.-MC*, **83**, 72–77.
54. Prashad, H. (1992). An approach to evaluate capacitance, capacitive reactance and resistance of tilted pads of a thrust bearing. STLE, *Tribol. Trans.*, **35**(3), 435–440.
55. Prashad, H. (1992). Analysis of the effects of shaft voltages on life span of pivoted pad thrust bearings. *BHEL J.*, **13**(2), 1–12.
56. Prashad, H. (1994). Theoretical model to analyze the minimum cycles before formation of craters due to leakage of charge energy on the liner surfaces of tilting-pads of a thrust bearing. *Inst. Eng*. (India), **75**, 82–86.
57. Prashad, H. (1993). An approach to determine minimum cycles before formation of craters due to leakage of charge energy on the liner surface of tilting-pads of a thrust bearing. NCIT-93, 24–26 March 1993, IIP, Dehradun.
58. Prashad, H. (1998). Analysis of pivoted pad thrust bearings on repeated start and stop of a machine operating under the influence of shaft voltages. Accepted for presentation in World Tribology Congress, London, 8–12 September 1997, published in *J. Inst. Mech. Eng. IE (I) J.-MC*, **79**, 42–47.
59. Prashad, H. (2001). Tribology in electrical environments. *Lubric. Sc.*, **13**(4), 359–369.

Chapter 2

SHAFT VOLTAGES AND THEIR ORIGIN IN ROTATING MACHINES AND FLOW OF ELECTRIC CURRENT THROUGH BEARINGS

2.1 Introduction

Shaft voltages and their origin in rotating machines is a very complex phenomenon. It can occur due to variety of causes including winding faults, unbalanced supplies, electrostatic effects, air-gap fields, magnetized shaft or other machine members, dissymmetry of magnetic fields, etc. These causes in the electric rotating machines result in a net flux or a current linking with the circuit consisting of shaft, bearing and frame. Shaft flux usually results in a localized current at each bearing rather than a potential difference between shaft ends. A current path along the shaft, bearings and frame results in a potential difference between shaft ends.

Certain important causes for the origin of shaft voltage in rotating machines are discussed in this chapter and highlighted as under.

2.2 Causes for the Origin of Shaft Voltages

2.2.1 Dissymmetry Effect

Even though the potential is not directly applied to a shaft, bearing current can still arise through induced effects. This occurs due to dissymmetry of the magnetic circuits developed in the electric machines, which closes over the yoke and induces the voltage on the shaft. The shaft voltage induced because of dissymmetry of the magnetic circuit is influenced by the rotor and stator arrangement of the machine.

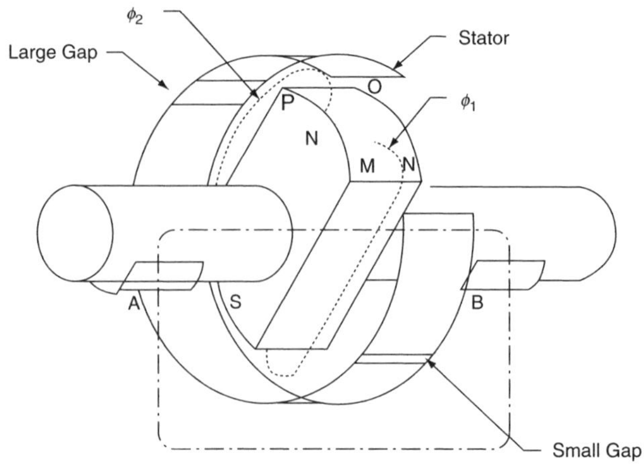

Fig. 2.1. Dissymmetry effect causing shaft voltage.

The principle, by which current is generated, is shown in Fig. 2.1. Figure 2.1 shows a two-pole machine with north and south poles as indicated with stator comprising of two semi-circular sections. If the construction of the machine is perfect, the reluctance of the magnetic circuit will be uniform around the periphery of the stator and hence there will not be induced shaft voltage and current through the bearings. If, however, there are some dissymmetries in the magnetic circuit, then a shaft voltage may be induced. Various sources of dissymmetry may crop up in the construction of the machine: difference in the space between the ends of laminations, non-symmetrical distribution of slots, etc., as shown in Fig. 2.1.

Voltage is induced in the shaft because the value of each of the two fluxes Φ_1 and Φ_2 (Fig. 2.1) varies during each revolution of the shaft. The magnitude of the flux Φ_1 or Φ_2 will decrease or increase depending on the position of the shaft, since, the flux flow is either through the small gap or the large gap depending on the location of the shaft. As the value of Φ_1 increases or decreases, it cuts the conducting material, causing an alternating emf to be set up between M and N (Fig. 2.1). The emf generated by the change in the flux Φ_2 reinforces the emf generated by the change in the flux Φ_1. This generates an alternating potential between the shaft ends.

2.2.2 Shaft Magnetization Effect

This is another source of bearing current in electric machines. This occurs due to the unbalanced ampere-turns, which surrounds the shaft. Such turns cause the shaft to become magnetized and set up a flux which flows from one bearing to another through the machine frame as shown in Fig. 2.2. The portion of the shaft that confines each bearing cuts the flux as the shaft rotates and generates the potential difference on that portion of the shaft. This tends to produce bearing current, which flows from one bearing to another through the shaft and the frame.

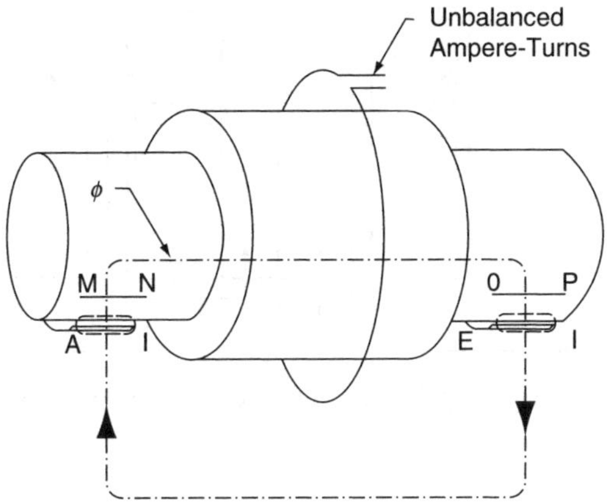

Fig. 2.2. Shaft magnetization effect

In addition, accidental grounding of a part of the rotor winding to the rotor core can lead to stray currents through the shaft and bearings. This may lead to the permanent magnetization of the shaft.

If the machine has a significant level of permanent magnetization, shaft voltage and current could be generated when the machine is rotated. The magnitude of these generated voltages and currents will depend on factors such as the strength of the residual magnetization, the relationship between various residual fields and the particular component that has been magnetized. It also depends on the magnitude of the air gap between rotor and stator, the current path and the variation in insulating properties of the oil films in seals and bearings during operation.

On many occasions where bearing damage occurs, it is found that the shaft of the machine is magnetized. In such cases, magnetic flux flows in the shaft from one bearing to another bearing as shown in Fig. 2.2. Hence, a magnetic circuit exists, which is formed by the shaft, bearing and casing. A local bearing current can flow due to the magnetic circuit causing shaft voltage through the oil films. The voltage over the bearing may be very low, but very high local bearing currents can flow if the resistance of oil film is very low.

Furthermore, homo-polar flux can result from an air gap or rotor eccentricities as shown in Fig. 2.3. The flux crosses the air gap and leads to generate a local

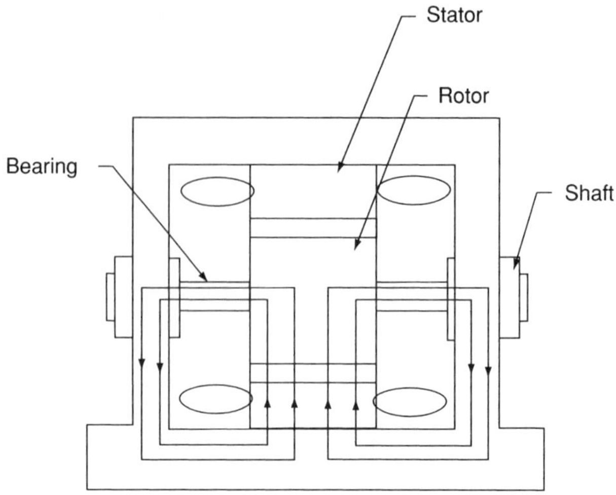

Fig. 2.3. Homo-polar fluxes between stator rotor and shaft

bearing current (Fig. 2.3). The homo-polar flux crossing the air gaps will generate an additional voltage causing the current to flow along the shaft, bearings, bedplate and the casing. At the inner edge of the bearings, the bearing currents due to the magnetized shaft and the shaft current combine. Hence, this part of the bearing and the inner oil ring will be subjected to more damage than the outer section of the bearings.

Another most important cause of the bearing currents is the linkage of an alternating flux with the shaft. The flux flows perpendicularly to the axis of the shaft and pulsates in the stator and rotor cores. It is caused by dissymmetry in the magnetic circuits of the machines, the details of these causes are listed in Section 1.1 of the Chapter 1.

The flux from each pole crosses the air gap and, if the magnetic path is symmetrical, will divide equally, half clockwise and half anti-clockwise as shown in Fig. 2.4. However, if there is a difference in reluctance of the core in one direction compared with the other direction, there will not be an equal division of the flux, and there exists a net flux linking with the shaft. This flux is alternating and will produce voltage in the circuit formed by the shaft, the bearing and the frame.

2.2.3 Electrostatic Effect

Due to the electrostatic effect, charges build up on the shaft when the resistivity of the lubricant used in the bearings is such that no current flows due to low voltage.

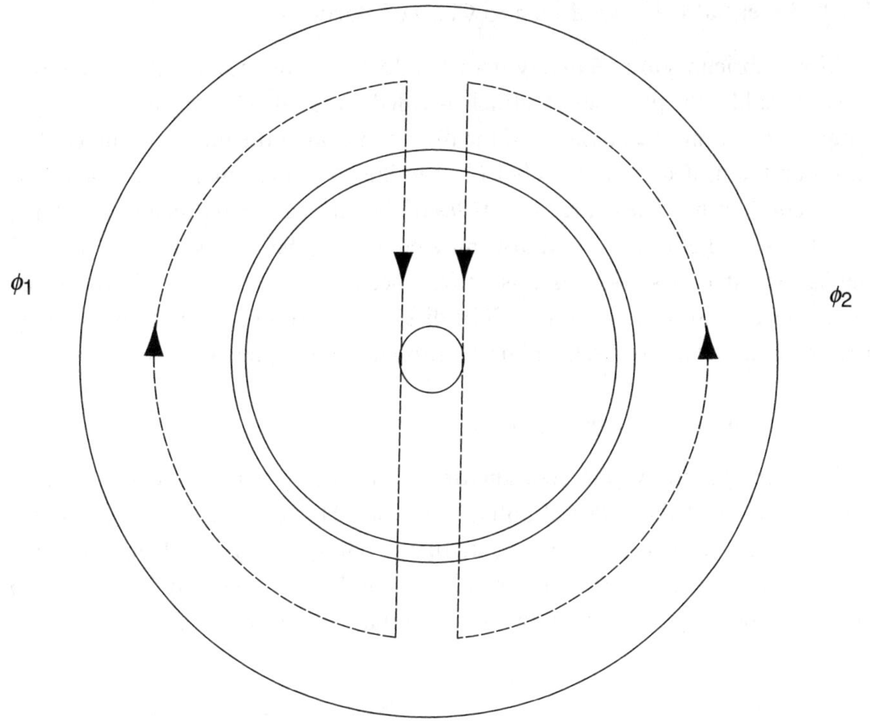

Fig. 2.4. Ring flux lining the shaft (if $\phi_1 \neq \phi_2$)

The minimum voltage, exceeding which, the momentary flow of current is established, is called threshold voltage. When the potential difference across the film thickness exceeds the threshold voltage, the accumulated charge is dissipated through the bearings causing flow of current. Some of the important sources of such charges are discussed as below.

2.2.3.1 Potential Developed by Impinging Particles

Each particle of a finely divided material has an equal number of positive and negative charges and this makes the material electrically neutral under equilibrium conditions. When such a stream of neutral particles strikes an object under certain conditions, the positive and negative charges of the particles on impact make the object also charged in the same fashion.

2.2.3.2 Potential Developed Due to Charged Particles

Most lubricants are relatively poor conductors. The individual molecules of the lubricant act like the particles of a finely divided material. The lubricant while passing through the certain fine passages of the filter can charge the lubricant molecules. The charges on the molecules are likely to be retained, because, in general, the lubricants are non-conducting. The charged particles of the lubricant can transmit the charges to the surfaces of the journal, and also may charge the bearing surface, provided, the latter is insulated. As soon as this charge becomes high or exceeds the threshold voltage, the accumulated charges will be dissipated through the oil film under certain operating conditions, as discussed in the subsequent chapters of this book.

2.2.3.3 Potential Developed by Belts

Belts are generally poor conductors of electricity. Charges originating at the point of contact of the belt and pulley are equal. However, if the condition of both the pulleys of the system is different, then a charge may build up on the shaft, provided, the shaft is insulated by the oil film, and the charge may not get dissipated through the bearings. This is shown diagrammatically in Fig. 2.5.

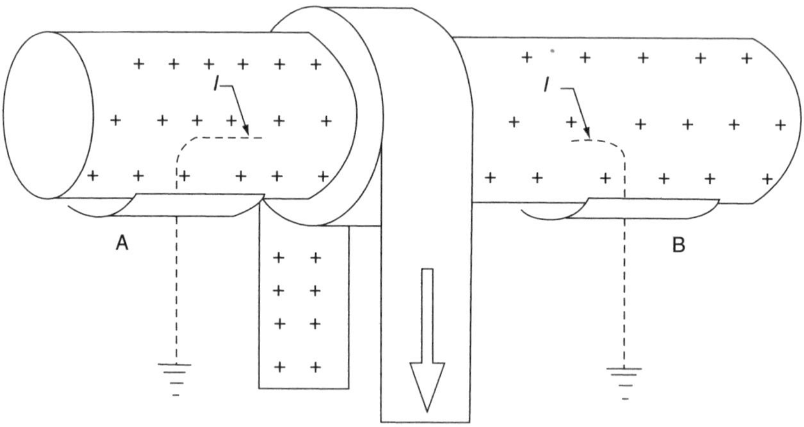

Fig. 2.5. Electrostatic effect-potential developed by charged belt

2.2.4 Shaft Voltage Caused by Permanent Magnetization of Casing, Shaft or Pedestals

In practice, shaft, bedplate, pedestals, casings and other components are made of the material, which possess good magnetic properties. In general, permanent magnetization is developed on the shaft, casing, etc., due to the heavy short circuit (like welding of components without proper grounding) or any other abnormal operation of the generator. Under certain conditions, bedplate or other such components can produce an axial residual flux in the shaft and this can produce self-excitation resulting in very large shaft and bearing currents. Such phenomenon leads to the damage of shaft, the bearings and other components. Failure of the oil pumps and shafts of the control devices in the generating units has been reported due to self-excitation phenomenon.

Shaft voltage has very close relevance with the electrical equipments. It is generated by residual magnetic fields in the stator and other components of a machine. The magnetic fields are the results of increased use of magnetic devices such as magnets used for lifting machines, magnetic chucks on machines, magnetic particle inspection and especially arc welding systems and various other reasons as explained in the subsequent chapters.

Every time the electric arc welding is carried out around a machine, current travels through the casing and then the rotor creates/develops a residual magnetic field on these components. When a machine with significant residual magnetism rotates, the electric voltage and current are generated in the same manner as in an electric generator or in an eddy current brake. The strength of these voltages and currents depends on the strength of the residual magnetic field. The path of the current flow and the voltage depends on the insulating properties of the oil films in the bearings and seals, the rotor speed and various other factors, which are investigated subsequently.

2.3 Factors Affecting the Shaft Voltages

The important factors affecting the magnitude and frequency components of the shaft voltage generated due to magnetic dissymmetry depend on the machine size, the stator core construction, the rotor alignment, the number of poles, the generator design and load conditions.

Larger units, in general, have higher magnitudes of the shaft voltage. This can be attributed to the higher magnitudes of air-gap flux, greater constructional dissymmetry and saturation of the core. Besides this, larger unit introduces non-uniformity in the air-gap length resulting in larger magnitudes of the shaft voltage. The excessive eccentricity distorts the magnetic field considerably and enhances the effect of the shaft voltage.

2.4 Reasons for Epidemic Shaft Current Problems

The significant reasons for the epidemic shaft current problems are:

(1) The magnetic tools and magnetic particle inspection devices are those widely used with portable units. Rotor and stator components must be demagnetized after repair, manufacturing and inspection.

(2) The electric arc welding around a machine or use of 'arc torch' to burn off foreign matter on a base plate leaves a strong permanent residual magnetic field which must be taken care of.

(3) A new unit, in most cases, has less degree of residual magnetism. But it increases continually and steadily with time due to repairs, welding and use of magnetic tools. Finally, residual magnetism will reach a level where the fields are strong enough to reorient themselves in such a manner as to allow self-excitation of an area. However, there are no chances for the billions of minute magnets to align themselves. Yet due to shock, surge, oil whirl, vibrations, rubbing, etc., these small magnets may align themselves in such a manner and increase field strength considerably.

The above factors will generate high voltage as the rotating field reacts with the stationary fields. This may allow the current passage through the bearings depending on the resistance of oil film thickness.

2.5 Passage of Current Through Bearings

The principal effect of a bearing current is the damage caused by arcing across the bearing surfaces. As an electric current passes through the surface in contact, the flow of current concentrates through the asperity contact points, and the local current density increases to a very high value. A flow of current through the asperity contacts generates heat at the contact area. Based on the resistance of the oil film and the resistivity of the lubricant used, the slow passage of current or arcing takes place at the contact zone. The main consequence of current passage and the arcing leads to the wear of bearing surfaces due to the removal of fused metals in the arc, resulting in closely pitched marks and burned craters. Due to this, roughening of the surface and mechanical wear is accelerated. The arcing, the formation of craters, the mechanical roughing of the active surfaces of bearings, and the liberated metallic particles cause the oxidation of the lubricant and loss of purity of the lubricant. Thus, deterioration of the lubricant gradually increases bearing destruction.

2.6 Bearing Electrical Parameters

The bearing current mainly depends on the magnitude of the shaft voltage and the bearing impedance. The magnitude of the shaft voltage is usually estimated as a function of the ring flux. The ring flux results from dissymmetry whose magnitude cannot be readily measured/estimated. The bearing impedance, resistance and capacitance depend on many factors such as lubricant resistivity, lubricant permittivity/dielectric constant, bearing temperature, load and speed of operation. In general, the bearing impedance reduces with the increase in temperature and increases substantially with the increase in speed. The characteristics of the bearings depend on their design and type. The impedance of a ball bearing differs from a roller bearing. Hydrodynamic journal and thrust bearings used in large machines have different electrical parameters and characteristics as compared to rolling-element bearings. The literature on the values of a bearing impedance, resistance, capacitance, capacitive reactance is very scarce and a better understanding of the bearing electrical parameters is needed to fulfill the know-how and know-why gap.

In the absence of bearing insulation, a shaft voltage appears across the bearing oil film. The oil film thickness in a rolling-element bearing ranges from 0.001 to 0.02 mm, and the area of contact depends on local surface roughness. The oil film will cease to be an insulator when the shaft voltage reaches the threshold value and then the flow of current through a bearing takes place.

In addition, presence of dirt, metallic particles and irregular oil film thickness makes the impedance of a bearing circuit so low that small shaft voltages may cause substantial bearing currents. If these currents are suddenly reduced by the reestablishment of the non-conducting film in the bearing, the self-inductance of the single loop of the shaft, the bearing and casing may then cause a relatively high-induced voltage across the oil film. This induced voltage may breakdown the oil film depending on the lubricant characteristics and other operating parameters. The study of these phenomena in different bearings is inadequate, and needs to be undertaken conceptually, theoretically and experimentally.

2.7 Conclusions

To fill the above know-how gap in the literature, different chapters of this book have been developed so that various bearings and their performance, behavior, etc., could be understood more precisely under the influence of electric currents and shaft voltages. This book on 'Tribology in Electrical Environments' is the first of its kind to probe into this hitherto unprobed area in many aspects and dimensions.

Chapter 3

BEHAVIOR OF LUBRICANTS IN ROLLING-ELEMENT BEARINGS UNDER THE INFLUENCE OF ELECTRIC CURRENT

3.1 Introduction

It is known that lubricating greases, like their base oils, are capable of forming elasto-hydro-dynamic films thick enough to prevent the metal-to-metal contact between asperities of the opposite surfaces. The films formed by greases are initially thicker than those formed by the corresponding base oils but, after continuous running, become thinner than the latter. There is evidence that the differences observed between oils and greases are related to their viscoelastic properties [1, 2]. Efforts were made to understand the role of the metal-soap and base oil in the lubricating process, for which purpose the friction coefficient was measured and also to explain the lubricating behavior of mineral greases (by using optical and electrical measurements) to find out the film thickness of the lubricant and to control the lubricating regime in which the greases are working [3, 4]. However, it is not properly understood how the greases, as bodies of rheologically complex behavior, would behave under the influence of electric field in statically bounded conditions and when used as lubricants in non-insulated rolling-element bearings. Amongst the factors defining functional interaction between the metallic surfaces, in the presence of lubricating grease, it is at present of basic importance to know the structure and the properties of the lubricant molecule, how it adapts to the metallic surfaces and the special conditions for the existence of boundary layer [5]. This interaction becomes even more complex under the influence of an electric field.

The behavior of greases in the presence of an electric field needs investigation. Greases are compounds of oil, soap, additives and derivatives of the homologous series of hydrocarbons, which have electric moments. The soap present in the grease is itself produced from a carboxylic acid or their glyceride (fat and oil) and alkali or alkaline earth hydroxide and alcoholate. The behavior of these constituents under the applied electric potential drop may be quite complex. A film of metal-soap of strong bonding to the crystal lattice of the metal, and its high breaking strength may also add to the complexity. Contamination of grease with water or solutions of acids, salts, etc. make its performance rather unpredictable. Under the influence of the electric field, polar-chain molecules present in the grease may tend to orient towards the field and try to adhere to the surface of metal, while non-polar hydrocarbons would behave indifferently. The grease may exhibit a change in its resistivity under the electric field with time, and may 'recoup' it when the electric field is withdrawn or may behave indifferently.

Despite the wide use of lubricating greases, most of the studies are concerned with their structure, manufacturing, processing, mechanical testing and evaluation [6, 7]. A few papers deal with the effect of soaps and base oils on the frictional behavior of greases [8, 9]. Greases, because of their high consistency and lack of mobility have a tendency to work in starved conditions. This diminishes the film thickness of grease compared to flood lubrication, which has been reported by various authors in experimental and theoretical studies [10–12]. Desired properties that lubricating grease must possess are summarized as follows:

- to form a film of lubricant,
- to prevent corrosion,
- to resist temperature rise without becoming unduly softened in the process, and
- to withstand shear without a breakdown of the structure.

The greases have an effective yield stress and they resist the deformation elastically until the shear stress greater than the yield stress is applied. Quite a few investigations have been reported related to the electrophysical and electrochemical phenomena in friction, lubrication, contact interaction and fretting corrosion [13–16]. The understanding of viscous properties of grease in relation to shear rate, temperature and pressure is still not complete.

Although the above factors have an influence on the behavior and deterioration of the greases, it is not easy to quantify them, because friction between bearing surfaces in a lubricating media is accompanied by complicated physicochemical processes taking place between the plastically deformed surface layers and surrounding medium. As a result of this interaction, secondary structure of lubricating medium may form which may eliminate the adhesion of conjugate surfaces. The composition of the products obtained by interacting surfaces with the lubricating medium under friction depends on the medium, bearing surfaces and conditions of interaction [17]. Investigations of these changes may give valuable information to diagnose the mechanism of the medium-metal interaction during friction.

The chapter explains the above unpredictable complex electrochemical and electrophysical behavior of the lubricating greases in statically bounded conditions and when used in non-insulated rolling-element bearings, under the influence of electric fields [18, 19].

Also, study was undertaken to understand the net effect on the change of structure of greases used in rolling-element bearings after being operated separately under the influence of electric fields and under pure rolling friction, and also to detect newly formed compounds in the medium. XRD technique has been applied in this work to diagnose the changes in chemical composition of soap residues of used greases apart from the analysis by XRFS and atomic absorption spectrophotometry techniques [20, 21].

3.2 The Experimental Facilities

3.2.1 Test Set-Up for Determining Resistivity of Greases in Statically Bounded Conditions

The test set-up, shown in Fig. 3.1, was used to study the effect of potential drop on the resistivity of three different greases which are used as lubricants in rolling-element bearings. Two cylindrical brass electrodes, having diameter of 2.5 cm, a length of 2.5 m and an area of 5 cm, were used. The gap between the electrodes, fitted in the cuboid, could be adjusted from 2.5 to 4 mm. The supply voltage to the electrodes could be varied between 10 and 500 V (DC).

A megohm meter, Model RM 160 MK 111 A, was used for measurement of resistances upto 400 million megohm, and of current down to 10^{-12} A. The instrument is completely self-contained and can be set at DC test voltages of 10, 50, 100, 250, 400 and 500 V.

The sensitive measuring circuitry, as well as the indicating meter, has complete over load protection, preventing any damage that might be caused by the breakdown of the test sample (grease). A signal lamp is provided for indicating over load.

Before making any measurement, the instrument was stabilized and the multiplier adjusted to get a convenient reading of the meter. The resistance (in megohm), R, of the test sample is given by:

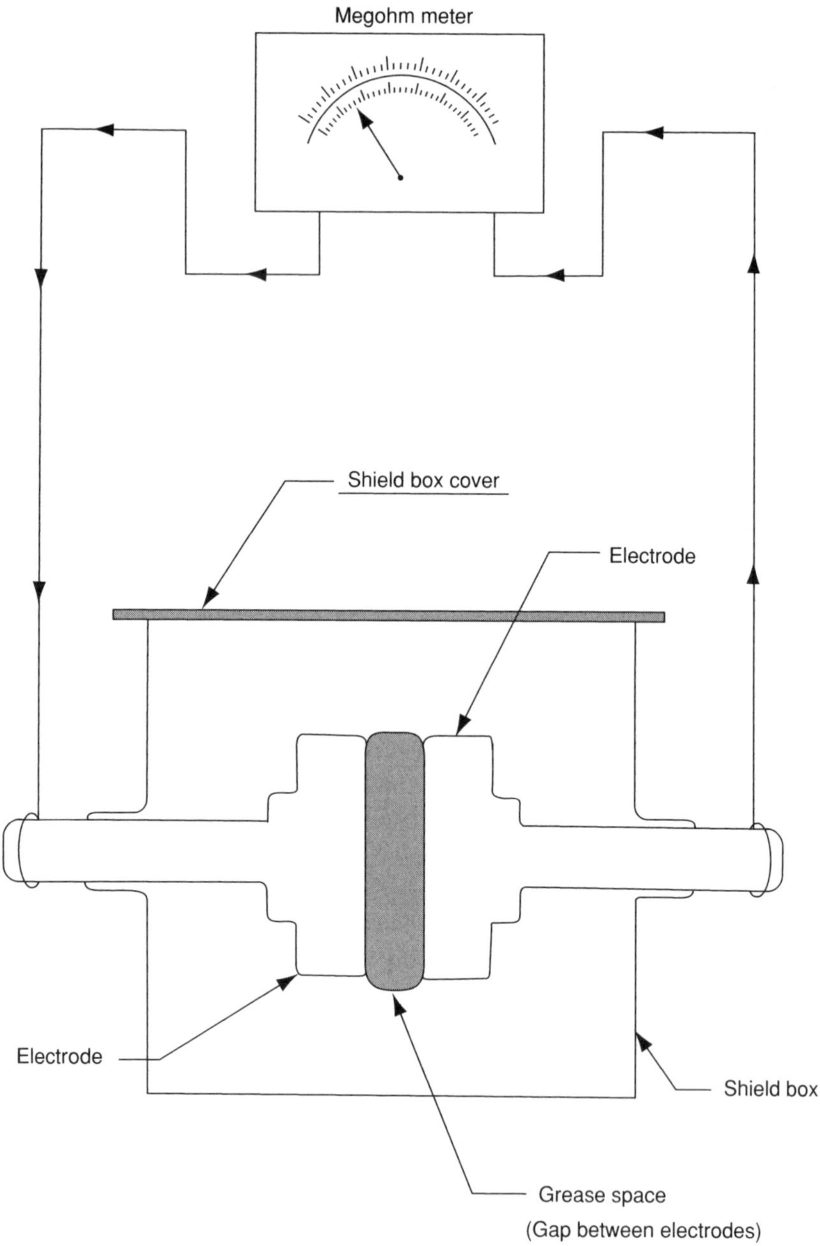

Fig. 3.1. Test set-up for measurement of resistivity

$$R = \text{Meter reading} \times \text{Multiplier} \times \frac{\text{Test Voltage}}{500}$$

The resistivity of the test sample can be calculated by using the formula

$$\sigma = \frac{R \times S}{L} \tag{3.1}$$

σ is restivity, ohm cm, R is resistance, ohm, S is area of the electrodes cm^2, L is gap between electrodes cm.

3.2.2 Roller Bearing Test Machine

The Roller Bearing Test Machine, shown in Fig. 3.2, was used to evaluate the performance of rolling-element bearings rotating at speeds of up to 3000 rpm under different combinations of radial, horizontal and axial loads in static and dynamic modes of operation. Bearings of different sizes (of inner diameter ranging from 44 to 150 mm) can be tested, and investigations in various tribological areas can be programmed. The test machine is driven by a rectifier-controlled 17 kW DC motor, type AG 2503 AZ. The condition of the bearing under test is monitored by means of various thermocouples and vibration pick-ups mounted on the bearing outer race and the end-shield. For evaluating the bearing performance, under the effect of rotor currents, a specially designed silver-lined slip-ring assembly is mounted in between the support bearings. The support bearings and the other relevant structures are fully insulated to avoid any current leakage. It is thus ensured that the current flows through the shaft, inner race of the bearing, rolling-elements, outer race of the bearing and the housing, to complete the electrical circuit.

In the present studies, bearings type NU 326 lubricated with lithium base grease 'A' has been tested at 1100 rpm under 1000 kgf and 500 kgf of radial loads acting at two different directions (90° to each other for a duration of 250 h by passing 50 A (AC)). During the test period voltage across the bearing varied from 1.12 to 2.3 V between points A and B (Fig. 3.2). Corrugated patterns on bearing surfaces and effect of operating parameters on the threshold voltages and impedance response were also studied. Similarly, NU 326 bearings were tested under identical conditions, without the passage of current under pure rolling friction.

3.2.3 X-Ray Diffraction (XRD) Unit

A Philips PW 1140 X-ray diffraction unit, complete with generator and goniometer, was used for analysis of soap residues of different greases. Instrument parameters are listed in Table 3.1. By X-ray diffraction, symmetry and regularity in arrangement of atoms was made visible when a monochromatic X-ray beam irradiates soap residues. In

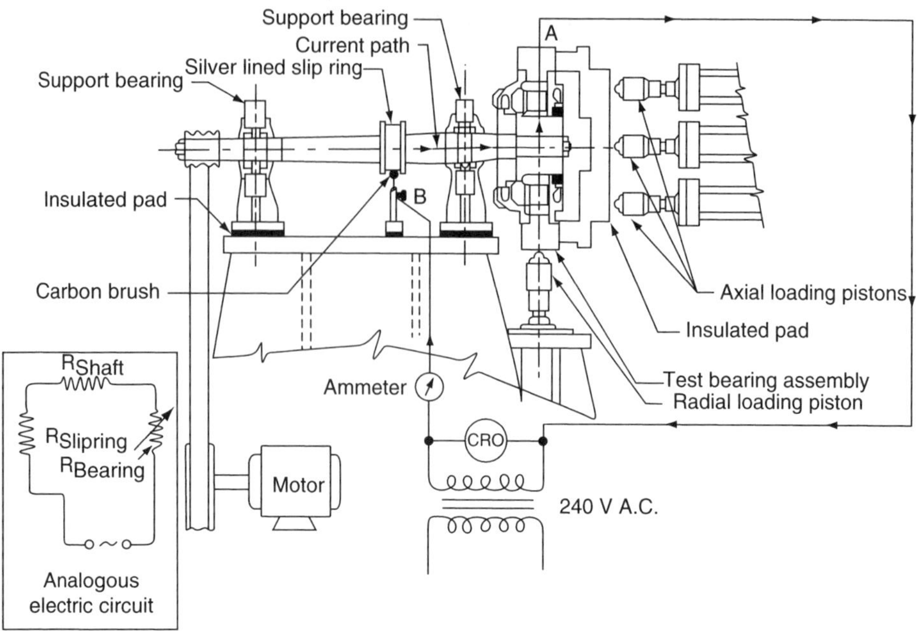

Fig. 3.2. Bearing test machine with electric current

Table 3.1. X-ray diffractometry instrument parameters

Model	Philips PW 114θ
Generator	3 Kw, 100 Kv Max, 100 mA Max
Horizontal goniometer	PW 138θ Philips
Detector	Xenon filled proportional counter
Channel control	PW 1390 with recorder PW 8203 (single pen recorder)
Debye scherrer camera	2 numbers (114 mm)
Powder sample	5 g fine powder
Solid sample	1.5 cm × 1 cm × 0.5 cm approximately (with smooth surface)

this way an unknown compound could be identified since XRD shows the typical lines for each compound separately, satisfying Bragg's law. It is given as:

$$n\lambda = 2d \sin\theta \tag{3.2}$$

where
n is order of diffraction, λ is wave length, θ is angle, d is inner planner distance (Å).

The particular advantage of diffraction analysis is that it discloses the presence of a substance as that substance actually exists in the sample, and not in terms of its constituent chemical elements [20].

3.2.4 X-Ray Fluorescence Spectrometry (XRFS)

This is an analytical tool for rapid quantitative and qualitative determination of elements present in a sample. When a beam of X-rays is directed onto the surface of the specimen, secondary fluorescent radiation is emitted by the specimen, which contains wavelength characteristics of each element present. The secondary radiation is directed onto an analyzing crystal separating the wavelengths, which are recorded by a detector. The whole arrangement is in the form of a vertical goniometer. Since the wavelength is a function of angular position of the crystal and goniometer, the elements in the specimen can be identified. A Philips model PW 1140 with channel control PW 1390 has been used for the elements present in the soap residues of the greases.

3.3 The Experimental Procedure

3.3.1 Grease Application to Bearings and Sampling Methodology

A permissible measured quantity of grease was filled in inner and outer bearing caps (about 250 g each), and a small quantity was applied on the working surfaces of the rollers as per the standard procedure of the bearing lubrication. Periodically, after a few hours of operation, varying between 41 and 250 h, bearing caps and outer race with rollers were dismantled and a small quantity of grease (5–7 g) was collected using sterilized glass rod from the working surfaces of the rollers for analysis, and then the bearings were reassembled without addition of fresh grease. Thus, the deterioration of the grease, once filled, was studied for a complete test duration without creating starvation of lubricant on the roller surfaces. Similarly, grease samples from different motor bearings, collected from the working surfaces of the rollers (after about 6000 h of operation) were also analyzed.

3.3.2 Determination of Resistivities and Infrared Spectra of Greases in Statically Bounded Conditions

Three different greases – designated as ‘A’, ‘B’ and ‘C’ – were taken. A fresh sample of grease was pressed between the electrodes kept 4 mm apart. On applying a potential drop of 10 V across the electrodes, the resistance of the grease was measured by the megohm meter within 1 min and subsequently at intervals of 15 min. Following this, the electrodes were thoroughly cleaned, and similar measurements were taken at potential drops of 50, 100, 250, 400 and 500 V (DC). This experimental procedure was adopted for each of the three types of greases. Resistivity was calculated by using Eqn. (3.1). The change in resistivities of the greases with time, at varying potential drops, is presented in Tables 3.2–3.4 and illustrated in Figs. 3.3–3.5. Tables 3.2–3.4 show the changes in resistivity with time, an average ratio of resistivity after 90 min to the initial value, and the ratio of resistivity at 500 V to the value at 10 V in certain periods, for greases 'A', 'B' and 'C', respectively.

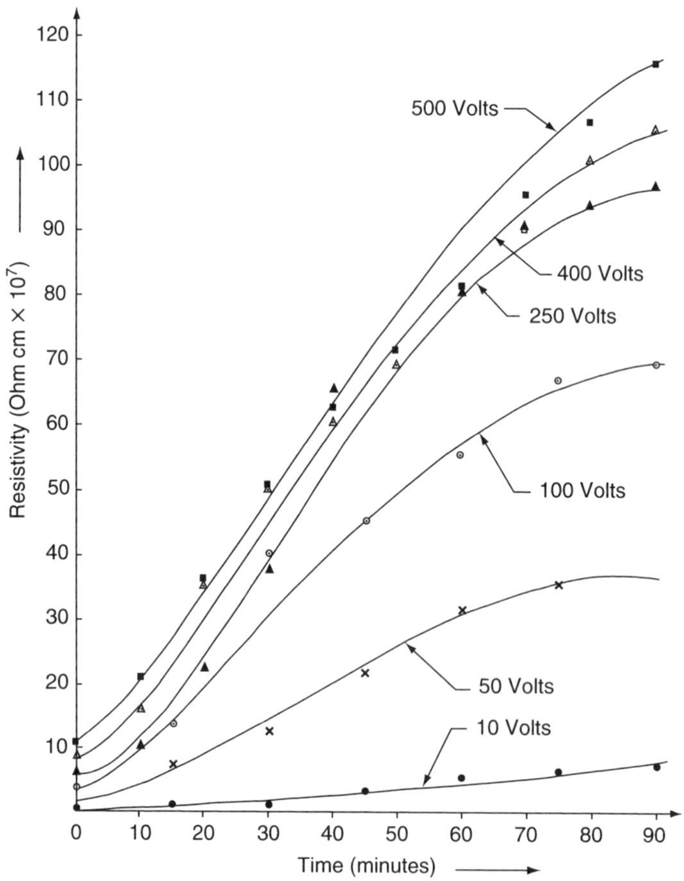

Fig. 3.3. Variation of resistivity of grease 'A' with time at different potential drops across electrodes

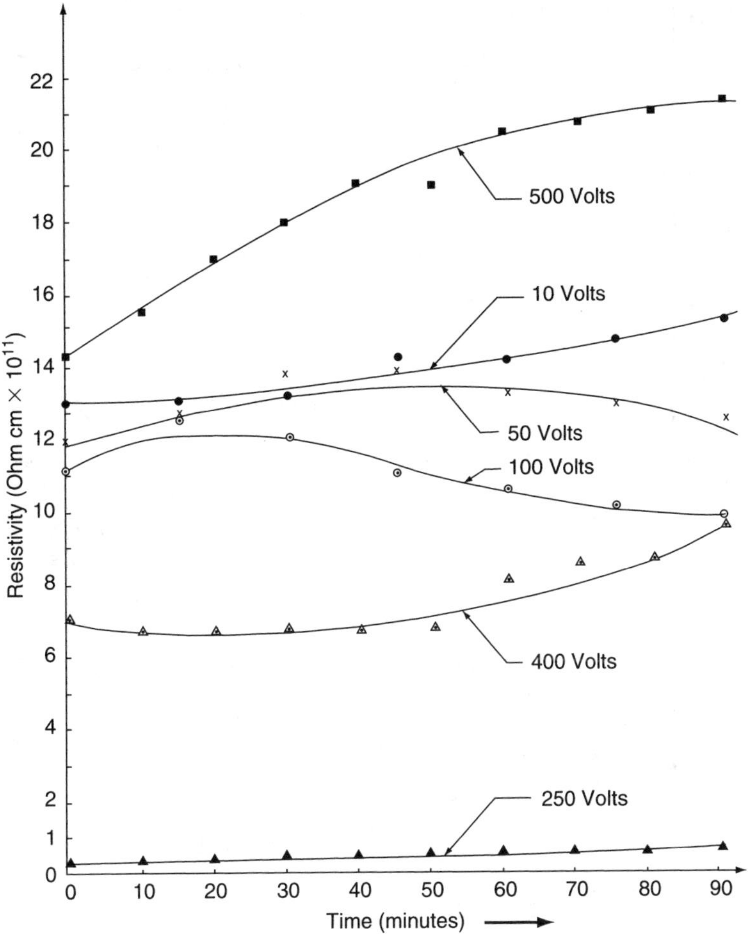

Fig. 3.4. Variation of resistivity of grease 'B' with time at different potential drops across electrodes

Subsequently, for grease 'A', which is normally used in AC and DC motors, restoration of resistivity with time, at 50 V (DC) potential drop across the electrodes kept 2.5 mm apart, was recorded. The distance between the electrodes was kept at 2.5 mm – the minimum available – to obtain a better 'recouping' behavior of the grease. The sequence of the records and the results are shown in Table 3.5 and Figs. 3.6 and 3.7. The 'recouping' behavior with time, and the percentage restoration of resistivity in consecutive records, for grease 'A', are clearly brought out [22].

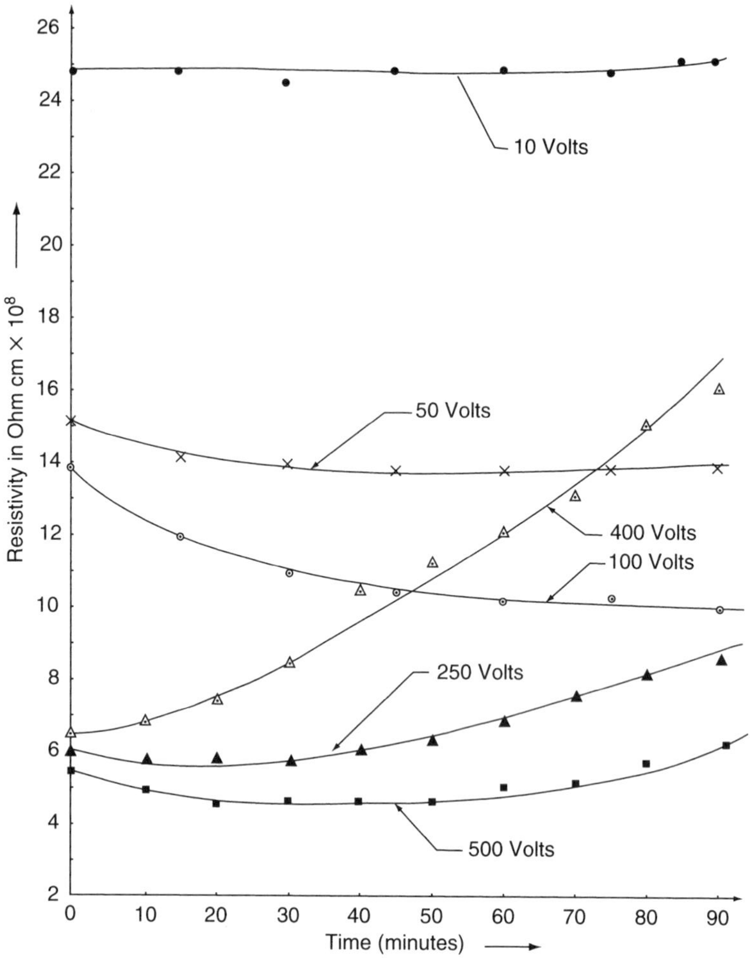

Fig. 3.5. Variation of resistivity of grease 'C' with time at different potential drops across electrodes

Functional performance characteristics and physicochemical properties of greases 'A', 'B' and 'C' were partly extracted from the sponsored work carried out at the Indian Institute of Petroleum, Dehra Dun (India) , and the results are given in Table 3.6. Infrared (IR) spectra of a sample of fresh grease 'A', and those of a sample of statically-bounded and electrically exposed grease 'A', taken after the completion of the eighth record, are shown in Figs. 3.8 and 3.9 respectively. Soap and oil were separated from these samples by dissolving them in petroleum ether. The IR spectra of soap and oil contents of these samples are shown in Figs. 3.10–3.13.

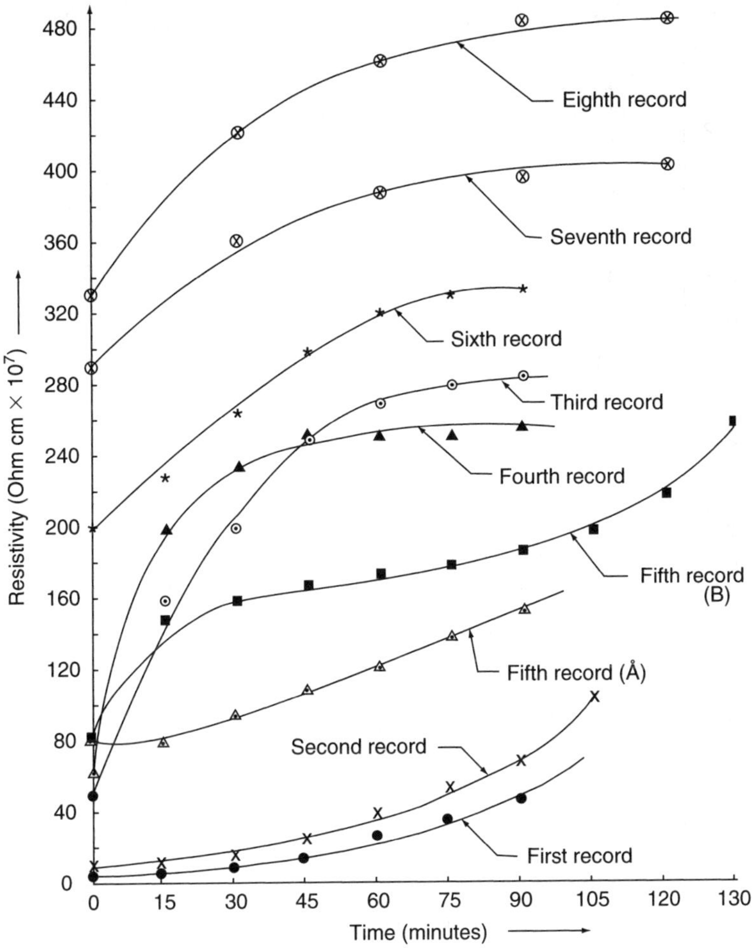

Fig. 3.6. Variation and recovery of resistivity of grease 'A' with time under 50 Volts potential drop across electrodes

3.3.3 Determination of Behavior of Grease when Used in Non-Insulated Bearings

The vibration levels and temperature rise of the test bearing were periodically monitored during operation. The bearing impedance was monitored at the rated operating conditions; it was measured between points 'A' and 'B'' (Fig. 3.2). After 41 h of operation, the bearing cap was removed, and a sample of grease collected from the rollers. The ash content of the sample was determined. The IR spectra of the sample of grease, its soap residue and oil content were taken, and are shown in Figs. 3.14–3.16 respectively. The bearing surfaces were found to have corroded after the

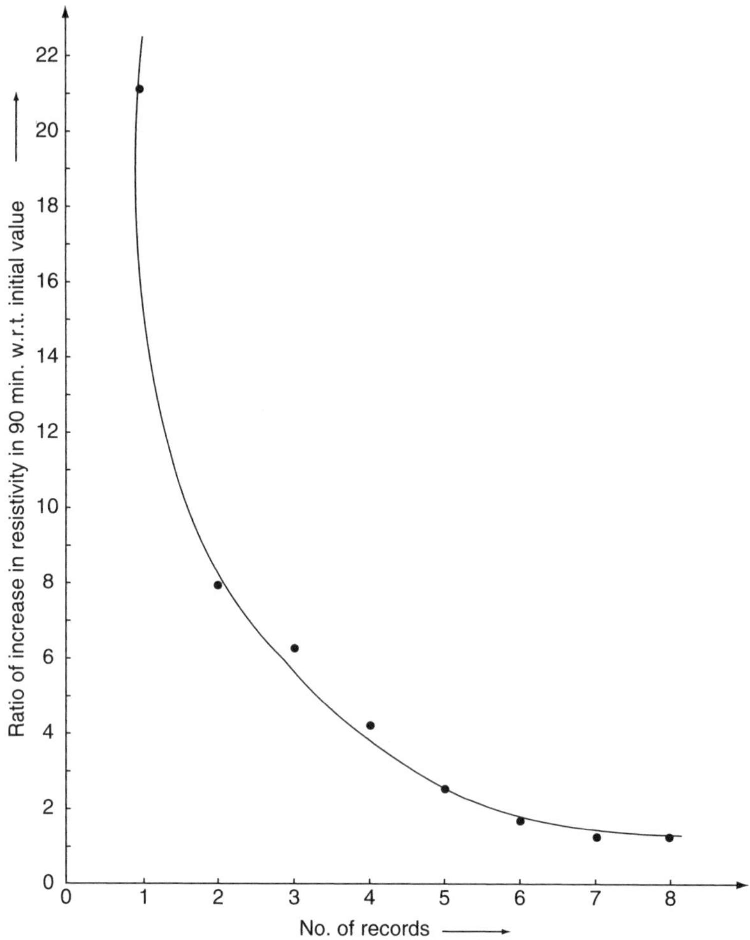

Fig. 3.7. Ratio of resistivity of grease 'A' after 90 minutes to the initial value vs. number of records under 50 V potential drop across electrodes

passage of electric current, as shown in Fig. 3.17. The relative percentage of the lithium metal present in the sample was determined, and compared with those present in fresh and statically-bounded samples.

3.3.4 Investigation by X-Ray Diffractometry (XRD)

Soap and oil contents of fresh grease and the grease samples from the test bearings and from motor bearings were separated by dissolving them in petroleum ether. The soap residues, thus obtained from different grease samples, were subjected

Table 3.2. Variation of resistivity of grease 'A' with time at different potential drops across electrodes kept 4 mm apart

Potential drop across electrodes (in volts)	Resistivity, σ (in ohm cm $\times$ 10^7) Time (in minutes) 1	10	15	20	30	40	45	50	60	70	75	80	90	Ratio of resistivity after 90 min to the initial value	Average ratio of resistivity after 90 min to the initial value
10	0.63	–	1.0	–	1.75	–	3.75	–	5.5	–	6.25	–	6.5	10.32	–
50	1.5	–	7.5	–	12.5	–	21.75	–	31.25	–	35.0	–	35.5	23.67	–
100	4.0	–	13.75	–	40.0	–	45.0	–	55.0	–	66.25	–	68.75	17.2	14.70
250	6.38	10.63	–	22.5	37.5	65.6	–	71.8	80.0	90.63	–	93.75	96.8	15.17	–
400	9.0	16.0	–	35.0	50.0	60.0	–	69.0	80.0	90.0	–	100.0	105.0	11.67	–
500	11.25	21.25	–	36.2	50.0	62.5	–	71.25	81.25	95.0	–	106.2	115.0	10.22	–
Ratio of resistivity at 500 V to that at 10 V	17.88	–	–	–	28.57	–	–		14.77		–	–	17.7	–	–
Average ratio of resistivity at 500 V to that at 10 V	–	–	–	–	–	–	19.73	–	–	–	–	–	–	–	–

Table 3.3. Variation of resistivity of grease 'B' with time at different potential drops across electrodes kept 4 mm apart

Potential drop across electrodes (in volts)	Resistivity, σ (in ohm cm $\times 10^{11}$) – Time (in minutes): 1	10	15	20	30	40	45	50	60	70	75	80	90	Ratio of resistivity after 90 min to the initial value	Average ratio of resistivity after 90 min to the initial value
10	13.0	–	13.0	–	13.75	–	14.25	–	14.0	–	14.5	–	15.0	1.5	–
50	12.0	–	12.60	–	13.63	–	13.75	–	13.0	–	12.63	–	12.38	1.03	–
100	11.25	–	12.5	–	12.0	–	11.00	–	10.5	–	10.0	–	9.75	0.87	–
250	0.033	0.0438	–	0.0469	0.05	0.0663	–	0.0625	0.071	0.073	–	0.0756	0.084	2.56	1.41
400	7.0	6.8	–	6.8	6.81	6.8	–	6.8	8.0	8.5	–	8.6	9.5	1.38	–
500	14.3	15.6	–	16.9	17.8	18.8	–	18.9	20.3	20.5	–	20.75	21.0	1.47	–
Ratio of resistivity at 500 V to that at 10 V	1.1	–	–	–	1.29		–	–	1.45	–	–	–	1.4	–	–
Average ratio of resistivity at 500 V to that at 10 V	–	–	–	–	–	–	1.31	–	–	–	–	–	–	–	–

Table 3.4. Variation of resistivity of grease 'C' with time at different potential drops across electrodes kept 4 mm apart

Potential drop across electrodes (in volts)	Resistivity, σ (in ohm cm × 10^8) Time (in minutes)													Ratio of resistivity after 90 min to the initial value	Average ratio of resistivity after 90 min to the initial values
	1	10	15	20	30	40	45	50	60	70	75	80	90		
10	24.8	–	24.8	–	24.5	–	24.8	–	24.8	–	24.8	–	25.0	1.01	–
50	15.2	–	14.25	–	14.0	–	13.88	–	13.75	–	13.88	–	13.75	0.91	–
100	13.75	–	12.0	–	11.0	–	10.5	–	10.25	–	10.25	–	10.0	0.73	–
250	6.0	5.93	–	5.93	5.81	6.06	–	6.43	6.93	7.50	–	8.12	8.62	1.44	1.29
400	6.5	6.8	–	7.5	8.5	10.5	–	11.20	12.00	13.00	–	15.0	16.0	2.46	–
500	5.5	5.0	–	4.62	4.75	4.62	–	4.75	5.00	5.37	–	5.75	6.37	1.16	–
Ratio of resistivity at 500 V to that at 10 V	0.22	–	–	–	0.19	–	–	–	0.20	–	–	–	0.25	–	–
Average ratio of resistivity at 500 V to that at 10 V	–	–	–		–	–	0.22	–	–	–	–	–	–	–	–

Table 3.5. Variation and 'recouping' of resistivity of grease 'A' with time under 50 V potential drop across electrodes kept 2.5 mm apart

Test record	Resistivity, σ (in ohm cm $\times 10^7$)										Ratio of resistivity after 90 min to the initial value	Percentage restoration of resistivity in consecutive records
	Time (in minutes)											
	1	15	30	45	60	75	90	105	120	135		
First record	2.32	6.0	11.0	17.0	29.9	38.5	50.0	–	–	–	21.16	–
Second record after 24 h from first record	9.0	11.0	15.0	27.0	40.0	56.0	72.0	–	–	–	8.0	82.0
Third record after 48 h from second record	45.0	160.0	200.0	250.0	270.0	280.0	284.0	–	–	–	6.31	37.5
Fourth record after 48 h from third record	60.0	200.0	236.0	250.0	250.0	252.0	254.0	–	–	–	4.23	78.8
Fifth record (A) after 20 h from fourth record	80.0	80.0	98.0	110.0	124.0	140.0	158.0	–	–	–	1.98	68.5
Fifth record (B) after 2 h from fifth record (A)	80.0	150.0	160.0	170.0	176.0	180.0	190.0	200.0	220.0	260.0	2.38	49.4
Sixth record after 13 days from fifth record (B)	200.0	230.0	266.0	300.0	320.0	330.0	332.0	–	–	–	1.66	23.0
Seventh record after 17 h from sixth record	290.0	–	360.0	–	385.0	–	395.0	–	402.0	–	1.36	12.7
Eighth record after 17 h from seventh record	330.0	–	421.0	–	460.0	–	461.0	–	481.0	–	1.40	17.9

Table 3.6. Physicochemical properties and functional performance characteristics of greases

S. No.	Properties/characteristics	Grease type		
		'A'	'B'	'C'
	Lubricating oil, %	86.6	86.5	86.5
	Thickener content	13.4	13.5	13.5
	Viscosity (in centi-stokes)			
	At 40°C	99.2	127.33	117.33
	At 100°C	9.68	10.43	11.93
	Alkali (free), %	0.03	0.05	0.07
	Free fatty acid			
	Free fact and filler, %	Nil	Nil	Nil
	Additives (type)	Amino	Amino	Amino
	Ash, %	0.86	0.98	1.27
	Structure modifiers, %			
	Glycerol	Nil	Nil	Nil
	Water	0.2	0.2	0.2
	Texture (color)	Smooth and brown	Smooth and brown	Smooth and brown
	Consistency (unworked)	236	260	260
	Dropping point, °C	185	187	185
	Oxidation stability (drop in pressure after 100 h at 99°C, in psi)	6	7	5
	Oil separation, %			
	At 80°C	Nil	Nil	Nil

(*Continues*)

Table 3.6. (*Continued*)

S. No.	Properties/characteristics	Grease type 'A'	'B'	'C'
	At 130°C	0.44	0.21	0.21
	Evaporation loss, %			
	At 105°C	0.43	0.50	0.65
	At 130°C	0.93	1.01	1.10
	Shell roll stability at standard temperature at 31°C as per ASTM–1831 (penetration difference)	8	8	9
	Low-temperature torque at 5°C as per IP–186 (in g cm)			
	Maximum	50	200.0	212.5
	Minimum	50	187.5	200.0
	Water resistance (water wash-out) test as per ASTM–1264 wt. loss (in g)			
	38°C	Nil	0.3	0.7
	30°C	0.3	0.1	0.3
	Wear and frictional characteristics determined by 4-ball wear tester at 2000 rpm at 100 kg load (time taken to initiate wear scar on the balls at the minimum setting of the vibration transducer and tripping of the machine)	7.7 min	34.83 min	22 min

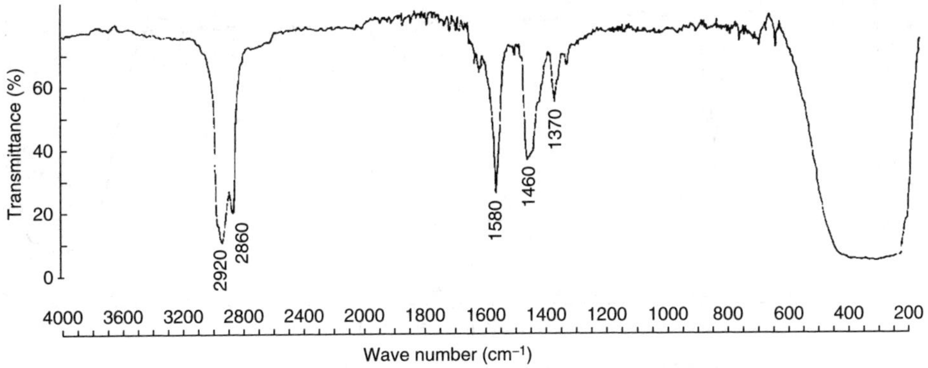

Fig. 3.8. IR spectrum of fresh grease 'A'

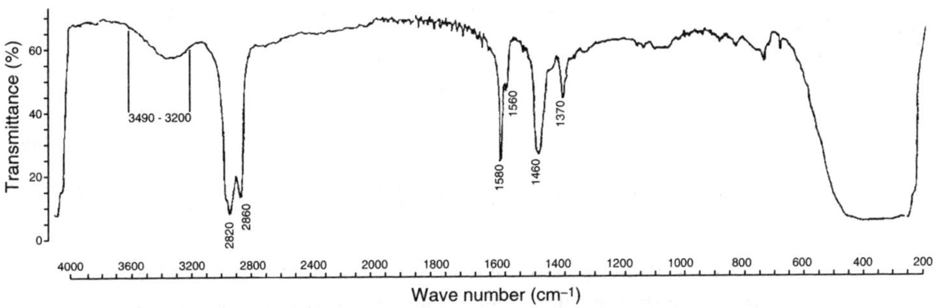

Fig. 3.9. IR spectrum of grease 'A' taken after **eighth** record on exposure to 50 V potential drop across electrodes

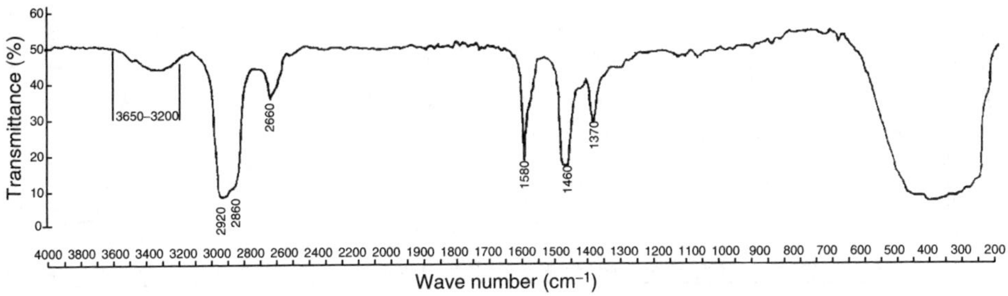

Fig. 3.10. IR spectrum of soap residue extracted from fresh grease 'A'

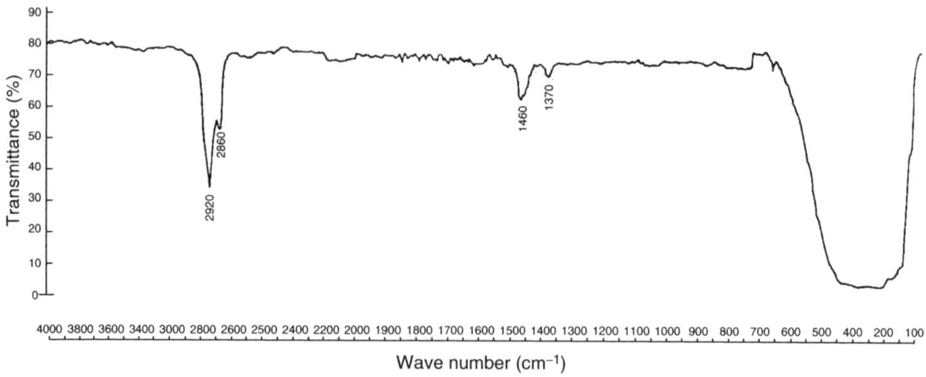

Fig. 3.11. IR spectrum of oil extracted from grease 'A'

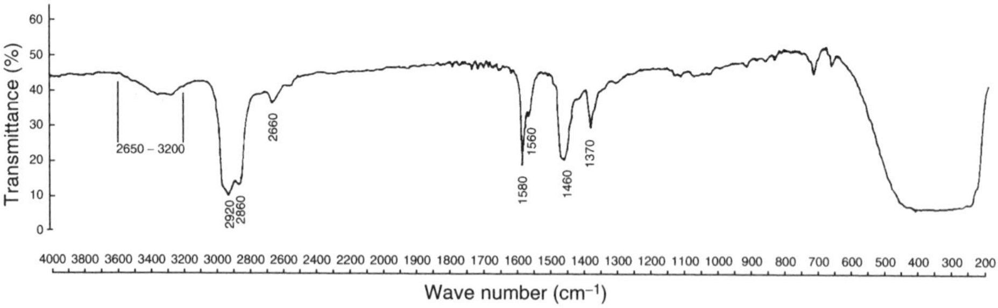

Fig. 3.12. IR spectrum of soap residue extracted from grease 'A' taken after **eighth** record on exposure to 50 V potential drop across electrodes

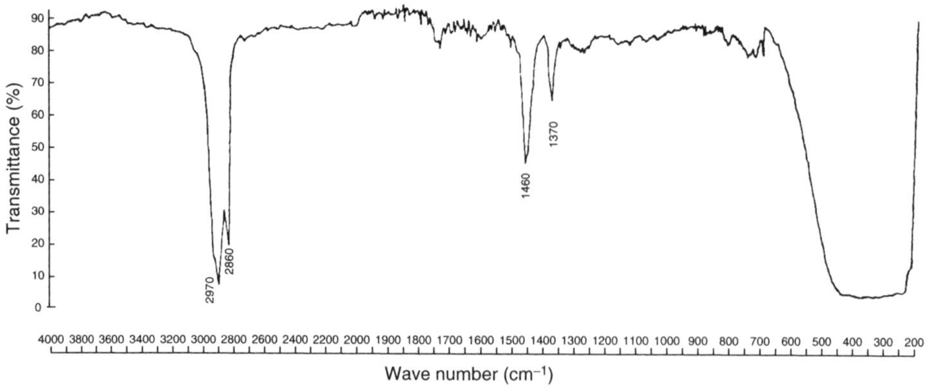

Fig. 3.13. IR spectrum of oil extracted from grease 'A' taken after **eighth** record on exposure to 50 V potential drop across electrodes

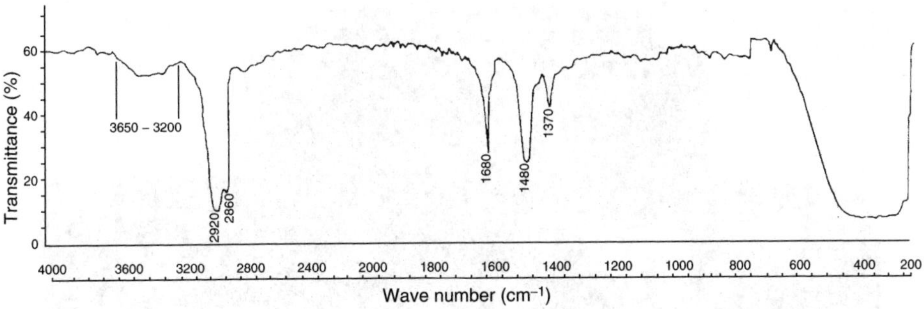

Fig. 3.14. IR spectrum of grease 'A' taken from NU 326 bearing after exposure to 50 A current (AC) for 41 h

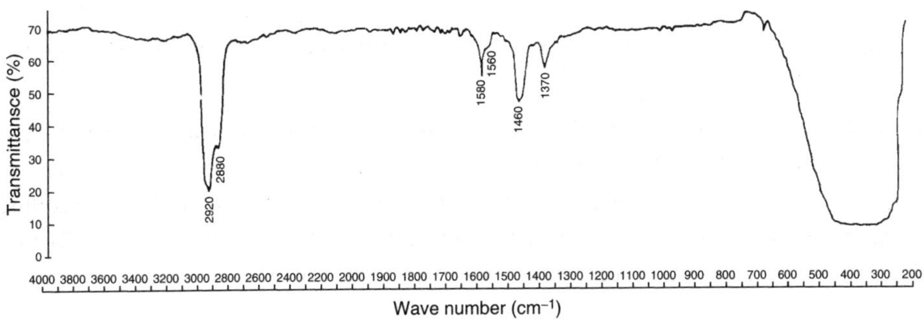

Fig. 3.15. IR spectrum of soap residue extracted from grease 'A' taken from NU 326 bearing after exposure to 50 A current (AC) for 41 h

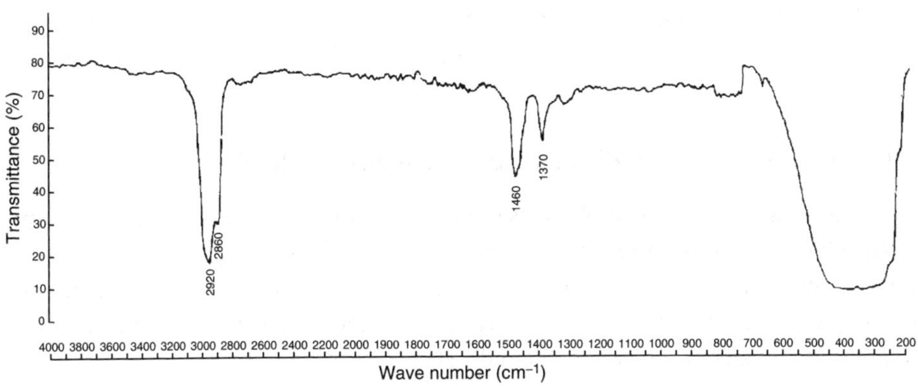

Fig. 3.16. IR spectrum of oil extracted from grease 'A' taken from NU 326 bearing after exposure to 50 A current (AC) for 41 h

Fig. 3.17. Inner race of NU 326 bearing after exposure of 50 A current (AC) for 101 h

to XRD investigations. Figure 3.18 shows XRD analysis of the soap residue of fresh grease. Figures 3.19 and 3.20 indicate XRD analysis of the soap residues of grease samples from the bearings after operation under electric fields and under pure rolling friction on the bearing test machine for duration of 250 h. Similarly, Figs. 3.21 and 3.22 illustrate analysis of the grease samples from AC motor bearing I and II, respectively.

XRD plots of different samples, thus obtained, were compared with standard cards to diagnose their chemical composition. Tables 3.7 and 3.8 show the experimental matching peaks d (Å) of the soap residues with those of the standard cards and their chemical compositions. The standard cards were chosen from the reference standards [23] to match closely with the minimum three peaks of the experimental XRD plots (Figs. 3.18–3.22).

3.3.5 Analysis by X-Ray Fluorescence Spectrometry (XRFS)

The soap residues of different samples were also analyzed by XRFS. Output from XRFS of the soap residue of the grease from the NU 326 bearing (operated under electric fields for a duration of 250 h) indicates Cu, and W peaks which are due to sample holder and the source, whereas Zn, Fe and Ni have been identified as the major constituents in the grease [20 and 21].

Table 3.7. X-ray diffraction (XRD) analysis of soap residues of different grease samples

Particulars	Fresh grease		Grease from NU 326 bearing (after operation under electric fields)				
			Matching standard cards			After 41 h	After 250 h
	Matching standard cards (4-0352)	Experimental analysis (XRD)	4-0385	17-937	24-684	Experimental analysis (XRD)	Experimental analysis (XRD)
Matching peaks (dA)			4.53				4.51
			4.34				4.39
	4.46	4.44					
	4.23	4.23	4.22			4.23	4.21
						4.20	
	4.11	4.11	4.12				4.12
					4.03	4.08	4.08
	3.97	3.97	3.97		3.97		3.95
		3.90				3.90	
		3.71	3.77			3.75	
				3.67			3.67
		3.56	3.54				
					3.18	3.16	3.16
	2.46	2.46	2.46	2.363			2.46
		2.34					
		2.24					
				2.184			2.16
	2.02	2.03	2.06	2.025		2.04	2.03
	1.75	1.75		1.771		1.78	
				1.602			
						1.54	
	1.45	1.43		1.481			
Chemical composition	Lithium stearate ($C_{18}H_{35}LiO_2$)	$C_{18}H_{35}LiO_2$	Lithium palmitate ($C_{16}H_{31}LiO_2$)	Gamma lithium iron oxide γ-$LiFeO_2$	Lithium zinc silicate $Li_{3.6}Zn_{0.2}SiO_2$	Transitional (unidentified)	(i) $C_{16}H_{31}LiO_2$ (ii) γ-$LiFeO_2$ (iii) $Li_{3.6}Zn_{0.2}SiO_2$
		Fig. 3.18					Fig. 3.19

Table 3.8. X-ray diffraction (XRD) analysis of soap residue of different grease samples

Particulars	Grease from NU 326 bearing (after operation without electric fields)			Grease from motor bearing I	Grease from motor bearing II
	Matching standard cards	After 41 h	After 250 h	After 6000 h	After 6000 h
Matching peaks (dA)	24-623	XRD analysis	XRD analysis	XRD analysis	XRD analysis
			4.46	4.39	
		4.26	4.27	4.27	4.26
		4.13	4.13	4.19	
	4.12	4.05	4.05	4.08	4.07
					3.96
		3.91	3.93	3.90	3.90
					3.67
	3.77	3.71	3.73	3.74	
			3.30		3.16
	2.65		2.5	2.5	2.46
		2.25	2.26	2.16	2.24
		2.03		2.03	2.03
			1.98		
		1.757			
		1.43			
Chemical composition	Li_5FeO_4 lithium iron oxide	$C_{18}H_{35}LiO_2$	(i) $C_{18}H_{35}LiO_2$ (ii) Li_5FeO_4 Fig. 3.20	(i) $C_{18}H_{35}LiO_2$ (ii) Li_5FeO_4 Fig. 3.21	$C_{18}H_{31}LiO_2$ γ-$LiFeO_2$ $Li_{3.6}Zn_{0.2}SiO_2$ Fig. 3.22

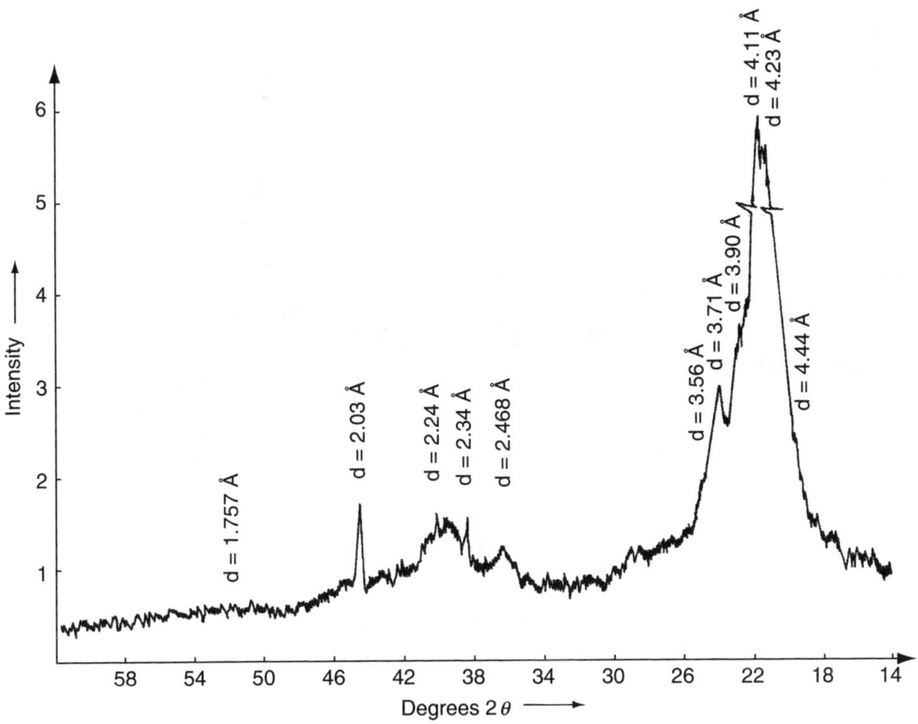

Fig. 3.18. X-ray diffraction pattern of soap residue of fresh grease

3.3.6 Analysis by Atomic Absorption Spectrophotometry (AAS) and Schmierstoff - / Lubricant - / Material (SRV) Test

To determine the change in the composition of the grease samples, the percentage of lithium metal present as lithium hydroxide and lithium carbonate were determined by an atomic absorption spectrophotometer of Pye Unicam-make. The percentage of lithium in an aqueous solution of a fixed quantity of different samples was compared with that of the fresh grease (Table 3.9).

The presence of metals Fe, Zn, Li, Ni, Cr, Si and their percentage in the ash extracted from different grease samples were also analyzed by AAS. The percentage of various metals in the ash extracted from the grease samples obtained after 101, 151 and 250 h were determined and given in Table 3.10.

Also, on SRV fresh and degraded grease samples from NU 326 bearing were evaluated under a 100 N load and a 1000 μm amplitude at 50 Hz frequency by running ball on flat for 2 h duration at 50°C. Minimum/ maximum friction coefficient, profile depth point and ball scar diameter were evaluated for grease samples

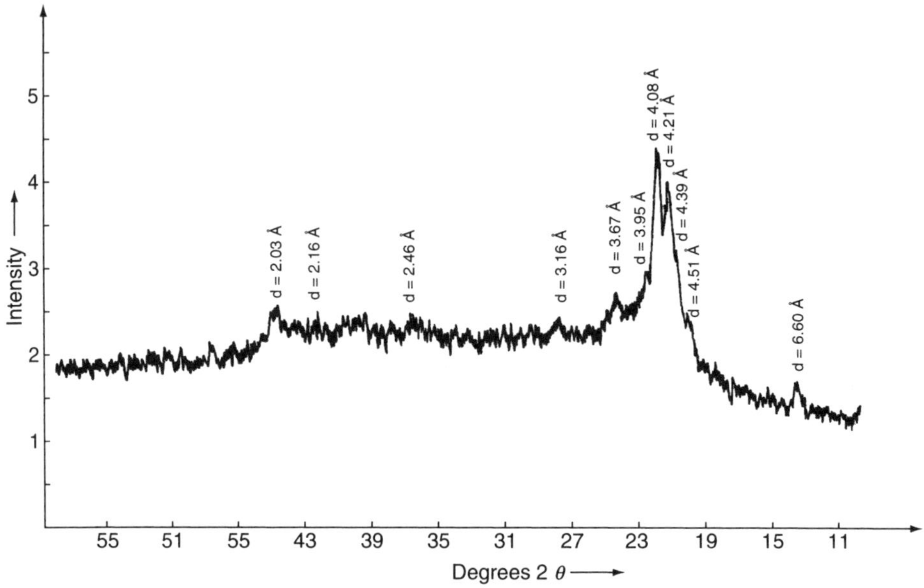

Fig. 3.19. X-ray diffraction pattern of soap residue of grease from NU 326 bearing after operation under electric fields for 250 h

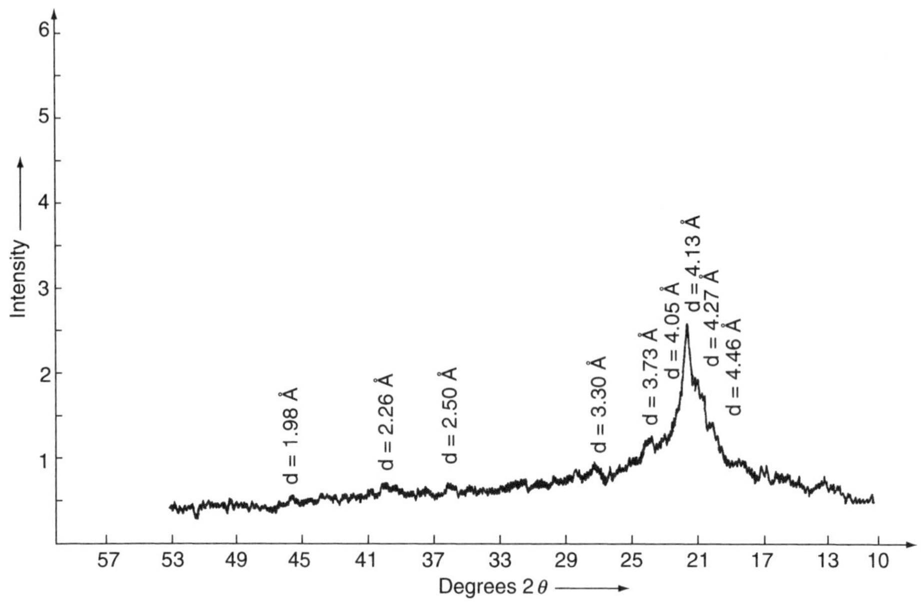

Fig. 3.20. X-ray diffraction pattern of soap residue of grease from NU 326 bearing after operation without electric fields for 250 h

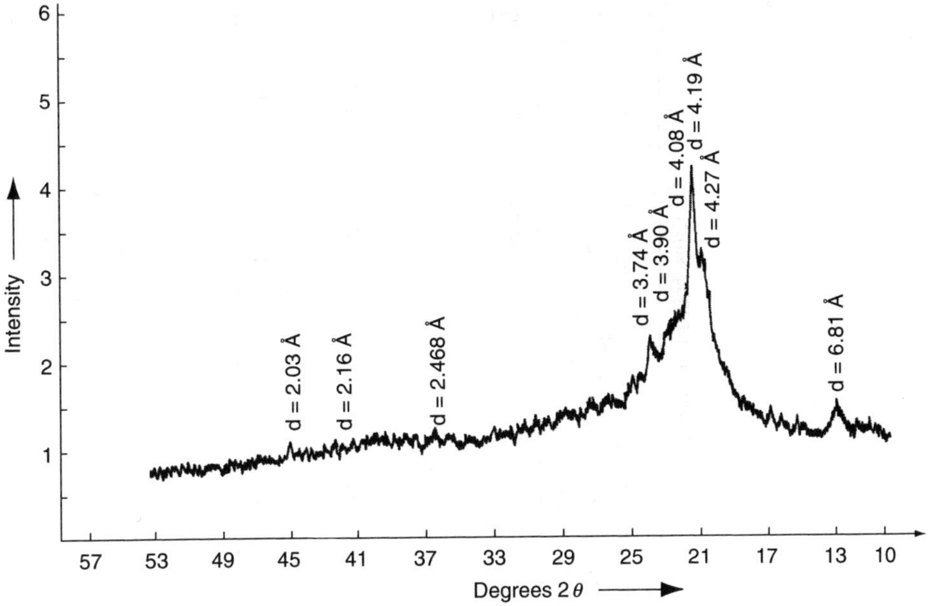

Fig. 3.21. X-ray diffraction pattern of soap residue of grease from NU 230 motor bearing I

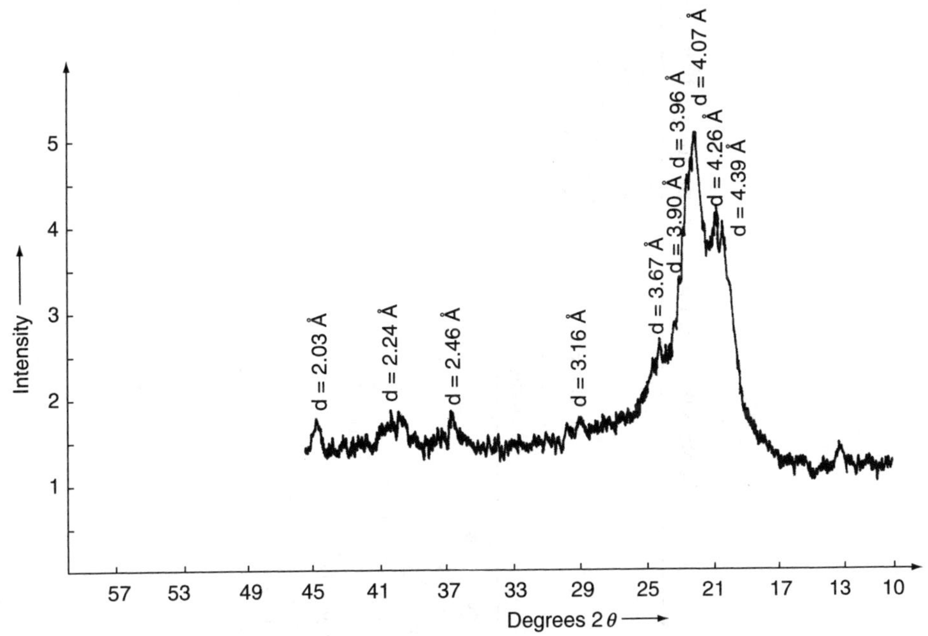

Fig. 3.22. X-ray diffraction pattern of soap residue of grease from NU 230 motor bearing II

Table 3.9. Atomic absorption spectrophotometry analysis of different grease samples

Test details	Fresh grease in NU 326 bearings	Duration (hours) after operation under electric fields in NU 326 bearing				Duration after operation without electric fields in NU 326 bearing				Fresh grease in motor bearings	From motor bearing I	From motor bearing II
		41	101	151	250	41	110	205	250		After about 6000 h of operation	
Lithium percentage as lithium hydro-oxide and lithium carbonate in aqueous solution of grease	0.0125	0.05	0.06	0.061	0.065	0.031	0.037	0.036	0.035	0.0067	0.014	0.026
Increase in lithium percentage as compared to fresh grease sample		4.0	4.8	4.9	5.2	2.48	2.96	2.88	2.80		2.1	3.8

Table 3.10. Analysis of gradual deterioration of the grease used in NU 326 test bearing

Typical test data	Fresh grease in NU 326 test bearing	Grease taken out from (in between) rollers			
		After 41 h	After 101 h	After 151 h	After 250 h
Ash content (%)	0.88	2.5	2.57	2.62	2.9
Increase in ash content as compared to fresh sample	–	2.84	2.92	2.96	3.30
Lithium (%) in aqueous solution of grease sample determined by atomic absorption spectrophotometry	0.0125	0.05	0.06	0.061	0.065
Increase in lithium (%) relative to that in fresh sample	–	4.0	4.8	4.9	5.2
Elements present in ash (wt. %)					
Fe	0.0045	–	0.5	0.52	0.66
Cu	0.0033	–	0.06	0.11	0.17
Zn	0.08	–	0.036	0.07	0.10
Ni	–	–	0.0014	0.073	0.0033
Al	0.065	–	0.00012	0.04	0.017
Li	0.15	–	0.135	0.13	0.12
Cr	Nil	–	0.011	0.04	0.04

(*Continues*)

Table 3.10. (*Continued*)

Typical test data	Fresh grease in NU 326 test bearing	Grease taken out from (in between) rollers			
		After 41 h	After 101 h	After 151 h	After 250 h
Si	Present	–	Present	Present	Present
SRV analysis at 100 N load (point contact) 1000 μm amplitude, 50 Hz frequency at 50° C for 2 h					
μ (minimum)	0.105	0.11	0.120	0.122	0.132
μ (maximum)	0.1125	0.115	0.125	0.188	0.40
Profile depth point (μm)	0.5	1.4	1.5	1.4	1.6
Ball scar diameter (mm)	0.52	0.64	0.65	0.62	0.66
Maximum increase in coefficient of friction relative to that for fresh grease	–	1.1	1.2	1.8	3.81
Increase in profile depth relative to that for fresh grease	–	2.8	3.0	2.8	3.2
Increase in ball scar diameter relative to that for fresh grease	–	1.23	1.26	1.2	1.27

after 41 h and 250 h of operation under the influence of electric current and given in Figs. 3.23 and 3.24, respectively. Table 3.10 gives the analysis of the gradual deterioration of the grease by AAS, SRV and elements present in ash used in NU 326 test bearing.

3.4 Discussion on the Results

3.4.1 Resistivity

It is evident from Tables 3.2–3.4, and Figs. 3.3–3.5 that the range of variation of resistivity of grease 'A' with respect to time (from 1 to 90 min) and potential drops (from 10 to 500 V) is 0.63×10^7 to 115×10^7 ohm cm, as against 0.033×10^{11} to 21×10^{11} ohm cm for grease 'B', and 4.62×10^8 to 25×10^8 ohm cm for grease 'C'. Variations of resistivity of grease 'A', at potential drops of 10, 50, 100, 250, 400 and 500 V, with time indicate similar trend, and show increase in resistivity at different rates (the maximum being 23.67 times at 50 V), as shown in Table 3.2. On the contrary, greases 'B' and 'C' indicate negligible variation (the maximum being 1.15 times) at potential drops of 10, 50 and 100 V. Only beyond 100 V, the variations show a significant increase in resistivity at different rates (the maximum being

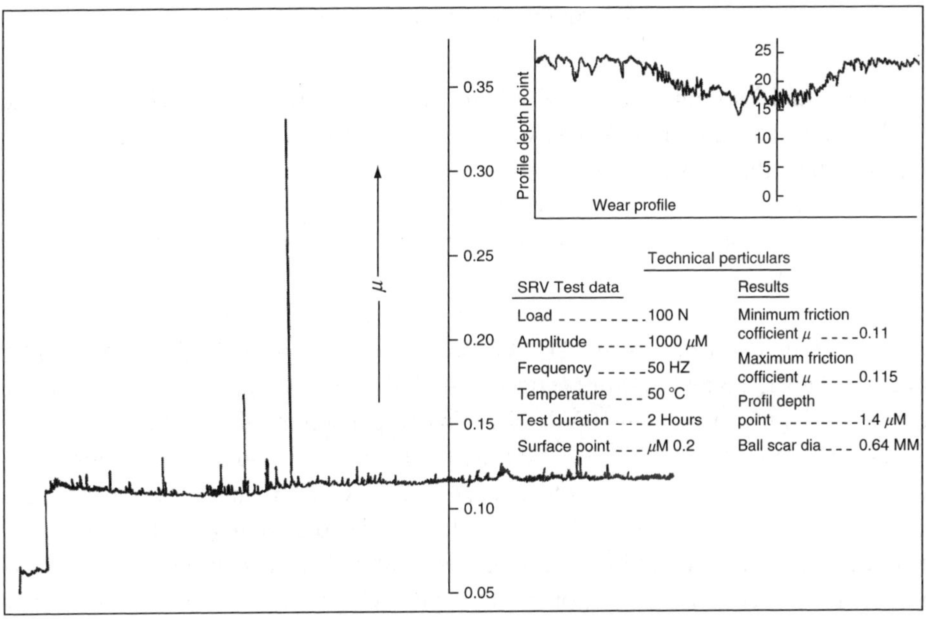

Fig. 3.23. SRV chart for grease 'A' taken from NU 326 bearing after exposure of 50 A current

(AC) for 41 h

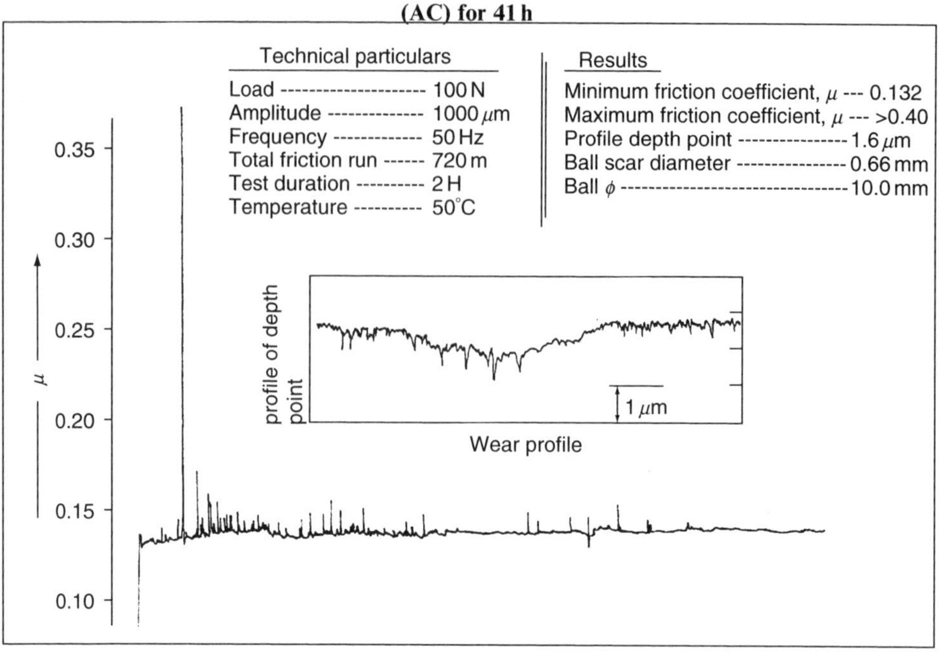

Fig. 3.24. SRV chart for grease 'A' taken from NU 326 bearing after exposure of 50 A current (AC) for 250 h

2.56 times and 2.46 times for grease 'B' and grease 'C', respectively). Greases 'B' and 'C' both indicate drop in resistivity with increase in potential drop from 10 to 250 V, but at 400 V, the resistivity rises. At 500 V, grease 'B' shows an increase, whereas grease 'C' shows a decrease in resistivity (Figs. 3.4 and 3.5). On the contrary, resistivity of grease 'A' always increases with increasing potential drop, as shown in Table 3.2 and Fig. 3.3.

The average ratio of resistivity of grease 'A' after 90 min to the initial value, at different potential drops, is 14.70 (Table 3.2) as against 1.41 and 1.29 for greases 'B' and 'C' (Tables 3.3 and 3.4) respectively. Similarly, the average ratio of resistivity of grease 'A' at 500 V to that at 10 V is 19.73, in comparison to 1.31 and 0.22 for greases 'B' and 'C', respectively. The above difference in resistivity and in its variation under the electric field is attributed to the greases 'B' and 'C' being much more stable than grease 'A'. Greases 'B' and 'C' show instability beyond potential drop of 100 V, whereas the grease 'A' exhibits instability even at 10 V. The continuous increase in resistivity of grease 'A' with applied potential drop and time is due to the electrochemical effect on the soap structure, which is much more significant than that of increase in density and the compressibility of the grease due to the forces exerted by the applied potential drop. This has been shown by the infrared and atomic absorption spectrophotometer analysis of the grease 'A' (Figs. 3.8–3.16 and Table 3.9).

Reduction in the resistivity of grease 'B' at 100 V (by 1.5 ohm cm) and that of 'C' at 50 V (by 1.45 ohm cm) and at 100 V (by 3.75 ohm cm) after 90 min may be explained on the basis of the nature of the impurities/byproducts, formed by the arc effect, which remained dissolved or in suspension causing a reduction of electric strength and resistivity, thus making it relatively more conductive. The reduction in resistivity of greases 'B' and 'C' at 250 V, as compared to the resistivity at 10 V, may also be attributed to the above effects. On the other hand, in the case of grease 'A', resistivity increases due to electrochemical decomposition (Figs. 3.3 and 3.6). Also, electric forces stretch the molecule structure till the critical stage, and transform the structure, affecting the resistivity of the grease. The change in the 'recouping' property of grease 'A' in statically bounded conditions confirms this behavior (Table 3.5) [22].

3.4.2 Recouping Properties

3.4.2.1 Recouping Properties under Statically Bounded Conditions

Variation of resistivity and percentage restoration of the original resistivity ('recouping' property) of grease 'A' with time, under 50 V (DC) potential drop across electrodes kept 2.5 mm apart, are shown in Table 3.5, and Figs. 3.6 and 3.7. It is evident that the resistivity increases from 2.32×10^7 to 481×10^7 ohm cm at the end of eighth record, and that on disconnecting the applied voltage and keeping the system idle for a number of hours, the resistivity of the grease tends to approach the original value (resistivity of 50×10^7 ohm cm attained after 90 min in the first record, the original resistivity being 2.32×10^7 ohm cm, reduced to 9×10^7 ohm cm in the second record taken after 24 h). If the grease is reexposed to the same potential drop, its resistivity again increases, but with different rates, as is evident from Table 3.5 which shows an increase in resistivity from 9×10^7 to 72×10^7 ohm cm in the second record, and from 60×10^7 to 254×10^7 ohm cm in the fourth record in 90 min. The 'recouping' property is found to be non-uniform. Even on keeping the grease unexposed to the potential drop for as long as 13 days, the original resistivity could not be restored; in fact, the resistivity of 260×10^7 ohm cm as observed in the fifth record, reduced to only 200×10^7 ohm cm in the next record taken after 13 days, as against the original resistivity of 2.32×10^7 ohm cm observed during the very first record. Initially, and after 90 min, resistivity in successive records is more than that in previous records: the initial resistivity varies from 2.32×10^7 to 330.0×10^7 ohm cm and, after 90 min, from 50×10^7 to 461×10^7 ohm cm, from the first record to the eighth record. The ratio of resistivity after 90 min to its initial value reduces with increase in the number of records (Table 3.5 and Fig. 3.7), and varies between 21.16 and 1.40. This indicates that the increase in resistivity with time is reduced if the grease is exposed to the potential drop for sufficient time. This may be attributed to the fact that the rate of stretching of the molecules reduces with time. The decrease in percentage restoration of resistivity, from 82 to 17.9 per cent (Table 3.5), confirms that the stretching of the molecules becomes increasingly permanent with the duration of exposure [18, 22].

The variation in ratio of resistivity after 90 min to the initial value at 50 V potential drop for greases 'A', 'B' and 'C' – 23.67 (Table 3.2), 1.03 (Table 3.3) and 0.91 (Table 3.4) respectively – indicates that molecule stretching phenomenon and 'recouping' property are prominent in case of grease 'A', and negligible in case of greases 'B' and 'C'. This is so, as the electric force at this potential drop is not sufficient to disrupt the insulant properties of greases 'B' and 'C', contrary to grease 'A'. That is why experiments pertaining to 'recouping' property were conducted only on grease 'A'.

3.4.2.2 Bearing Impedance and 'Recouping' Phenomenon of Grease 'A' in the Bearing

'Recouping' phenomenon, similar to 'under statically bounded conditions', was observed on the Roller Bearing Test Machine in the first few hours of operation during testing of NU 326 bearing. The impedance of the bearing was found to have increased from approximately 20 ohms to 3000 ohms, on switching off the current after the bearing had been operated at the rated parameters for about 10 h. After keeping the bearing idle for 24 h, the impedance reduced to 250 ohms. Subsequently, after again passing current for some time, the impedance of the bearing was found to have reduced to 15 ohms, on switching off the current to the bearing. It is obvious that the 'recouping' phenomenon ceased to exist due to the decomposition of the grease. Under these conditions, the bearing impedance was found to decrease with the rise in bearing temperature, and increase with the decrease in the grease temperature, on keeping the bearing idle. However, the bearing impedance significantly reduced to 0.024 ohm on passage of current through the bearing. Thus, the change in the impedance of the bearing indicates stretching of the molecules and decomposition of the grease under the electric fields [19, 22].

The considerable difference between the impedance of the bearing under operation (15–3000 ohms) and the resistance of the grease in the statically-bounded condition (of the order of 10^7 ohms) may be attributed to the asperities of the opposite metal surfaces in the bearing, small film thickness, and the shear of the grease in between the rotating interacting surfaces. However, the bearing impedance changes with respect to the film thickness and electrochemical behavior of the rheologically complex lubricating greases, although the resistance of the bearing in static conditions was found to be about 1 ohm, and the impedance, on passing AC or DC current in the presence of grease 'A' in the bearing, was measured to be around 0.024 ohm.

3.4.3 Properties and Performance Characteristics of Lubricants

3.4.3.1 Physicochemical Properties and Functional Performance Characteristics

Comparison of physicochemical properties and functional performance characteristics of the greases (Table 3.6) shows that the values of viscosity and oxidation stability for grease 'B' are higher than those for greases 'A' and 'C'. Grease 'A' exhibits the lowest temperature torque (50 g cm) and unworked consistency (236). Oil separation for grease 'B' is 0.21 per cent at 130°C as compared to 0.44 per cent and 1.21 per cent for greases 'A' and 'C', respectively. The average time taken to initiate wear scar on balls of a 4-ball wear tester at 2000 rpm at 100 kgf load, for grease 'A', is only 7.7 min, as compared to 34.83 min and 22.0 min for greases 'B' and 'C', respectively, which indicates that the extreme pressure (EP) properties of grease 'A' are much inferior to those of grease 'B' or grease 'C' (Table 3.6).

It has also been observed that the ash content of grease 'A', when used in NU 326 bearing, increased from 0.86 to 2.5 per cent even after 41 h of operation of the bearing under the electric fields.

3.4.3.2 Correlation between functional performance characteristics and resistivity of greases

On analyzing the functional performance characteristics and resistivity of greases, it may be deduced that resistivity of the greases is mainly influenced by EP properties alongwith viscosity, torque characteristics and consistency. The resistivity of grease 'A' is the lowest, compared to either grease 'B' or grease 'C', which may be related to these functional performance characteristics, apart from the difference in structure, behavior of additives and the processes used for producing these greases.

3.4.3.3 IR Spectra

The IR spectra of a sample of fresh grease 'A', and of a sample of grease 'A' taken after the eighth record on exposure to 50 V (DC) potential drop, and of a sample of grease 'A' from NU 326 bearing after exposure to 50 A (AC) current for 41 h (shown in Figs. 3.8, 3.9 and 3.14 respectively) indicate strong C–H bending absorption at 1460 cm^{-1} and two absorptions at 1580 cm^{-1} and 1370 cm^{-1}. These absorptions are caused by symmetric and asymmetric >C = O stretching vibrations of the resonance-stabilized carboxylate anion group. Two bands indicating weak stretching vibrations are also observed at 2920 cm^{-1} and 2860 cm^{-1}, due to symmetric and asymmetric stretching of C–H and C = O bonds. Comparison among Figs. 3.8, 3.9 and 3.14 clearly shows that the percentage transmittance due to absorption at 1580 cm^{-1}, related to carboxylate anion stretching, as against at 1460 cm^{-1} related to C–H bending, is the lowest in the sample from NU 326 bearing, as compared to the statically bounded sample and the fresh grease sample. O–H stretching vibrations including free O–H, inter-molecular hydrogen-bonded single-bridge compounds and polymeric association are observed in the region 3650–3200 cm^{-1} in electrically exposed greases (Figs. 3.9 and 3.14). This shows that the carboxylic group, in both statically-bounded grease and grease from a bearing, is decomposed under the influence of an electric field. The appearance of an additional peak at 1560 cm^{-1}, as observed in Fig. 3.9, suggests a change in grease structure of statically bounded and electrically exposed grease 'A', under 50 V potential drop, and this might have resulted in increase in its resistivity (Tables 3.2 and 3.5).

The IR spectra of oil extracted from these three samples (shown in Figs. 3.11, 3.13, and 3.16 respectively) indicate the absence of O–H stretching vibration in the region 3650–3200 cm^{-1} and absorption due to bonded O–H stretching in the region near 2600 cm^{-1}. Only C–H bending at 1370 cm^{-1} and 1460 cm^{-1} is indicated, thereby suggesting that oil in the greases under these conditions has not been affected at all.

The IR spectra of the soap residue extracted from these samples, shown in Figs. 3.10, 3.12 and 3.15 respectively, indicate C–H bending at 1370 cm^{-1} and 1460 cm^{-1}. It also reveals that the carboxylate anion stretching at 1580 cm^{-1} is the least in the sample of grease from the bearing (Fig. 3.15), as against C–H bending at 1460 cm^{-1}. The peaks observed at 1370 cm^{-1} and 1460 cm^{-1} are also contributed by Nujol used for smearing on the KBr (Potassium Bromide) window for taking the IR spectra.

The appearance of peak at 2660 cm^{-1} in soap residues from the samples of fresh grease and statically-bounded electrically exposed grease, as indicated in Figs. 3.10 and 3.12 respectively, is the diagnostic of a carboxylic acid functional group. It is evident that the percentage transmittance of carboxylic acid for fresh grease is higher than that for the statically-bounded grease, while it is negligible in the sample taken from the bearing (Fig. 3.15). This suggests complete decomposition of carboxylic acid in the soap residue of the sample from the bearing rollers, partial decomposition in the soap residue of the sample of statically bounded grease, and the presence of carboxylic acid function group in the soap residue of the sample of fresh grease. The complete decomposition of carboxylic acid led to corrosion of the surface of the NU 326 bearing, as shown in Fig. 3.17. In case of the statically bounded grease, the byproducts of the partial decomposition of carboxylic acid remained unutilized and dissolved or suspended, which might have resulted in the change in its resistivity. The presence of peak at 1560 cm^{-1}, as seen in Figs. 3.12 and 3.15, suggests a change in the soap structure, which also might have contributed to the change in resistivity [18, 19].

The variation in O–H group in the region 3650–3200 cm^{-1} observed in IR spectra of soap residues, may be related to the process of separation of oil and soap contents from the respective greases.

3.4.3.4 Condition of Bearings after Test

The bearings NU 326, after testing separately, under electric fields and pure rolling friction for duration of 250 h, were examined. The inner race of the bearing tested under the electric field was found to be fully corroded with corrugations, whereas corrosion and corrugations were not observed on the bearing tested under pure rolling friction without the passage of current. The bearing I, NU 230, used in the motor was found to be corroded after 6000 h of operation but was not corrugated. On the contrary, the bearing II, NU 230, used in another motor was fully corroded and corrugated.

3.4.3.5 XRD and XRFS Analysis of Greases from NU 326 Bearings

From the XRD analysis as illustrated in Table 3.7 and Fig. 3.18, it is established that chemical composition of the fresh grease – before filling in the bearings – is lithium stearate ($C_{18}H_{35}LiO_2$). After operation of the bearing under electric fields for duration of 41 h, the chemical composition of the grease undergoes changes from lithium stearate (intermediate changes). Faint corrosion on the bearing surfaces indicates decomposition of the grease, which is also supported by four times increase in the relative percentage of free lithium (Table 3.9).

After 250 h of operation, the chemical composition of the fresh grease in the bearing changes to lithium palmitate ($C_{16}H_{31}LiO_2$) and peaks of gamma lithium iron oxide (γ-$LiFeO_2$) and lithium zinc silicate ($Li_{3.6}ZnO_{0.2}SiO_2$) are found (Table 3.7 and Fig. 3.19). Besides this the bearing surfaces were found to be fully corroded. On the contrary, no change in the chemical composition of the grease was detected after 41 h of operation of the NU 326 bearing under pure rolling friction without passage of current. However, after 250 h of operation the formation of lithium iron oxide (Li_5FeO_2) was detected, but chemical composition still remained as lithium stearate (Table 3.8). Moreover no corrosion on the bearing surfaces was found, which suggests that the grease was not decomposed. However, crystalline structure of the fresh grease (Fig. 3.18) changes to amorphous structure under pure rolling friction as well as in a bearing operating under electric fields (Figs. 3.19–3.22).

It may be noted that during interaction of surfaces under rolling friction, individual components of the lubricating medium react with bearing metal to form a layer of secondary structures, which may get destroyed by normal and tangential stresses. That is why the number of components taking part in the reaction or formed in the process of friction will be changed. Thus, the exact final compound in the grease will not be easy to identify (as in case of the grease from the bearing after 41 h of operation under electric fields). And also, it might be difficult to identify the exact matching composition of the formed compounds in presence of a number of components of the lubricating media and overlapping/close existence of peaks of various compounds (Tables 3.7 and 3.8) [20–22].

The role of oxygen is very significant in contact interacting surfaces. It has been found that oxygen takes an active part in the process of friction. The contact stresses, temperature, composition of lubricating medium and impurities greatly influence the rate of oxygen consumption during friction. The increase of contact stresses leads to a corresponding increase in the rate of oxygen consumption [17]. That is why oxides of the metals i.e. lithium iron oxide is formed easily in rolling friction as shown in Tables 3.7 and 3.8.

The decomposition of the zinc additive, i.e. zinc dithiosphosphate or zinc dialkyl dithiosphosphate under the influence of electric fields has lead to the formation of lithium zinc silicate in presence of high relative percentage of free lithium and silica impurity in the grease under high temperature in the asperity contacts (Table 3.10). On the contrary, lithium zinc silicate compound has not been detected in the grease from the bearing operated without the influence of an electric field (Table 3.8) [20, 21].

The formation of lithium iron oxide and lithium zinc silicate also contribute to medium-metal interaction, rate of chemical reaction, affinity of lubricating medium components for the metal, availability of the free metal and the temperature. Besides this, reactions having fast rates will be more energetic and carried out to a higher extent. The rate of reaction also determines the end composition of interacting products and their characteristics.

The detection of Fe, Zn, Ni, Li, Cr, Si by XRFS and AAS in the grease samples also suggests the formation of lithium iron oxide and lithium zinc silicate in presence of oxygen (Table 3.10).

3.4.3.6 XRD Analysis of Greases from Motor Bearings

XRD analysis of the grease from bearings I and II of the AC motors indicate that the original chemical composition of lithium stearate of fresh grease is not changed in the bearing I, but it is changed to lithium palmitate in the bearing II (Table 3.8).

Furthermore, corrosion and corrugation patterns on the surfaces of bearing II are found, and peaks of gamma lithium iron oxide and lithium zinc silicate are detected. However, peaks of lithium iron oxide have been detected from the grease of bearing I (Fig. 3.21). Comparison of Fig. 3.22 with that of Fig. 3.19 indicates that the motor bearing II was exposed to electric fields, and the comparison of Fig. 3.21 with that of Fig. 3.20 suggests that motor bearing I was not exposed to electric current.

3.4.3.7 Analysis of Data from Atomic Absorption Spectrophotometry (AAS) and Schmierstoff - / Lubricant - / Material (SRV)

The AAS analysis illustrates that change in relative percentage of free lithium is from 4 to 5.2 for a corresponding duration of 41 and 250 h of operation under electric fields, whereas without the influence of electric fields this change is only from 2.48 to 2.88 (Table 3.9). This indicates that the change in the chemical composition of the greases is more significant under the influence of electric fields than in pure rolling friction. The corrosion of bearing surfaces also justifies this conclusion (Fig. 3.17). Furthermore, it was found that the relative percentage of lithium in aqueous solution of a fixed quantity of grease is 1.6 times higher in the case of statically-bounded grease as compared to fresh grease. The relative percentage of free lithium would have been much higher than 5.2, if the bearing surfaces were not corroded [25].

The absence of corrosion on the surfaces of the bearing (tested without current) and the low value of relative percentage of lithium (maximum 2.96) in the grease show the effect of pure rolling friction. However, the bearing surfaces would have been corroded after prolonged operation, due to the significant change in the composition of the grease. Relative percentage of free lithium (2.1), corrosion of bearing surfaces and XRD analysis of the grease from motor bearing I suggest the effect of rolling friction after 6000 h of operation (Fig. 3.21). Whereas higher relative percentage of free lithium (3.8), XRD analysis of the grease from motor bearing II and complete corrosion of the surfaces indicate the exposure of bearing to the current (Table 3.9 and Fig. 3.22).

A similar trend is observed in the grease samples taken from different NU 230 motor bearings. This indicates a rapid and extensive deterioration and ionization of the low resistivity lithium-base grease in the initial hours of operation of the bearing. A similar trend is shown by the increase in ash content by 2.84–3.20 times after 41 and 250 h of operation, respectively, compared with the fresh grease sample (Table 3.10). SRV analysis of the greases (Figs. 3.23 and 3.24) shows an increase in the coefficient of friction by 1.1–3.8 times after 41–250 h of operation compared with that of the fresh grease sample. SRV analysis also shows an increase in profile depth by 2.8–3.2 times and in ball scar diameter by 1.23–1.27 times during this period compared with the fresh grease sample.

The presence of moderate levels of silicon in the grease, as indicated in Table 3.10, suggests dirt or dust contamination. It may be due to a silicone-base fluid or clay thickener. The increase of 0.5–0.66 per cent in iron levels after 101–250 h of operation suggest not so much wear but rust or fretting corrosion (Figs. 3.7, 5.1–5.5). Metals such as chromium and nickel, but not zinc, present in the grease samples can be attributed to wear metals (Table 3.10). The presence of zinc may be related to the low level of lubricant additive, e.g. zinc dithiophosphate or carbonate. A reduction in the percentage of lithium (from 0.15 to 0.130 to 0.12) in the ash form of the used grease relative to that of fresh grease after 151–150 h of operation may be attributed to amalgamation of the lithium to the bearing surface.

3.4.4 Decomposition of Lubricants and Bearing Failure

The capability of a lubricant to carry the load at the point of contact of the friction surfaces in a bearing depends on the degree of lubricant molecule orientation and by the bond of the molecules to the bearing interacting surfaces [25]. When this bond is disturbed – under the influence of electric current, rolling friction or shear – the lubricating film is destroyed. Under these conditions the lubricating grease loses its property to resist high pressures and capability to separate friction surfaces. This leads to changes in the structure of the grease, change of higher hydrocarbon molecules to low hydrocarbon molecules, increase in percentage of free lithium in the lubricating medium and formation of other compounds as shown in Tables (3.7 and 3.8). The 'recouping' of the resistivity of the grease under electric field suggests the stretching of the molecules. Zero 'recouping' indicates the change of molecules of higher hydrocarbons to low hydrocarbons. This finally leads to contact of mechanical surfaces, which causes a sharp increase of the coefficient of friction and heating of the bearing surfaces. The outer race temperature as high as 90°C has been detected when the bearing was operated under electric field. And the bearing surfaces under high contact pressure were found to be corrugated.

The chemical process leading to the formation of lithium iron oxide, lithium hydroxide, lithium carbonate, lithium zinc silicate and low hydrocarbon molecules is explained as under:

$$RCOOLi \rightarrow Li^{+}RCOO^{-}$$
$$RCOOLi \rightarrow Li^{+} + R_1COO^{-}$$

(R is $C_{17}H_{35}$ or $C_{15}H_{31}$, and R_1 is $C_{15}H_{31}$)

$$H_2O \rightarrow H^{+} + OH^{-}$$
$$RCOO^{-} \rightarrow CO_2 + \text{Organic Compound}$$
$$CO_2 + H_2O \rightarrow H_2CO_3$$
$$H_2CO_3 \rightarrow 2H^{+} + CO_3^{2-}$$
$$2Li^{+} + CO_3^{2-} \rightarrow Li_2CO_3$$
$$Li^{+} + OH^{-} \rightarrow LiOH$$
$$Li^{+} + Fe^{+} + O_2 \rightarrow \gamma LiFeO_2$$
$$LI^{+} + Fe^{+} + O_2 \rightarrow Li_5FeO_2$$
$$Li^{+} + Zn^{+} + SiO_2 \rightarrow Li_{3.6}Zn_{0.2}SiO_2$$

The formation of lithium hydroxide (LiOH) and lithium carbonate (Li_2CO_3) make the dielectric alkaline in nature and corrode the bearing surfaces. It also increases the percentage of lithium in the aqueous solution (Table 3.9). The percentage increase in lithium content is correlated to the corresponding decrease in carboxylic group as indicated in infrared spectra in contrast to the fresh grease and the soap residue.

It may be noted that the process leading to the damage of the bearing is usually caused by the presence of oxygen and water. Corrosion on the bearing surfaces develops as an array of pits, which often lowers the fatigue life of a bearing causing corrugations on the surfaces under the influence of electric fields. Similar corrugation patterns are observed on the motor bearing II.

In view of the above, the mechanism of a bearing failure under the influence of electric current is explained as follows:

When current passes through a bearing, silent discharge occurs through the inner race, rolling-elements and outer race that causes to electrochemical decomposition of the low-resistivity lithium grease (10^7 ohm cm) and subsequently corrosion of the bearing surfaces, which finally leads to increased wear and high local temperatures on asperities, before pitting starts on the surfaces and ultimately resulting failure of the bearings. This effect is accompanied by an increase in the lithium content of the grease and change in its chemical composition accompanied by the formation of lithium iron oxide and lithium zinc silicate compounds.

On the contrary, a bearing under pure rolling friction does not corrode the surfaces in the initial stages, and also change in lithium content of the grease is much lower. Furthermore, no change in chemical composition takes place. However, formation of lithium iron oxide, corrosion of the surfaces and change of crystalline structure of the grease to amorphous leads to bearing failure after prolonged operation.

3.5 Conclusions

From the various investigations and analyses reported in this Chapter, the following conclusions are drawn [18–22]:

(1) The resistivity of grease depends on its EP properties, viscosity, torque characteristics and consistency.

(2) Resistivities of greases vary with respect to the applied voltages and time. The difference in resistivity among greases can be as high as 10^5 times.

(3) The change in resistivity with time and applied voltage may depend on the nature of the impurities or byproducts and the types of additives in the grease, besides its density, compressibility and structure.

(4) The low-resistivity greases tend to 'recoup' their resistivities when the applied electric field is switched off. In the case of grease 'A', the percentage of 'recouping' varies from 18 to 82, and depends upon the stretching of the molecules.

(5) Under the influence of the electric field, carboxylic group at $2660\,cm^{-1}$ in the low-resistivity greases decomposes and lithium metal concentration in an aqueous solution relatively increases. When grease is used in a bearing, the decomposition of carboxylic acid leads to corrosion of the bearing surfaces before the pitting process is initiated, whereas, in statically bounded greases, the byproducts of decomposition remain dissolved/suspended, which affect their resistivities.

(6) The oil content of the grease is not affected by the electric field, but the carboxylate anion stretching and carboxylic group, present in the soap residue, undergo changes.

(7) X-ray diffraction analysis indicates that the structure of the fresh lithium grease is lithium stearate ($C_{18}H_{35}LiO_2$), which changes to lithium palmitate ($C_{16}H_{31}LiO_2$); the lower fraction of hydrocarbons after operation of a bearing under electric fields. Whereas the original structure of the grease is not changed under pure rolling friction.

(8) After prolonged operation, crystalline structure of a fresh lithium grease in a bearing changes to amorphous structure.

(9) When the bearing is operated under electric fields, the lithium grease gets decomposed and gamma lithium iron oxide (γ-$LiFeO_2$) and lithium zinc silicate ($Li_{3.6}Zn_{0.2}SiO_2$) are formed in presence of Zn, Fe and SiO_2.

(10) Under pure rolling friction, lithium iron oxide (Li_5FeO_2) is formed in the presence of free Li and Fe in the grease.

(11) A percentage increase of free lithium in aqueous solution is about two fold in the grease from a bearing operated under electric fields than under pure rolling friction.

(12) During bearing operation the free lithium percentage in the grease increases and then the formation of lithium iron oxide takes place.

(13) The formation of lithium hydroxide and lithium carbonate make the dielectric alkaline and corrode the bearing surfaces, which finally leads to increased wear and failure of a bearing.

(14) The coefficient of friction, profile depth, ball scar diameter, ash content and percentage of lithium in grease increase rapidly in the beginning under the influence of electric fields and later the process with time is slowed down.

The present studies of XRD, XRFS, IR, SRV and AAS of greases, alongwith the study of bearing surfaces, have a potential to diagnose the bearing damage under the influence of electric fields, and also to establish the deterioration of the lithium greases used in the bearings. In the case of bearings using grease of low-resistivity, bearing failure occurs under the 'silent' electric discharge due to chemical decomposition, while with grease of high-resistivity, the failure may occur through accumulation of charges and their subsequent breakdown.

REFERENCES

1. Dyson, A. and Wilson, A. R. (1969–70). Film thickness in elasto-hydrodynamic lubrication of roller by greases. *Proc. Inst. Mech. Engrs.*, **184**, Part 3F, 1.
2. Jonkisz, W. and Krezeminski-fred, H. (1979). Pressure distribution and shape of an elasto-hydrodynamic grease film. *Wear*, **55**, 81.
3. Horth, A. C., Sproule, L. W. and Patterden, W. C. (1968). Friction reduction with greases. *NLGI Spokeman*, **32**, 155.
4. Polacios, J. M. (1980). Elasto-hydrodynamic films of lithium greases, *Macanique, Materiaux, Electricite (GAME)* 365–366, 176, (also published in *NLGI Spokeman*, March 1981).
5. Akhmatov, A. S. (1966). Molecular physics of boundary friction. Jerusalem: Program For Scientific Translations.
6. Langborne, P. L. Grease lubrication: A review of recent British papers. *Proc. Inst. Mech. Engrs.*, **184**, Part 3F, 82.

7. Barnett, R. S. (1970). Review of recent USA publications on lubricating grease. *Proc. Inst. Mech. Engrs.*, **184**, Part 3F, 87. Also in *Wear*, **16**, 87.
8. Godfrey, F. (1964). Friction of greases and grease components during boundary lubrication. *ASLE Trans.*, **7**(24).
9. Horth, A. C. Sproule, L. W. and Pattenden, W. C. (1968). Friction reduction with greases. *NLGI Spokeman*, **32**, 155.
10. Lander, W. (1965). Hydrodynamic lubrication of proximate cylindrical surfaces of large relative curvature. *Proc. Inst. Mech. Engrs.*, **180**, Part 3B, 101.
11. Dowson, D. and Whitaker, A. V. (1965). The isothermal lubrication of cylinders. *ASLE Trans.*, **8**, 224.
12. Dowson, D. and Whomes, T. L. (1966). Side leakage factors for a rigid cylinder lubricated by an iso-viscous fluid. *Proc. Inst. Mech. Engrs.*, **181**, 165.
13. Pastnikov, S. N. (1978). *Electrophysical and Electrochemical Phenomena in Friction, Cutting and Lubrication*, VNR.
14. Matveyevsky, R. M., Markov, A. A. and Buyanovsky, I. A. (1973). *Proc. of Conf. on physico-chemical mechanics of contact interaction and fretting corrosion* (in Russian), 46, Kiev.
15. *Proc. of Conf. on electrochemical proceses in friction and their use against wear* (in Russian). Odessa, 1973.
16. Postnikova, S. N., Avdonin, I. M., Gromyko, G. G., Zinkin, Yu. I. and Polunichev, A. I. (1973). In friction, lubrication, wear (in Russian). Translated by Gork. Politech. Inst. **29**(9) 29.
17. Aksyenoy, A. F., Beliansky, V. P. and Shepel, A. Y. (1981). On some aspects of the interaction mechanism between hydrocarbon liquid and metals during friction. 3rd Inter. Trib. Cong. Eurotrib 81, Lubricant and their applications, III, 7–16.
18. Prashad, H. (1996). Experimental study on influence of electrical fields on behaviour of grease in statically bounded conditions and when used in non-insulated bearings. *BHEL J.* **7**(3), 18–34.
19. Prashad, H. and Murthy, T. S. R. (1988). Behaviour of greases in statically bounded conditions and when used in non-insulated anti-friction bearings under the influence of electrical fields. *Lubric. Eng.*, **44**(3), 239–246.
20. Prashad, H. (1989). Diagnosis of lithium greases used in rolling-element bearings by X-ray diffractometry. *STLE Trans.*, **32**(2), 205–14.
21. Prashad, H. and Murthy, T. S. R. (1998). Deterioration of lithium greases under the influence of electrical current – An investigation. France, 10–4 August J. *Lubric. Sc.*, **10**, 323–342.
22. Prashad, H. (1998). Variation and recovery of resistivity of greases – An experimental investigation. 11–1, France, November J. *Lubric. Sc.*, **11**, 73–103.
23. *Powder diffraction File, Search Manuals, Organic/Inorganic Compounds*, Publication SMO-27, Published by JC PDS, International Centre for Diffraction Data, 1601 Parklane, Swarthmore, PA 19081, USA (1977).

24. Prashad, H. (1994). Investigations of damaged rolling-element bearings and deterioration of lubricants under the influence of electrical fields. *Wear*, **176**, 151–161.
25. Gersimov, M., Novacheva-Ranheva, A. and Staynov, S. T. (1981). Study of fatty and oily base effect on the lubricating properties of lithium lubricants. 3rd Inter. Trib. Cong. Eurotrib 81, Lubricant and their application, III, 102–113.

Chapter 4

THRESHOLD VOLTAGE PHENOMENON AND EFFECT OF OPERATING PARAMETERS ON THE THRESHOLD VOLTAGE IN ROLLING-ELEMENT BEARINGS

4.1 Bearing Threshold Voltage

Threshold voltage phenomenon exists in the bearings operating under the influence of shaft voltages, similar to the voltage-current characteristics of an electric arc [1]. The current does not begin to flow in the bearings until a threshold voltage is reached. There occurs first and second threshold voltages in the bearings under the effect of shaft-voltage. At the first threshold voltage (the increase in voltage from zero to a specific maximum value with zero current), the increase in flow of current is momentary. After the first threshold voltage is reached, even a further slow increase in voltage results in a very high increase in current. This is termed as the second threshold voltage in the bearings.

This chapter discusses the studies undertaken to understand the behavior of impedance of the rolling bearings, under the influence of varying levels of electric current, using lubricants having different resistivities. The effect of the operating parameters and lubricant characteristics on the threshold voltages and threshold coefficients of the bearings to determine the safe levels of voltage, is analyzed in order to preclude damage of the bearings. A method for the determination of the lubricant film thickness on the basis of the aforementioned investigations is suggested.

4.2 Test Set-Up and Investigations

4.2.1 The Experimental Set-Up and Electric Circuits

The bearing test machine shown in Fig. 3.2 is used for studying the threshold voltage phenomenon in bearings using two different grades of lubricants 'A' and 'B'. The detailed performance characteristics and physicochemical properties are given in Table 3.5. For generating different levels of potential difference across the test bearing, a variable transformer, type 160, no.738, is used. The transformer is connected to a 240 V, 50 cycles AC supply to get an output voltage of 24 V. The output from the transformer is fed to the input of a dimmerstat, type 100 RM (2 KV A). A variable output, ranging between 200 mV and 2 V, at variable current intensity (upto 1000 mA), thus achieved is fed to the bearing test machine shaft through the slip ring and carbon brush assembly (Fig. 3.2). The peak-to-peak voltage across the test bearing is monitored by a cathode ray oscillograph, and the current (rms) through an ammeter. The average contact resistance of the bearing is ascertained by monitoring the current through the bearing at varying voltages fed to the shaft. When metal-to-metal contact occurs in the bearing, the low resistance of the bearing indicates a short-circuit condition. A complete open circuit is obtained when a lubricant film exists between the bearing surfaces and the rolling-elements.

4.2.2 The Experimental Tests Conducted on the Bearings under the Influence of Electric Current

The roller bearing of type NU 330, with lubricant 'B' having a resistivity of 10^{11} ohm cm, has been tested at different speeds (450, 750, 1000 and 1200 rpm) and at various radial loads (450, 750 and 1000 kgf). During the test, a maximum temperature rise of 8° C has been observed on the bearing outer race. Overall radial vibration levels between 10 and 15 μm have been noticed. The voltage across the bearing between points 'A' and 'B' (Fig. 3.2) during each test varied between 100 mV and 1.4 V. No sign of damage was noticed on removing the inner race of the bearing. Based on the data thus obtained, a set of plots showing the variation of the different parameters is drawn for the bearing of type NU 330 (Figs. 4.1–4.5).

Figures 4.1 and 4.2 show the variations in bearing current with different voltages at different operating speeds, at 750 kgf and 1000 kgf loads, respectively. Figure 4.3 indicates variation in bearing impedance with current at different operating speeds, at 450 kgf of radial load. Figure 4.4 shows the variation in bearing impedance with the operating speed at different levels of bearing current at 750 kgf of radial load. Similarly, Fig. 4.5 shows the variation in bearing impedance with load at an operating speed of 1200 rpm, at different levels of current. A similar behavior, as shown for

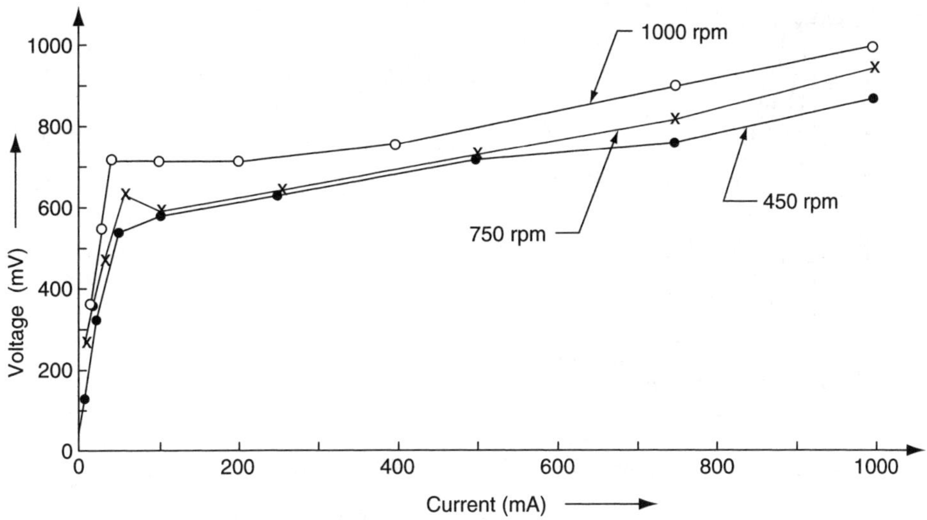

Fig. 4.1. Variation of bearing current with voltage of NU 330 bearing using lubricant 'B' at different speeds at 750 kgf load

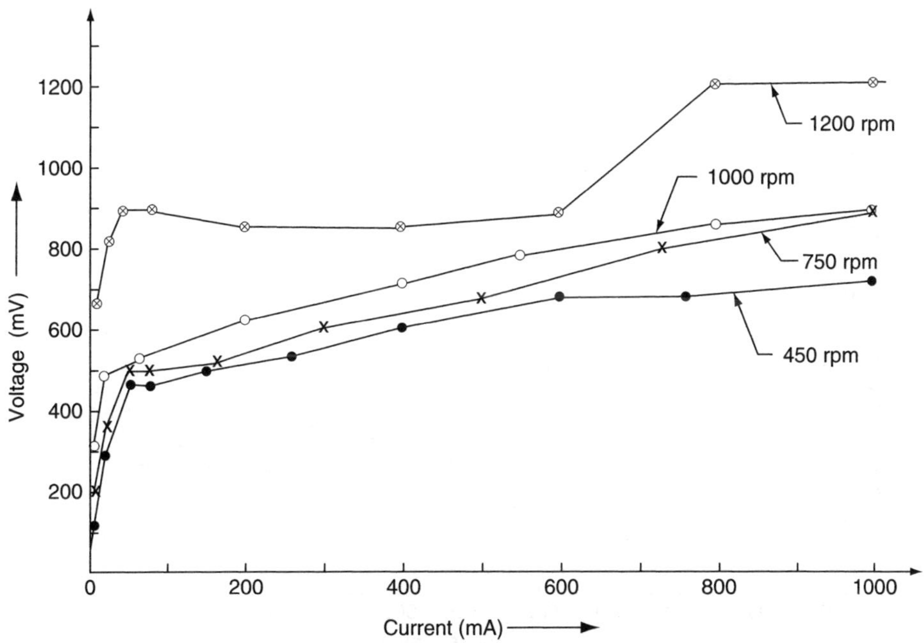

Fig. 4.2. Variation of bearing current with voltage of NU 330 bearing using lubricant 'B' at different speeds at 1000 kgf load

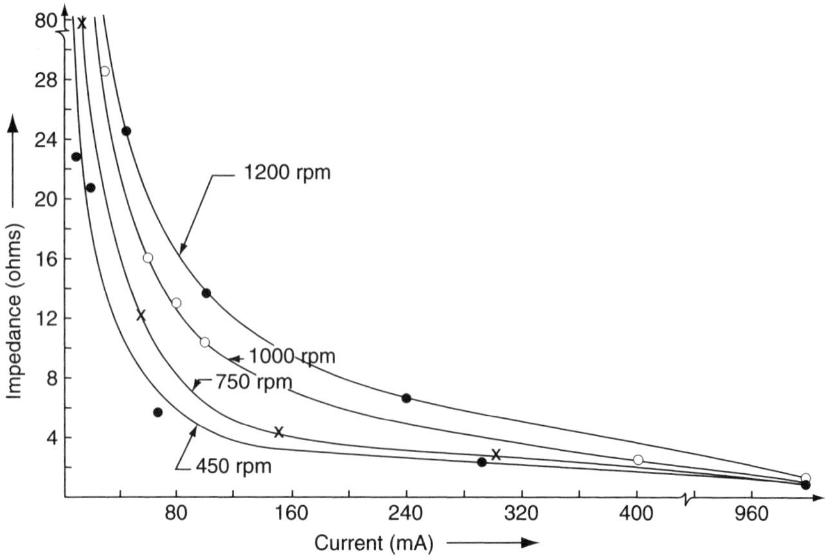

Fig. 4.3. Variation of bearing impedance with current at different speeds and 450 kgf load – for NU 330 bearing (using lubricant 'B')

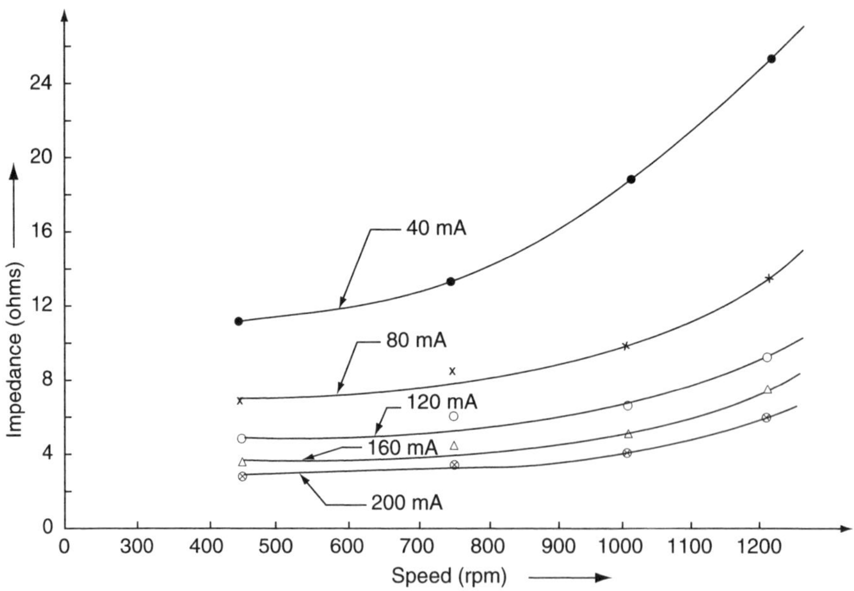

Fig. 4.4. Variation of impedance of bearing NU 330 using lubricant 'B' with rpm at different levels of bearing current at 750 kgf load

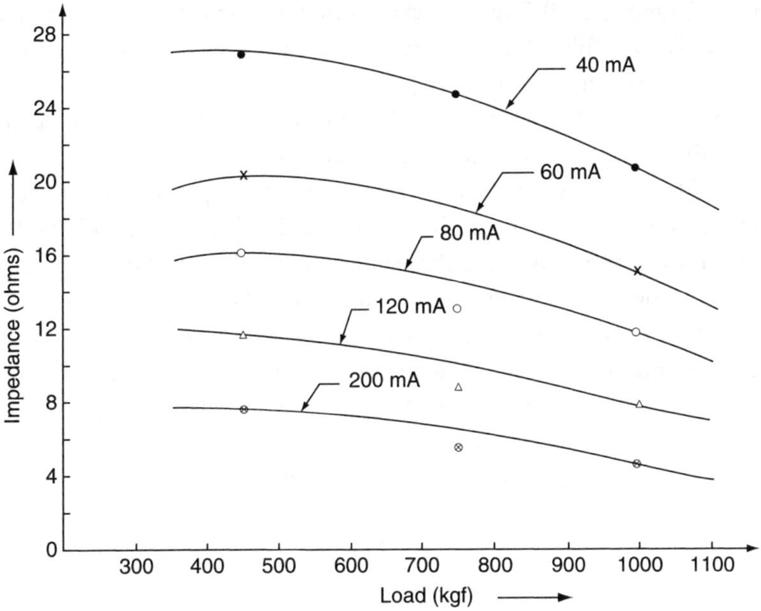

Fig. 4.5. Variation of impedance of bearing NU 330 using lubricant 'B' with load at various levels of bearing current at 1200 rpm

the NU 330 type of bearing, has also been exhibited by the NU 2215 type of bearing. The bearing of type NU 330 was operated for the study of impedance response and threshold voltage, and the bearing of type NU 2215 was also operated intermittently for a duration of 250 h to study the effects of electrical fields on the bearing surfaces.

When tested under identical conditions, the bearing of type NU 326, with low-resistivity lubricant 'A' (10^7 ohm cm), has not shown the voltage-current behavior pattern exhibited by the bearing of type NU 2215 as well as the bearing of type NU 330, both with lubricant 'B'. The bearing of type NU 326 has shown the bearing impedance to be as low as 0.024 ohm even at 100 mV potential difference existing across the bearing as against 0.3–0.5 ohm bearing impedance measured under static conditions without the passage of current through the bearing.

4.3 Theoretical

4.3.1 Determination of Minimum Thickness of Lubricant Film

The relatively large actual area of contact in stationary bearing shows that the lubricant film is approximately 50 Å (0.005 μm) thick, and the quantummechanical tunnel effect enables the current to pass through the contact zone with little additional resistance (0.3–0.5 ohm). Hence, a high temperature does not occur in the contact zone. As a result, a strong electric current may pass through non-rotating bearings without causing any damage [2].

On the other hand, in a rotating bearing the actual surface area in contact depends on the lubricant film thickness, surface roughness and 'asperity contacts'. The average film thickness, in turn, depends upon the operating parameters, lubricant characteristics, and bearing radial clearance (load sharing of rollers) [3, 4].

Based on the Grubin formula, the minimum film thickness of the bearings using lubricant 'B', under different operating parameters, has been evaluated [3, 5]. The minimum film thickness (h_0) depends on diameter of rolling-element (d), pitch diameter of bearing (D), contact angle (α), rpm (n), lubricant viscosity at atmospheric pressure (η_0), length of rolling-element (L), modulus of elasticity (E), maximum roller load at the inner raceway ($Q_{\max}$) and the pressure coefficient of viscosity (λ). The minimum film thickness (in inches) is given by

$$h_o = 0.22\left[\frac{d}{2}(1-y)\right]^{1.09} \times \left[\left(\frac{1+y}{y}\right)\eta_0\lambda n\cos\alpha\right]^{0.727} \times \left[\frac{\mathrm{p}EL}{2Q_{\max}}\right]^{0.09}$$

where

$$E = E(1-\mu^2)$$
$$Y = d\cos\alpha/D$$
$$Q_{\max} = \frac{5P}{Z}$$

P being the radial load on a bearing and Z the number of rollers in a bearings ($Z = 14$ in NU330 as well as NU 2215 type of bearings) [5].

$$\mu = \text{the Poison's ratio}$$

The minimum film thickness, thus evaluated, has been studied against the impedance of the bearings at different levels of bearing current, determined under similar operating parameters (Figs. 4.3 and 4.5). The variation in minimum film thickness of NU 330 and NU 2215 type of bearings with bearing impedance at different levels of current is shown in Figs. 4.6 and 4.7.

4.3.2 Phenomenon of Threshold Voltages

When the bearing is under operation in an electric field, current does not begin to flow until the first threshold voltage is reached. At the first threshold voltage, the

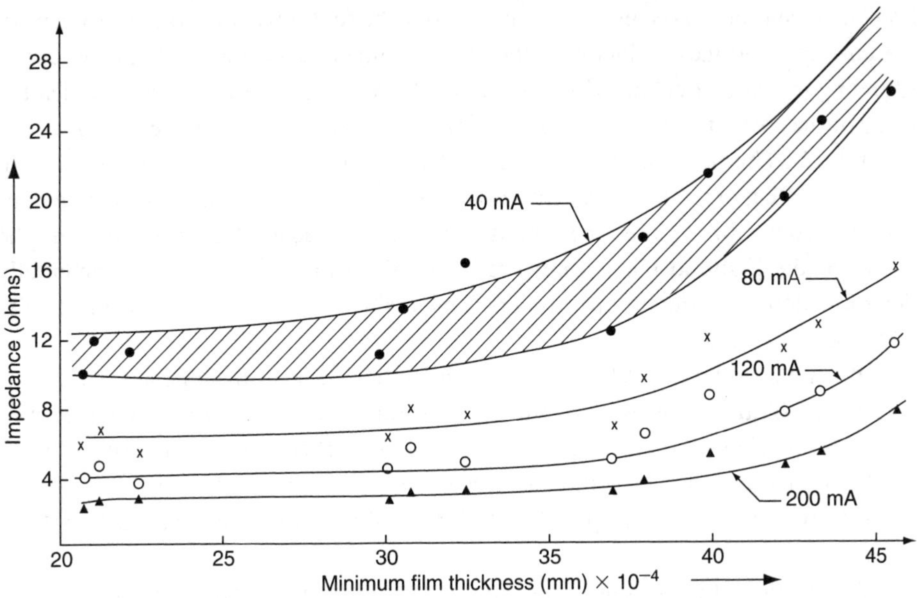

Fig. 4.6. Variation of minimum film thickness with impedance of bearing NU 330 using lubricants 'B' at different levels of bearing current

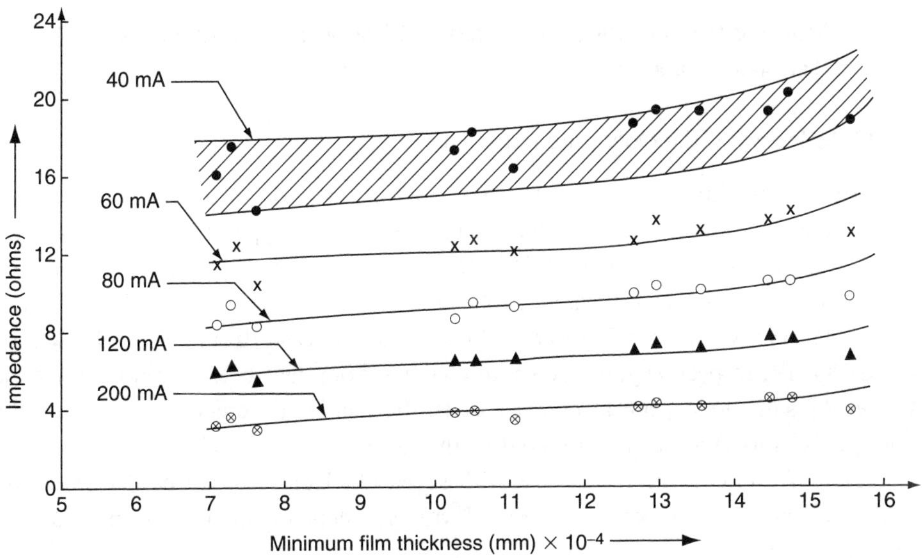

Fig. 4.7. Variation of minimum film thickness with impedance of bearing NU 2215 using lubricant 'B' at different levels of bearing current

flow of current increases instantaneously. After the first threshold voltage is reached, the current continues to increase, though to a much lesser degree than the corresponding increase in voltage (Figs. 4.1 and 4.2). A point is then reached at which the value of the current changes sharply with only a marginal change in the voltage (Figs. 4.1 and 4.2). This junction showing the sudden change in voltage-current relationship due to the effect of temperature, vibration, operating parameters and phenomenon of cage and roller slip [6] in the bearing is termed as the second threshold voltage. The values of the first and the second threshold voltages (V_{T1}, V_{T2}), as experimentally determined under different operating parameters, are shown in Table 4.1 for the NU 330 type of bearing.

When the voltage rises from zero, a reduction in the bearing impedance occurs at threshold voltage. This is apparent by a sudden change in the voltage-current relationship (Figs. 4.1 and 4.2). In the zone of higher current intensity (100 to 1000 mA), the increase in voltage is much slower as compared to the zone of low current intensity (less than 100 mA), at all the different operating parameters (Figs. 4.1 and 4.2).

At certain voltages, sometimes during the intermediate zone, the current varies between two values. This phenomenon may be explained in terms of the circuit having been broken down at high frequencies because of arcing in the contact area.

4.4 Influence of the Operating Parameters and Lubricant Characteristics on Threshold Voltages

4.4.1 Influence of the Operating Parameters

Both the first and the second threshold voltages (V_{T1}, V_{T2}) increase with speed at a particular load, as shown in Table 4.1 for the NU 330 type of bearing. A similar trend is clearly observed for the NU 2215 type of bearing using the same lubricant 'B'. It is observed that at 450 kgf and 1000 kgf radial loads, the first threshold voltage of the NU 330 type of bearing using lubricant 'B', increases from 50 to 850 mV and 40 to 580 mV, respectively, as the speed changes from 450 to 1200 rpm (Table 4.1). Under the same loads, but at higher speeds, the relative difference in the threshold voltages is more significant compared to that at lower speeds. This is similar to the behavior of lubricant film thickness with speed [3]. Hence, it can be deduced that both the first and the second threshold voltages depend on the thickness of the lubricant film.

Also, the two threshold voltages decrease as the load on the bearing is gradually increased at a constant speed. However, both the threshold voltages increase as the speed changes from 450 to 1200 rpm at different loads on the bearing.

Table 4.1. First and second threshold voltages (V_{T1} and V_{T2}) and threshold coefficients (V_{T1K} and V_{T2K}) of NU 330 bearing at different operating parameters

Speed of operation (rpm)	Load on bearing (kgf) 450		750		1000		
	V_{T1} (mV)	V_{T1K}	V_{T1} (mV)	V_{T1K}	V_{T1} (mV)	V_{T1K}	V_{T1K} (Average)
450	50	0.623	45	0.654	40	0.634	0.635
750	–	–	200	2.90	150	2.38	2.64
1000	270	3.36	210	3.05	200	3.17	3.19
1200	850	10.6	–	–	580	9.19	9.90
	V_{T2} (mV)	V_{T2K}	V_{T2} (mV)	V_{T2K}	V_{T2} (mV)	V_{T2K}	V_{T2K} (Average)
450	–	–	535	7.78	464	7.35	7.57
750	678	8.45	625	9.1	500	7.93	8.49
1000	785	9.79	714	10.38	–	–	10.09
1200	1160	14.47	–	–	892	14.13	14.30

It is found that the threshold coefficients (V_{T1K} and V_{T2K}) are almost constant at a particular operating speed: the change of load does not significantly affect the threshold coefficients (Table 4.1). The threshold coefficients are given by

$$V_{T1K} = V_{T1}P^{0.3}$$
$$V_{T2K} = V_{T2}P^{0.3}$$

where P is the load on the bearing.

As shown in Table 4.1, the average value of the first threshold coefficients varies between 0.635 and 9.90 and that of the second threshold coefficient between 7.57 and 14.30 for the NU 330 type of bearing using lubricant ‘B’ and operating in the speed range of 450 to 1200 rpm. In the case of the NU 2215 type of bearing using the same lubricant, the average value of the first threshold coefficient is 9.90 and that of the second threshold coefficient is 12.44 for an operating speed of 1200 rpm. The values for these two bearings (NU 330 type and NU 2215 type) are close enough to deduce that the threshold coefficients depend upon the type of bearing characteristics of the lubricant used, and the speed of operation.

The values of threshold coefficients are useful to predict the levels of threshold voltage under different loads at a particular speed of operation. They could help to predict the safe limit of the allowable potential drop – lesser than the first threshold voltage – across the bearing to prevent damage of bearings under different operating parameters. However, the temperature rise of the bearings also affects the threshold voltages and, hence, the safe limit of the allowable potential drop.

4.4.2 Influence of Lubricant Characteristics

The investigations on passage of current through the bearings show that the threshold voltages vary with the lubricant characteristics, particularly with the resistivity of the lubricants. The lubricant ‘B’ (resistivity 10^{11} ohm cm, viscosity 127.33 centistokes at 40 °C and 10.43 centistokes at 100 °C) allows a current of 1 A (max.) at 1.36 V to pass, whereas lubricant ‘A’ (resistivity 10^{7} ohm cm, viscosity 99.92 centistokes at 40 °C and 9.68 centistokes at 100 °C) gets disintegrated even at 1.2 V and allows a current of the order of 50 A (AC) to pass. The very low order of the threshold voltages (mV range) on account of using the low-resistivity lubricant ‘A’ could not be detected.

4.5 Phenomenon of Bearing Impedance

When electric current passes through the bearing surfaces in contact, separated by high-resistivity lubricant of type ‘B’, the current does not flow across the surfaces and, consequently, the current intensity on the bearing surfaces increases. This affects the insulation resistance of the lubricant. It is apparent from Figs. 4.3 to 4.5 that the increase in intensity of current through the bearing reduces the bearing impedance very significantly. The average contact resistance for current level of 10 mA for the NU 330 type of bearing operating at 450 rpm and 450 kgf load, is as high as 22.85 ohms (Fig. 4.3). Further increase in the current intensity to 1 A reduces the contact resistance to as low as approximately 1 ohm. Similarly, a still higher resistance of 75.38 ohms for current level of 9 mA at 1200 rpm and 1000 kgf load reduces to 1.214 ohms at 1 A.

It has also been observed that the bearing impedance does not return to the previous high value in the reverse cycle – on decreasing the flow of current in the bearing – but, instead, tends towards a comparatively smaller value, because of temperature rise, change in viscosity, deterioration of lubricant and change in bearing conditions with time.

At any value of current intensity, the bearing impedance is larger at higher operating speeds and under a given load (Fig. 4.3). However, the difference in impedance under different operating speeds at lower current intensity (<40 mA) is very much larger than that at higher current intensity (>900 mA), as shown in Fig. 4.3. This pattern may be attributed to the behavior of the lubricant molecules and their bonding strength under different levels of current intensity. Exposure of the bearing to higher current intensity for a considerable length of time may lead to breakdown and ionization of the insulating lubricant film apart from the arcing phenomenon occurring in the contact area due to make and break of contacts of the surfaces because of bearing vibrations.

Furthermore, the breakdown may occur anywhere within the bearing [7]. But it may be more significant in the contact area due to the negative cage and roller slip, which leads to a close interaction of the rolling-elements with the outer race at moderate speed under load and no-load operation, and also, due to the positive cage and roller slip, leading to a close interaction of the rolling-elements with the inner race at high speed under no-load operation of the bearing [6]. Besides this, the edge loading becomes more significant in a highly loaded roller bearing because of the depression of the raceway outside the roller ends. Under these conditions, the roller end compressive stresses become higher than even those in the centre of contact, which may also lead to breakdown of the lubricant film.

Impedance of bearings is distinguishable at different operating speeds till the potential drop has not exceeded 600 mV and the current intensity 300 mA (Fig. 4.3). At higher current intensities, the bearing impedance decreases rapidly, and a slight increase in the potential drop results in flow of large currents through the bearing. There is a tendency for the impedance to approach the same value (less than 1 ohm) at about 1 A (AC), irrespective of the operating speed, for the bearing using high-resistivity lubricant.

4.5.1 Instantaneous Versus Average Electrical Impedance

Electric voltage as examined on the oscilloscope under various loads and speeds shows instantaneous high or low value and oscillates rapidly between these extremes. The duration of the low voltage is found to be as short as one millisecond. Under normal operating conditions, it may be due to the intermittent nature of metallic contact caused by the surface irregularities and the metallic particles of the bearing surfaces entangled in the grease. Furey [8] has shown the similar metallic contacts.

Peak-to-peak voltage was recorded at different operating parameters, and this was used for determining the rms voltage and subsequently the 'average' bearing impedance using the rms value of current. Even when the 'average' impedance is fairly high, a considerable amount of metallic contact can be taken place. The change of impedance of the bearing indicates the relative change in the metallic contacts in the bearing. The bearings have been tested under different operating conditions for short time, to avoid the effect of temperature rise and other factors on the impedance of these bearings.

4.6 Influence of the Operating Parameters and Minimum Film Thickness on Bearing Impedance

4.6.1 Influence of the Operating Parameters

Increase in speed reduces the extent of metallic contacts and thus increases the bearing impedance. The bearing, when operated at different loads and speeds with various levels of current intensity, exhibited differences in the slopes of the curves (Figs. 4.4 and 4.5). Increase in the operating speed from 450 to 1200 rpm at 750 kgf load and 40 mA current resulted in the impedance varying from 11.0 to 24.5 ohms, as against the variation from 3.2 to 5.5 ohms at 200 mA current intensity (Fig. 4.4).

The variation of load at constant bearing speed has very little effect on the bearing impedance at different levels of current intensity (Fig. 4.5), as against the effect of change in speed at fixed loads except at high current intensity (200 mA). The increase in current intensity reduces the bearing impedance considerably with increase in load from 450 to 1000 kgf at 1200 rpm, the impedance varied from 27.0 to 20.5 ohms at 40 mA current intensity, as against the variation from 8 to 4.5 ohms at 200 mA (Fig. 4.5).

From the above, it is evident that changes in load and speed of the bearing have much less effect on the bearing impedance than the current intensity which reduces the bearing impedance drastically.

4.6.2 Influence of Minimum Film Thickness

At current intensity between 40 and 200 mA, there is a negligible variation in the impedance with film thickness upto 37×10^{-4} mm, for the NU 330 type of bearing (Fig. 4.6). But beyond this, the impedance increases considerably with increase in film thickness. However, the pattern of variation in impedance is identical at different current intensities. It is observed that the impedance is around 3, 4.5 and 6.5 ohms at 200 mA, 120 mA and 80 mA current intensities with film thickness of 32×10^{-4} mm, whereas the impedance varies between 10 and 14.5 ohms at 40 mA with the same film thickness (Fig. 4.6). The zone of variation in impedance at lower intensity of current (shown by the shaded area in Fig. 4.6) may also be attributed to intermittent polar and non-polar behavior of the different molecules of the lubricant, apart from the influence of film thickness.

As the film thickness varies from 37×10^{-4} to 45×10^{-4} mm, the impedance changes from 3 to 6 ohms, 5 to 10 ohms and 8 to 15 ohms at current intensities of 200, 120 and 80 mA, respectively (Fig. 4.6). Similarly, at 40 mA current intensity, the impedance varies in the range of 11–17 ohms to 25–32 ohms, with the same variation in film thickness. This trend in variation of impedance with film thickness is also found in the case of NU 2215 type of bearing using the same lubricant (Fig. 4.7); only the absolute values of the impedance and film thickness are observed to be different.

It may be noted that after the initial breakdown of the lubricant, increase in current intensity (40 to 200 mA) causes enlargement of the electric contact region by fritting [9], and the bearing impedance reduces further at a constant film thickness (Figs. 4.6 and 4.7). Increase in film thickness beyond a certain value leads to a significant increase in impedance only at lower current intensity (40 mA). This behavior, in particular, can be used to determine the minimum thickness of the film, by measuring the impedance of the bearings at different levels of current intensity.

The low value of bearing impedance at high current (200 mA) is probably due to the complete disruption of the film (Figs. 4.6 and 4.7). This would cause the electrical contact area to approach a constant value, perhaps a value close to the area of mechanical contact. The recouping property of the lubricant increases the bearing impedance if the electric field is withdrawn [10]. This may continue to happen until the stretchings in the hydrocarbon molecules are not permanently set in. An increase in current, or load, also results in increased plastic flow of the metal, which may be responsible for increase in wear rate provided the bearing is not seized and the surfaces are not damaged by the arcing effect resulting in the welding of the surfaces. It appears that small filaments of current start flowing when high points on the asperities of the races and rolling-elements come close or when conducting particles bridge the lubricant film. When these conducting paths are broken as the asperities become separated either by higher thickness of the lubricant film, vibration effects, or a combination of both, or by the presence of high-resistivity lubricant, arcing results. Arcing may also result as conducting particles in the film are rearranged by the relative motion of the bearing elements.

4.7 Different Types of Bearing Failure

Two types of bearing failure can be distinguished by the phenomenon of the passage of current through a bearing under electric field: firstly, in bearings using lubricants having high-resistivity lubricant (10^{11} ohm cm or higher), charges accumulate till the voltage reaches the threshold critical value, at which the breakdown takes place. The damage is caused by arcing when the passage of current is interrupted. This is accompanied by mass transfer, elevated local temperature on asperities of the friction surfaces. (Fig. 4.8 shows the inner race of NU 2215 bearing – using high-resistivity lubricant type 'B' – damaged under these conditions). Rehardening of the bearing surfaces also takes place during the subsequent rapid cooling. As the bearing of type NU 330 was not exposed to electric field for a long duration, no damage of the bearing was detected.

On the other hand, bearing using low-resistivity lubricants (10^7 ohm cm), no arcing takes place. 'Silent' discharge occurs through the inner race, rolling-elements and outer race of the bearing. The current passes between the surfaces in contact as a result of the quantum mechanical tunnel effect and because of fritting. The steel is thus heated and the hardness of the surface is reduced. This is also accompanied by electrochemical decomposition of the lubricant, leading to corrosion on the bearing surfaces as discussed in Chapter 3, due to formation of $LiOH/Li_2Co_3$ in the first few hours of operation [10]. After this, the accelerated formation of corrugations takes

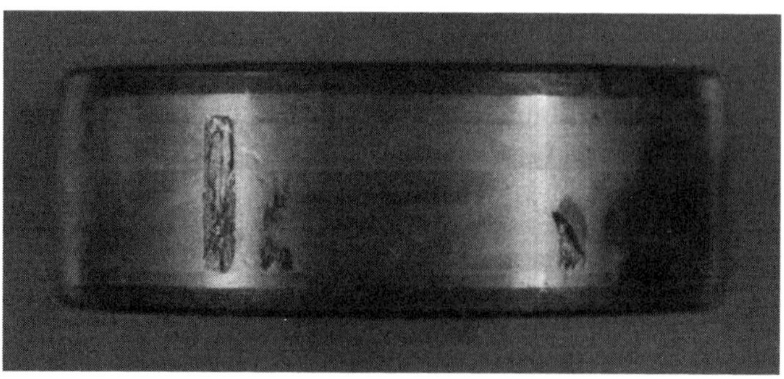

Fig. 4.8. Damaged inner race of NU 2215 bearing (using lubricant 'B') under electric fields

place due to increased erosion and deformation in the loaded zone of the corroded surfaces of the bearings, as shown on the inner race of NU 326 bearing using low-resistivity lubricant ‘A’ (Fig. 4.9). This finally leads to the bearing failure accompanied by an increase in ash content, coefficient of friction, profile depth, ball scar diameter, and depth/width of corrugations on the bearings [10, 11].

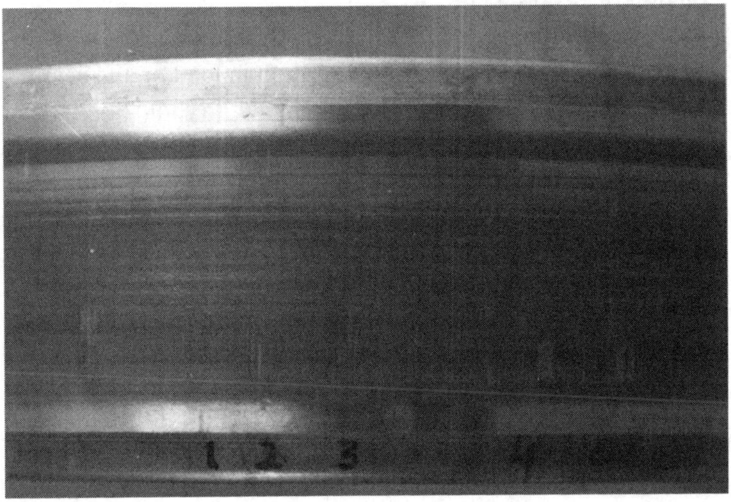

Fig. 4.9. Inner race of bearing NU 326 on exposure of 50 A (AC) for 250 h (using lubricant ‘A’)

4.8 Conclusions

From the analysis and investigations brought out in this chapter, the following conclusions are drawn [12, 13]:

(1) Impedance of the bearing varies with the operating parameters, current intensity and lubricant characteristics. It is as high as 75 ohms at 10 mA (AC) and as low as 1 ohm at 1 A current for the NU 330 type of bearing using lubricant having resistivity of 10^{11} ohm cm.

(2) Variation of load at constant speed has very little effect on the bearing impedance at the fixed levels of current intensity, as against the change in speed at fixed load.

(3) Increase in current intensity reduces the bearing impedance very significantly irrespective of the operating parameters.

(4) The phenomenon of threshold voltages is indicated by the bearings using high-resistivity lubricants (10^{11} ohm cm or more). Bearings using low-resistivity lubricants (10^7 ohm cm or less) do not exhibit such a phenomenon.

(5) Bearings using high-resistivity lubricant fail by the arcing phenomenon. On the other hand, bearings using low-resistivity lubricant fail by electrochemical decomposition and ionization of the lubricant under the action of electric field.

(6) Threshold voltages and threshold coefficients increase with the operating speeds and are less affected by variation in load on the bearings. Threshold coefficients can be used to determine threshold voltages and the safe limit of the potential drop across the bearing, to preclude damage of the bearings under different operating parameters.

(7) Safe limit of the potential drop across the bearing is lower than the first threshold voltage.

(8) The bearing impedance increases considerably with the minimum film thickness at low current intensity (40 mA). The minimum film thickness of a bearing can be evaluated by measuring the impedance and current intensity in the bearing.

The bearing impedance, threshold voltages and threshold coefficients are very critical parameters and are influenced by temperature, vibration, alignment, bearing quality, material and different maltreatment factors apart from the operating parameters and lubricant characteristics. With judicious adoption of the experimental techniques and extensive developmental efforts towards correlating the various factors influencing the bearing performance, it should be possible to bring about a considerable improvement in the assessment of the effect on the various parameters, as reported in this chapter.

REFERENCES

1. Rodstein, L. (1974). *Electrical Control Equipment*, Mir Publishers, Moscow.
2. Anderson, S. (1968). Passage of electric current through rolling bearings. *Ball Bearing J.*, **168**.
3. Harris, A. T. (1966). *Rolling Bearing Analysis*, John Willey and Sons, Inc., New York.
4. Tallian, T. E., Mc Cool, J. I. and Sibley, L. B. (1966) Partial elastohydrodynamic lubrication in rolling contacts. I. Mech. E., **180**, 169.
5. Kovalev, M. P. and Narogetski, M. Z. (1975). Design Calculation of Precision Rolling-Element Bearings. *Mashinostroenie*, Moscow.
6. Prashad, H. (1987). The effect of cage and roller slip on the measured defect frequency response of rolling-element bearings. *ASLE Trans.*, **30**(3), 360–367.

7. Garnell, P. and Higginson, G. R. (1966). The Mechanics of Roller Bearings. Elastohydrodynamic Lubrication Conference, Leeds, I. Mech. E. Publication.
8. Furey, J. N. (1961). Metallic contact and friction between sliding surfaces. *ASLE Trans.*, **4**, 1–11.
9. Holm, R. (1958). *Electric Contacts Hand Book*, Springer-Verlag, Berlin, 3rd Ed.
10. Prashad, H. (1986). Experimental study on influence of electrical fields on behaviour of greases in statically bounded conditions and when used in non-insulated bearings. *BHEL J.*, **7**(3), 18–34.
11. Prashad, H. and Murthy, T. S. R. (1988). Behaviour of greases in statically bounded conditions and when used in non-insulated anti-friction bearings under the influence of electrical fields. *Lubric. Eng.*, **44**(3), 239–246.
12. Prashad, H. (1987). Effect of operating parameters on the threshold voltages and impedance response of non-insulated rolling-element bearings operated under the action of electrical current. *Wear*, **117**, 223–240.
13. Prashad, H. (1987). Effects of operating parameters and lubricant characteristics on threshold voltages and impedance of non-insulated roller bearings under the influence of electric current. *BHEL J.*, **8**(2), 36–48.

Chapter 5

EFFECT OF ELECTRIC CURRENT ON THE TRACK SURFACES OF ROLLING-ELEMENT BEARINGS

5.1 A General Review

In Chapter 2 various causes of the origin of shaft voltages in electrical machines have been discussed. The asymmetry of faults in an electric machine causes net flux resulting in shaft voltage that leads to a current flow linking with the circuit consisting of shaft, bearing and frame. Shaft flux usually results in a localized current at each bearing rather than a potential difference between shaft ends [1]. A current path, however, along shaft, bearings and frame results in a potential between shaft ends. At a certain threshold potential difference, depending on resistivity of the lubricant, it creates an electrical breakdown in the lubricating film of a bearing causing an arcing effect or allowing a passage of 'silent' discharge through the lubricant.

Bearing current mainly depends on the magnitude of the shaft voltage and the bearing impedance. Bearing impedance in turn depends upon lubricant viscosity, bearing temperature, load, speed and shaft voltage [2]. The current flows within the shaft and the bearings through the oil films [3, 4]. The voltage over the bearing may be very low but local bearing currents of very high order might flow depending on the bearing impedance [5]. Dirt, metallic particles, surface roughness and irregular oil film thickness reduce the impedance of the bearing circuit such that the small shaft voltages induce substantial bearing currents, which finally can damage the bearing by surface pitting, arcing effects or by deterioration, contamination and aging of the lubricant [6].

In rotation, the actual surface area in contact is smaller than in a non-rotating bearing. The average film thickness is of the order of 0.2–2 microns, which primarily depends upon speed and lubricant characteristics, and to a lesser extent on the rolling-element load [7, 8]. In general, passage of electric current leads to the development of flutings and corrugations before final seizure of the bearings [9–11].

This chapter reports the studies undertaken to understand experimentally the mechanism of gradual formation of corrugations on roller and ball bearing surfaces under the exposure of an electric current, to evaluate depth, width and pitch of corrugations on various bearing surfaces under different conditions of operation, and also, to correlate pattern of corrugations, thus determined, with that of the results from derived theory. The theory is based on operating conditions, lubricant characteristics, kinematics, positions of the plane of action of loading, number of rolling-elements in the loaded zone and quality of bearings.

5.2 Process of Formation of Corrugations

At each revolution of the shaft, at an instant, part of the circumference of the inner race passes through a zone of maximum force, and Hertzian pressure between the rolling-elements and races (at the line/point contact) leads to a maximum shear stress. The maximum shear stress is taken as the criterion for yielding, and this occurs in the sub-surface at a depth equal to half the radius of contact surface [12]. It is generally at this point that the failure of material, if occurring will initiate. As soon as the fatigue spall appears on the surface, the actual area of asperity contact between the rolling-element and the race is reduced and there is a gradual increase in the width of corrugation by deformation at the asperity contacts due to an increase in contact pressure per unit area. Furthermore, corrugations are extended along the track width in a ball bearing due to axial oscillations of balls in the raceways.

When a bearing is significantly loaded in a particular direction, deformation is caused by rolling-elements (K) on the races in the loaded zone. The value of K is generally less than half of the total number of rolling-elements in a bearing ($K \leq N/2$). The process of deformation which leads to the formation of corrugation pattern on the surfaces is accelerated by: (1) the passage of current, (2) corrosion and oxidation on the surfaces, (3) resultant load, (4) quality of a bearing and (5) lubricant characteristics.

5.3 Theoretical Determination of Pitch and Width of Corrugations on Roller Track of Races and Rollers of Roller Bearings

5.3.1 Pitch of Corrugations

The rolling-element frequency of a roller bearing is determined assuming the outer race is stationary and the inner race is rotating, and is given as [13, 14]:

$$f_b = \frac{Df_s\left[1 - \left(\frac{d}{D}\right)^2\right]}{2d} \tag{5.1}$$

In a roller bearing the centers of rollers describe an 'orbit' with a radius $(r_i + r)$ or $D/2$ on the outer surface of inner race, and are loaded by different positions of plane of loading (p) by f_b/f_s times during one revolution. The resultant of p loads the rolling-elements, rolling in half of the circumference of a bearing. Thus, the rollers describe an arc of length πr_i on outer surface of the inner race, and similarly, an arc of length $\pi(r_i + 2r)$ on inner surface of the outer race in the loaded zone of a bearing during each revolution. Also, each roller element describes an arc of length $2\pi r$ by its own rotation.

As a result of mutual interaction of the rollers with the races, the arc length of corrugations on a bearing surface is given as:

$$C_{ir} = \frac{p(D - d)}{2} \tag{5.2}$$

$$C_{or} = \frac{p(D + d)}{2} \tag{5.3}$$

$$C_{re} = pd \tag{5.4}$$

5.3.1.1 Pitch of Corrugations on Inner Race

The pitch of corrugations on the inner race (Δ_{ir}) is determined by number of similar corrugations (M_{ir}) on C_{ir} and is calculated as:

$$\Delta_{ir} \times M_{ir} = \frac{p(D - d)}{2} \tag{5.5}$$

It is obvious from the section 5.2 that the M_{ir} depends on f_b, p, K and the bearing coefficient (S_1). Hence

$$M_{ir} = S_1 f_b K p \tag{5.6}$$

and

$$\Delta_{\text{ir}} = \frac{\text{p}Dd}{[S_1 f_s Kp(D + d)]} \tag{5.7}$$

The bearing coefficient (S_1) in seconds is arbitrarily used as the factor considering the bearing quality, maltreatment and lubrication characteristics. S_1 is unity when a roller with a unit frequency loaded by a single position of plane of action makes a unit corrugation on the inner/outer race of a bearing.

5.3.1.2 Pitch of Corrugations on Outer Race

Similar to Eqn. (5.7), the pitch of corrugation on the outer race is given as:

$$\Delta_{\text{or}} = \frac{\text{p}Dd}{[S_1 Kpf_s(D - d)]} \tag{5.8}$$

5.3.1.3 Pitch of Corrugations on Rollers

The pitch of corrugations on a roller is determined by the number of corrugations (M_{re}) on C_{re} and is given as:

$$\Delta_{\text{re}} \times M_{\text{re}} = \text{p}d \tag{5.9}$$

It is obvious that the number of corrugations on a roller depends on f_s, p and S_1. Hence:

$$M_{\text{re}} = S_1 f_s p \tag{5.10}$$

and

$$\Delta_{\text{re}} = \frac{\text{p}d}{S_1 f_s p} \tag{5.11}$$

During operation, the rollers are comparatively less heated than the inner race because of higher values of f_b than f_s ($f_b/f_s = 2.61$ for the bearing NU 326 and 3.7 for the bearing NU 230), and thus the possibility of tempering of roller surfaces is initially less than that of the races. This reduces the probability of earlier formation of ridges and corrugations on the rollers. Steady gyroscopic movements of the rolling-elements i.e. change of polarity (axis of rotation) between the races also add to it. Because of this, Δ_{re} will be affected by the corrugation pattern formed on the races, and also by surface distress due to asperity contact on the races. Cage and roller slip under load and no load conditions also affect it [15]. Thus, the pitch of corrugation on the rollers may appear either as Δ_{re}, $\Delta_{\text{re}} - \Delta_{\text{ir}}$, or $\Delta_{\text{re}} - \Delta_{\text{or}}$, or like a replica of Δ_{ir} or Δ_{or} after long operation.

5.3.2 Comparison of Pitch of Corrugations on the Surface of Bearings

The pitch of corrugations is compared by Eqns. (5.7), (5.8) and (5.11) and is given as:

$$\Delta_{ir} = \frac{\Delta_{or}(D - d)}{(D + d)} \tag{5.12}$$

$$\Delta_{re} = \frac{\Delta_{ir} K(D + d)}{D} \tag{5.13}$$

$$\Delta_{re} = \frac{\Delta_{or} K(D - d)}{D} \tag{5.14}$$

From this, it is evident that Δ_{or} and Δ_{re} is always more than Δ_{ir}, but Δ_{re} can be less than the Δ_{or} when $K = 1$.

5.3.3 Theoretical Determination of Width of Corrugations

Rollers of diameter d_1 and d_2 of the same material in contact under load form a rectangular surface of deformation of a length equal to the length of the rollers. Width of corrugation on these rollers is given by [12]:

$$W_w = 2.15\left[\frac{Pd_1 d_2}{EL(d_1 + d_2)}\right]^{0.5} \tag{5.15}$$

5.3.3.1 Width of Corrugations on Inner Race

Since the complete load acting on a bearing from p is taken up by K rollers, the width of corrugations on outer surface of the inner race is calculated as:

$$W_{ir} = 2.15\left[\frac{Pd(D - d)}{pKELD}\right]^{0.5} \tag{5.16}$$

where d_1 of Eqn. (5.15) is taken as d and d_2 as outer diameter of inner race $(D - d)$.

5.3.3.2 Width of Corrugations on Outer Race

Similar to Eqn. (5.16), the width of corrugations on inner surface of the outer race is given as:

$$W_{or} = 2.15\left[\frac{Pd(D + d)}{pKELD}\right]^{0.5} \tag{5.17}$$

5.3.3.3 Width of Corrugations on a Roller

The width of corrugations on a roller (W_{re}) will be affected by both W_{ir} and W_{or}, and is determined as:

$$W_{re} = W_{ir} \tag{5.18}$$

or

$$W_{re} = W_{or} + \beta W_{ir} \tag{5.19}$$

β varies between 0 and 1. $\beta = 0$ indicates overlapping of W_{or} and W_{ir}, and $\beta = 1$ shows maximum bandwidth of corrugations equivalent to $W_{ir} + W_{or}$.

5.4 Theoretical Determination of Pitch and Width of Corrugations on Ball Track of Races and Balls of Ball Bearings

5.4.1 Theoretical Determination of Pitch of Corrugations

The ball frequency f_b of a bearing is determined, assuming the outer race is stationary and inner race is rotating, as [14]:

$$f_b = Df_s \frac{\left[1 - \left(\frac{d}{D}\right)^2 \cos^2 \alpha\right]}{2d} \tag{5.20}$$

The centers of the balls describe an 'orbit' with a radius ($r_i + r$) (assuming $\alpha = 0$) or $D/2$ on outer surface (ball track) of the inner race and inner surface (ball track) of the outer race. The balls are loaded at the asperity point contacts by different positions of plane of action of loading (p) by f_b/f_s times, during one revolution. Furthermore, the frequency of stress application on the races is very high in a loaded zone, since the balls follow each other in rapid succession depending on f_b. Thus the time required for the stressed material on the surfaces of ball tracks of the races to reach the original state is not available when a bearing is in operation. This creates residual stresses in the material and may explain why a corrugation pattern is formed by passage of current under asperity contacts on the races in due course. However, it is accelerated by decomposition of the lubricant [16–18].

5.4.1.1 Radius of Curvature of Width of Corrugation along Bearing Width on Races

The repetitive contact of balls and their axial oscillation in the raceways form an impression at the asperity contact points on the ball tracks of races. The radius of curvature of the width of corrugation along the bearing width on the circumference of ball track of the inner race is given as [19]:

$$r_{ib} = \frac{2rr_i}{(r_i - r)} = \frac{d(D - d)}{D} \tag{5.21}$$

Similarly, the radius of curvature of the width of corrugation along the bearing width on the circumference of ball track of the outer race is given as:

$$r_{ob} = \frac{2rr_0}{(r_0 - r)} = \frac{d(D + d)}{(D + 2d)} \tag{5.22}$$

and the ratio of these radii of curvatures on outer to inner race is:

$$\frac{r_{ob}}{r_{ib}} = \frac{\left[1 + \left(\frac{d}{D}\right)\right]}{\left[\left(1 - \frac{d}{D}\right)\left(1 + \frac{2d}{D}\right)\right]} \tag{5.23}$$

5.4.1.2 Pitch of Corrugations

At each revolution of the inner race, the resultant of P_1, P_2 and P_3 loads passes through a certain position of the race, and at that instant some of the balls passing through this position of the race are in the loaded zone. That is why, for an instant, maximum of half of the circumference of a bearing is considered as a zone of maximum loading. Thus, balls describe an arc of length πr_i on ball track of the inner race, and similarly, an arc of length πr_o on ball track of the outer race in the loaded zone of a bearing, during each revolution. Also each ball describes an arc of length of $2\pi r$ by rotation about its own axis on the circumference of inner and outer races.

As a result of instant mutual close interaction of the balls with the races, the arc length of corrugations is given as:

$$C_{ib} = \frac{\mathrm{p}(D - d)}{2} \tag{5.24}$$

$$C_{ob} = \frac{\mathrm{p}(D + d)}{2} \tag{5.25}$$

$$C_{rb} = \mathrm{p}d \tag{5.26}$$

5.4.1.2.1 Pitch of Corrugations on Inner Race

The pitch of corrugations on the ball track of inner race is determined by number of similar corrugations (M_{ib}) on C_{ib} and is given as:

$$\Delta_{\text{ib}} \times M_{\text{ib}} = \frac{\text{p}(D - d)}{2} \tag{5.27}$$

From the above sections it may be deduced that M_{ib} increases with f_{b}, K, positions of plane of action of loading (p), and depends on the bearing coefficient S_2. Furthermore, number of corrugations is reduced with an increase in circumference of the contact area. Thus M_{ib} is taken as inversely proportional to the circumference of the circle (πW_{ib}) formed by the width of corrugations (W_{ib}) as diameter. Hence,

$$M_{\text{ib}} = \frac{S_2 f_{\text{b}} K p}{\text{p} W_{\text{ib}}} \tag{5.28}$$

and

$$\Delta_{\text{ib}} = \frac{3.14\text{p}(D - d) d W_{\text{ib}}}{\left\{ K p S_2 D f_S \left[1 - \left(\frac{d}{D} \right)^2 \cos^2 \alpha \right] \right\}} \tag{5.29}$$

The bearing coefficient S_2 (in mm sec) is arbitrarily used as a factor depending on the bearing quality, maltreatment and lubricant characteristics. S_2 is defined as unity when a ball with unit frequency, loaded by a single position of plane of action makes a single corrugation with a width of corrugation equal to a diameter of a circle having a unit circumference ($\pi W_{\text{ib}} \pi W_{\text{ob}} = 1$) on the ball track of race.

5.4.1.2.2 Pitch of Corrugations on Outer Race

Similar to Eqns. (5.27)–(5.29), the M_{ob} and pitch of corrugations on the ball track of outer race are given as:

$$\Delta_{\text{ob}} \times M_{\text{ob}} = \frac{\text{p}(D + d)}{2} \tag{5.30}$$

where

$$M_{\text{ob}} = \frac{S_2 f_{\text{b}} K p}{\text{p} W_{\text{ob}}} \tag{5.31}$$

and

$$\Delta_{\text{ob}} = 3.14\text{p}(D + d) d W_{\text{ob}} \left\{ K p S_2 D f_S \left[1 - \left(\frac{d}{D} \right)^2 \right] \cos^2 \alpha \right\} \tag{5.32}$$

5.4.1.2.3 Pitch of Corrugations on a Ball

The pitch of corrugations on a ball is determined by an arc length of corrugations M_{rb} on C_{rb} and is given as:

$$C_{rb} = \Delta_{rb} \times M_{rb} = \mathrm{p}d \tag{5.33}$$

It is evident that the number of corrugations on a ball depends on f_s, p and S_2. Also, it will be inversely proportional to the circumference of the circle (πW_{rb}) formed by the width of corrugations on a ball as a diameter (W_{rb}). W_{rb} is determined by Eqns. (5.42) and (5.43). Hence,

$$M_{rb} = \frac{S_2 f_s p}{\mathrm{p} W_{rb}} \tag{5.34}$$

and

$$\Delta_{rb} = \frac{3.14\mathrm{p}dW_{rb}}{S_2 f_s p} \tag{5.35}$$

The probability of earlier formation of corrugations on the surface of a ball as compared with the race is quite remote because $f_b > f_s(f_b/f_s = 2.13$ for bearing 6326). Thus, the pitch of corrugations on the balls may appear either as Δ_{rb}, $\Delta_{rb} - \Delta_{ib}$, or $\Delta_{rb} - \Delta_{ob}$ or like a replica of Δ_{ib} or Δ_{ob} after long periods of operation.

5.4.1.3 Comparison of the Pitch of Corrugations on the Surfaces of the Ball Bearings

The pitch of corrugations is computed by Eqns. (5.29), (5.32) and (5.35), and is given as:

$$\Delta_{ib} = \frac{\Delta_{ob}(D-d)W_{ib}}{[(D+d)W_{ob}]} \tag{5.36}$$

$$\Delta_{rb} = \frac{\Delta_{ib} KD\left[1 - \left(\frac{d}{D}\right)^2 \cos^2\alpha\right]}{(D-d)} \tag{5.37}$$

or

$$\Delta_{rb} = \frac{\Delta_{ob} KD\left[1 - \left(\frac{d}{D}\right)^2 \cos^2\alpha\right]}{(D+d)} \tag{5.38}$$

From the above eqns, it is evident that Δ_{ob} and Δ_{rb} are always more than the Δ_{ib}.

5.4.2 Width of Corrugations

Two balls of diameters d_1 and d_2 of the same material in contact under load form a surface of deformation due to load P arising from elastic deformation. The radius of contact area is given by [12]:

$$r_{12} = 0.88\left[\frac{Pd_1d_2}{E(d_1 + d_2)}\right]^{1/3} \tag{5.39}$$

5.4.2.1 Width of Corrugations on the Inner Race

The resultant load (P) acting on a bearing due to different positions of plane of action of loading (p) is taken by K balls, the width of corrugations on the ball track of the inner race is calculated by modifying Eqn. (5.39), and is given as:

$$W_{ib} = 1.76\left[\frac{Pd(D - d)}{pEK(D - 2d)}\right]^{1/3} \tag{5.40}$$

where d_1 of Eqn. (5.39) is taken as ball diameter (d), and d_2 as the ball track diameter of inner race i.e. ($D - d$). The radius of contact area is doubled to get the width of corrugations.

5.4.2.2 Width of Corrugations on the Outer Race

Similar to Eqn. (5.40), the width of corrugations on ball track of the outer race is given as:

$$W_{ob} = 1.76\left[\frac{Pd(D + d)}{pEDK}\right]^{1/3} \tag{5.41}$$

where d_1 of Eqn. (5.39) is taken as d and d_2 as ($D + d$).

The ball track of the outer race and ball track of the inner race in a 6326 ball bearing are concave at the interacting points of contact with the balls, so the value of d_2 has been taken with negative sign in the Eqns. (5.40) and (5.41).

5.4.2.3 Width of Corrugations on a Ball

The width of corrugations on a ball (W_{rb}) will be affected by the existing W_{ib} and W_{ob} and is determined as:

$$W_{rb} = W_{ob} \tag{5.42}$$

or

$$W_{rb} = W_{ib} + \beta W_{ob} \tag{5.43}$$

where β, the overlapping coefficient of corrugation width on a ball due to width of corrugations on ball track on inner and outer races, varies between 0 and 1. $\beta = 0$ indicates complete overlapping of W_{ob} and W_{ib}, and $\beta = 1$ shows maximum band width of corrugations equivalent to $W_{ib} + W_{ob}$ on a ball.

5.5 Experimental Investigations

5.5.1 Bearing Test Machine and Test Procedure

The bearing test machine, which was used for these experimental investigations, is described in Chapter 3 (Fig. 3.2). In the present study, roller bearings type NU 326 and ball bearing type 6326 was tested separately under the influence of electric currents and without the passage of current. The bearings were lubricated with lithium base grease (viscosity 99.91 cSt at 40 °C and 9.68 cSt at 100 °C) having a resistivity of 10^7 ohm cm. Testing was at 1000–1100 rpm under applied 1000 Kgf and 500 Kgf radial loads (P_1 and P_2), and 350 Kgf axial load (P_3), for a duration of 250 h. Current through bearing was 50 A (AC).

The voltage across the test ball bearing 6326 varied from 2.1 to 3 V during the test period to pass 50 A (AC), and the voltage across the NU 326 roller bearing during the same test period varied from 1.12 to 2.3 V between points 'A' and 'B' (Fig. 3.2). The impedance was high before switching on current to the bearing, but the impedance became significantly low immediately after the supply was on. The grease collected from the active zone of the bearings was analyzed and investigations are reported in the Chapter 3. The roller bearing surface condition was monitored periodically after 41, 101, 151 and 250 h of operation. Figures 5.1 and 5.2 show photographs of the bearing inner race after 41 and 250 h of operation, respectively. Magnified views of the bearing surfaces were taken, which indicate the initiation and development of corrugations with time (Figs. 5.3 and 5.4). Figure 5.5 illustrates the corrugation pattern on the inner surface of the NU 230 bearing after about 6000 h of operation in an AC motor. Figure 5.6 and 5.7 show photographs of the inner and outer races of the 6326 ball bearing, respectively after 250 h of operation.

The pitch and width of corrugations were measured on the circumference of races of the test bearings. Similar measurements were made on the different damaged bearings of AC motors after about 5000–6000 h of industrial operation. The data are tabulated in Tables 5.1 and 5.2.

Fig. 5.1. Inner race of NU 326 bearing on exposure of 50 A (AC) for 41 h

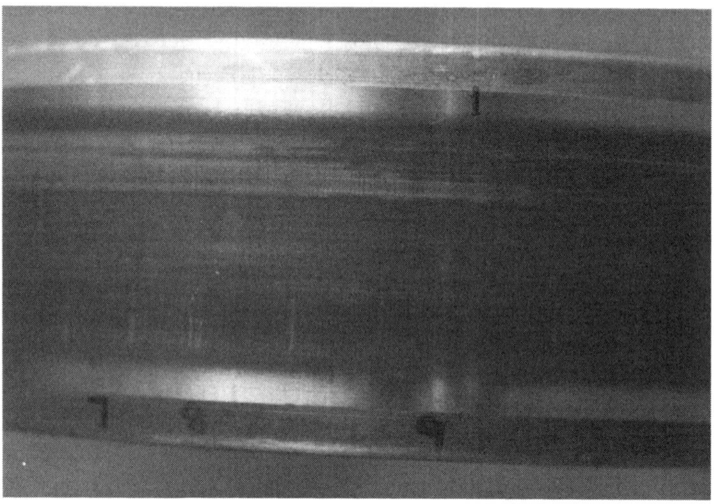

Fig. 5.2. Damaged inner race of NU 326 bearing using lubricant 'A' under electric field on exposure of a 50 A (AC) for 250 h

Fig. 5.3. Scattered corrugations on inner race of NU 326 bearing on exposure of 50 A (AC) for 41 h (enlarged view)

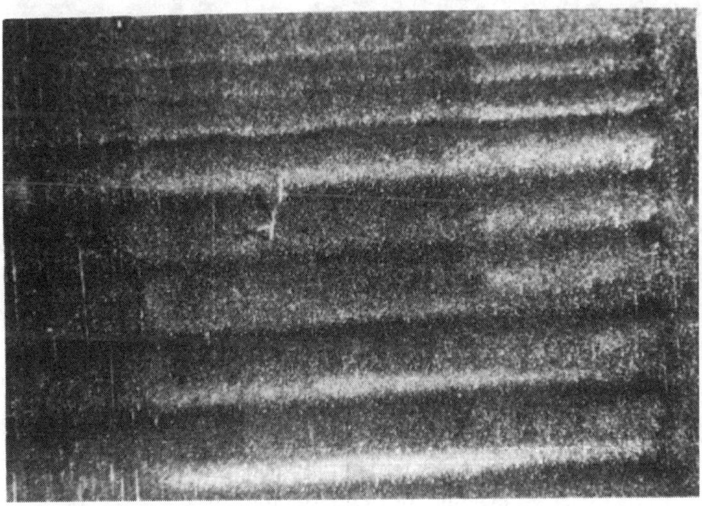

Fig. 5.4. Corrugations on inner race of NU 326 bearing on exposure of 50 A (AC) for 250 h (enlarged view)

Fig. 5.5. Corrugation pattern on the inner race of NU 326 motor bearing after about 6000 h of operation

Fig. 5.6. Inner race of 6326 ball bearing on exposure of 50 A (AC) for 250 h

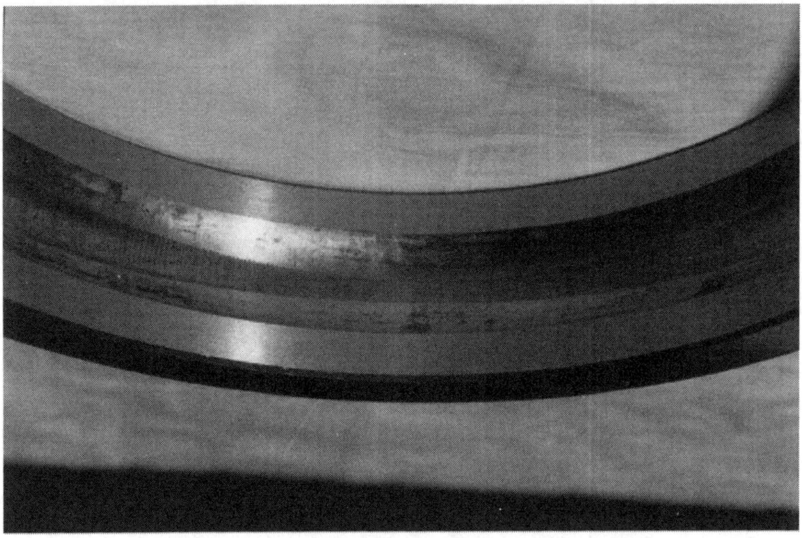

Fig. 5.7. Outer race of 6326 ball bearing on exposure of 50 A (AC) for 250 hours

Surface tempering on the races was detected in the test bearings subjected to an electric current. But no damage was observed on the bearings tested without the passage of current.

The bearing NU 330, lubricated with EP grease having resistivity of 10^{11} ohm cm was tested similar to other bearings. At 1.21 V potential drop, the current through the NU 330 bearing was around 1 A as against 50 A in the other bearing tested with low-resistivity lubricant (10^7 ohm cm).

5.5.2 Roughness Measurement

For measuring roughness on the curved surfaces of a bearing, a perthometer model M 3A has been used. The r_z, r_t and r_a were measured on the races and depth of various corrugations was established. A typical plot of the surface roughness of the inner race of NU 326 bearing around position 1 of Fig. 5.2 is shown in Fig. 5.8.

5.6 Effect of Various factors on Corrugations and on Bearing Surfaces

5.6.1 Effect of Radial Clearance on Pitch and Width of Corrugations

The theoretical derivation of the pitch and width of corrugations on the surfaces of a bearing is based on zero clearance. This could be modified for non-zero clearance conditions by adding the bearing clearance with factors $D + d$, $D - d$,

Table 5.1. Comparison of theoretical and measured values of the pitch of corrugations on races and rollers of various bearings

S. No. (1)	Bearings type (2)	Bearing parameters and operating conditions (3)	Place of corrugation (4)	Corrugation pitch (mm): Measured (average) (5)	Corrugation pitch (mm): Theoretical (6)	Parameters for calculation (7)	Difference in theoretically calculated and measured values (8)
1.	NU 326	$D = 205$ mm	δ_{ir}	0.381	0.392	$p = 2, k = 7$	0.011
		$d = 38$ mm		0.257	0.261	$p = 2, k = 7$	0.004
		$f_s = 18.99$					
		$N = 16$					
		$n = 1000$ rpm					
		$P_1 = 1000$ kgf					
		$P_2 = 500$ kgf					
		On testing NU 326 bearing for a duration of 250 h on exposure of 50 A(AC) at 1.12 to 2.30 V on bearing test machine	Δ_{or}	0.580	570	$p = 2, k = 7$	0.01
							No significant corrugation pattern detected
			Δ_{re}	–	3.25	$p = 2$	
					2.17	$p = 3$	

2.	NU 230	D = 210 mm	Δ_{ir}	0.640	0.621	$p = 1, k = 5$	0.619
		d = 28 mm		1.10	1.034	$p = 1, k = 3$	0.066
		f_s=25					
		N = 17					
		n = 1500 rpm	Δ_{or}	0.90	0.812	$p = 1, k = 5$	0.088
		Bearing taken after 6000 h of operation from motor		–	1.35	$p = 1, k = 3$	–
			Δ_{re}	–	3.52	$p = 1$	No significant corrugation pattern detected
3.	NU 230	Bearing taken after 6000 h of operation from motor operating at 1500 rpm	Δ_{ir}	0.75	0.7755	$p = 1, k = 4$	0.025
				1.12	1.034	$p = 1, k = 3$	0.086
			Δ_{or}	1.10	1.014	$p = 1, k = 4$	0.086
				–	1.35	$p = 1, k = 3$	–
			Δ_{re}	–	3.52	$p = 1$	No significant corrugation pattern detected
4.	NU 230	Damaged bearing from motor operating at 1500 rpm	Δ_{ir}	1.18	1.034	$p = 1, k = 3$	0.146

Table 5.2. Comparison of theoretical and measured values of width and pitch of corrugation on races and balls of various bearings

Ball bearing Type 6326	Bearing kinematics and operating parameters	Place of corrugation	Width of corrugation (mm) Measured (average)	Width of corrugation (mm) Theoretical	Pitch of corrugation (mm) Measured (average)	Pitch of corrugation (mm) Theoretical	Parameters for calculation	Difference in theoretical and measured values of pitch of corrugation (mm)
1.	$D = 206$ mm	W_{ib}, Δ_{ib}	–	1.30	–	3.15	$p = 3$, $k = 3$	
	$d = 46$ mm		1.10	1.20	2.10	2.18	$p = 3$, $k = 4$	0.08
	$f_s = 16.67$ Hz		1.00	1.10	1.55	1.60	$p = 3$, $k = 5$	0.05
	$N = 9$		0.92	0.99	1.10	1.08	$p = 4$, $k = 5$	0.02
	$p_1 = 1000$ kgf	W_{ob}, Δ_{ob}						
	$p_2 = 500$ kgf		–	1.25	–	4.76	$p = 3$, $k = 3$	
	$p_3 = 350$ kgf		–	1.14	–	3.26	$p = 3$, $k = 4$	
	$P = 1170$ kgf {After testing for a duration of 250 h at 50 A (AC) at 2.1–3 V on bearing test machine}		0.98	1.06	2.2	2.43	$p = 3$, $k = 5$	0.23
			0.88	0.94	1.5	1.61	$p = 4$, $k = 5$	0.11
		W_{rb}, Δ_{rb}	–	1.20	–	10.89	$p = 3$	–
			–	1.06	–	9.62	$p = 3$	–

2.	$P = 7000$ kgf (average) (60% of motor rating) $f_s = 25$ Hz	W_{ib}, Δ_{ib}	2.4	2.49	4.30	4.52	$p = 2, k = 4$	0.22
			2.28	2.31	3.30	3.36	$p = 2, k = 5$	0.06
	Bearing from AC motor after about 5000 h of operation	W_{ob}, Δ_{ob}	2.35	2.38	6.32	6.82	$p = 2, k = 4$	0.50
			2.10	2.20	4.8	5.04	$p = 2, k = 5$	0.25
		W_{rb}, Δ_{rb}	–	2.49	–	22.61	$p = 2$	–
3.	$P = 9500$ kgf (average) (75% of motor rating)	W_{ib}, Δ_{ib}	3.0	3.03	7.50	7.34	$p = 3, k = 3$	0.16
	$f_s = 25$ Hz		2.67	2.75	5.20	4.99	$p = 2, k = 4$	0.21
	Bearing from AC motor after about 5000 h of operation	W_{ob}, Δ_{ob}	2.58	2.64	7.4	7.56	$p = 2, k = 4$	0.16
		W_{rb}, Δ_{rb}	–	3.03	–	27.51	$p = 2$	Scattered corrugations
4.	$P = 12\,000$ kgf (100% of motor rating) and $f_s = 25$ Hz	W_{ib}, Δ_{ib}	–	2.97	–	5.9	$p = 2, k = 4$	–
		W_{ob}, Δ_{ob}	–	2.84	–	8.13	$p = 2, k = 4$	–
		W_{rb}, Δ_{rb}	–	2.97	–	26.97	$p = 2$	–

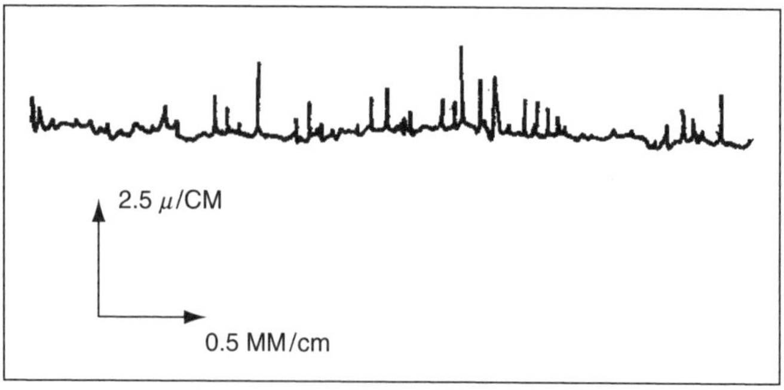

Fig. 5.8. Surface roughness profile of inner race of NU 326 bearing around position I of Fig. 5.2

$D - 2d$ and D in the respective equations of the above sections. This will change the pitch and width of corrugations. However during operation temperature rises, rings expand and non-zero clearance tends to approach to zero clearance conditions.

Based on the derived formulae, the pitch and width of corrugations have been calculated and compared with the measured values (Tables 5.1 and 5.2).

5.6.2 Effect of Electrical Current on Bearing Surfaces

The passage of current causes local heating which leads to low-temperature tempering and accelerates the formation of slip bands leading to the flutings/corrugations on the surfaces with time (Figs. 5.3 and 5.4).

As rolling continues, lubricant deterioration increases under the effect of electric current and small particles of material are pulled out from the craters formed on the tracking surfaces (Figs. 5.1 and 5.2). After long operation under electric current, the softer tempered surfaces of the races become harder; and thus harder/rehardened particles, due to localized high temperature and load, eject from the craters and intensify the depth of corrugations as shown in the Figs. 5.6 and 5.7. Also coalescence of ejected particles in the raceways continues.

5.6.3 Effect of Current on Corrugations

In the presence of low-resistivity grease (10^7 ohm cm), in the close asperity line/point contacts between rolling-elements and races, the current intensity is increased by short circuit (50 A at 1.12–2.1 V). This is accompanied by a gradual increase in voltage with time (2.3–3 V at 50 A in tested roller/ball bearings in 250 h). However, in certain instances, there occurs a momentary drop in resistance and voltage. This may happen during the formation of corrugation and removal of particles from a crater, which creates a better contact of interacting surfaces, and thus results in a drop of resistance. After coalescence of particles, the resistance and voltage build up again.

After formation of corrugations in one half of the races, the resistance become higher due to a decrease in contact area, the current leaks through the other half, and thus full-fledged corrugations are formed on the surfaces (Figs. 5.5–5.7). Corrugations slightly less than the full width of track surface of the races indicate misalignment and unequal corrosion due to lubricant decomposition (Figs. 5.2 and 5.8). However, with time, corrugations would have become uniform across the full width.

5.7 Discussion on Various Aspects of Corrugations

5.7.1 Corrugation Pattern on Bearing Surfaces

The passage of current through roller/ball bearings causes formation of corrugations – a system of transverse grooves – on the surface of races and corrosion due to electrochemical decomposition of the grease (Figs. 5.1, 5.2, 5.7 and 5.8). No such phenomenon is observed in bearings operated without the effect of electric current. However, after 41 h of operation a few scattered corrugations like wire string of a width approximately 30 microns, are observed on the inner race of the NU 326 bearing (Fig. 5.3). As a duration of exposure of current increases, more distant and closely spaced pronounced corrugations of relatively increased width are formed as shown in Fig. 5.4 (after 250 h of operation).

An increase in positions of plane of action of loading (p) and number of rolling-elements in the loaded zone increase the number of corrugations but diminishes their pitch [Equations (5.7), (5.8) and (5.28)–(5.32)]. Conversely, for a ball bearing an increase in width of corrugations with applied load diminishes the number of corrugations but increases the pitch. However, theoretically, the resultant of radial loads acts at one position. Additionally, the ratio of radius of curvature of width of corrugations along a bearing width on the outer to inner race is in general more than unity as shown in Eqn. (5.23) (r_{ob}/r_{ib} is 1.09 for the 6326 bearing). However, a corrugation pattern may appear on full/part bearing width depending on the oscillations of balls in the raceways (Figs. 5.6 and 5.7).

5.7.2 Measured and Theoretically Analyzed Width of Corrugations

The width of corrugations depends on p, K, P, E, d and D of a bearing and is not influenced by f_s [Eqns. (5.16)–(5.19) and (5.40)–(5.43)]. The width of corrugations on the races of the NU 326 test bearing is determined as 0.145 and 0.120 mm, which are found to match closely with the measured values. The width of corrugations on the inner race of a ball bearing is more than that on the outer race because of the concave raceways at the points of interaction with the balls. Theoretically, under identical conditions, the ratio of W_{ib} to W_{ob} is about 1.05 for the bearing 6326, which also matches closely with that of the measured values (Table 5.2).

5.7.3 Pitch of Corrugations

The measured and theoretically evaluated pitches of corrugations on the races match closely. The difference varies between 0.004 and 0.5 mm under the identical operating conditions. The pitch on the outer race is higher than on the inner race. For the NU 326 roller bearing the theoretical ratio of outer to inner race pitch is 1.46 and 1.31 as against measured ratio of 1.52 and 1.41 (Table 5.1). For the 6326 ball bearing, the theoretical ratio of outer to inner race pitch is 1.50 as against an average measured ratio of 1.43 (Table 5.2).

The positions of plane of action of loading (p) are taken as 2 on a ball bearing of a motor during its operation, considering that the bearing is subjected to both radial and axial loads. For the test ball bearing p is 3 since the load is applied from three different directions (two directional radial loads (P_1 and P_2) and unidirectional axial load (P_3). For the test roller bearing p is taken as 2 and for a roller bearing used in motors p is 1.

Different pitches of corrugations are detected at different locations on the ball track of the races of the test bearing (1.55 and 1.10 mm) and damaged bearings of motors (4.30 and 3.30 mm) as shown in Table 5.2. Also, the different pitches on the test roller bearing (0.381 and 0.257 mm) and motor bearings (1.10 and 0.640 mm), (1.12 and 0.75 mm) are detected (Table 5.1). This may be explained by periodically changing characteristics of coupled mechanism. In the corresponding theoretical calculations, this phenomenon is considered by assuming different values of p and K (Tables 5.1 and 5.2). The absence of corrugation pattern on the rolling-elements is due to 2–15 times higher calculated values of pitch than that on the races, which suggest remote possibility of formation of corrugation pattern. However, after a long operating period, a developed pattern on the rolling-elements of a few motor bearings has been observed.

On various locations, superimposition of different corrugation patterns takes place, which generates wear bands on the races (Figs. 5.2, 5.6 and 5.7). The resultant average pitch on these locations can be determined analytically or graphically by superimposition of different patterns of corrugations.

The difference in theoretical and measured values of width and pitch of corrugations may be related to the limitation of accurate assessment of f_s, p, K, P, S_1 and S_2 during long period of operation and changing characteristics of driven mechanism.

5.7.4 Depth of Corrugations

The average measured depth of corrugations on the races of different damaged bearings of AC motors is about 8–10 times higher than that of the test bearings. The average depth on the test bearings is 3 μm. The longer duration of operation of a bearing under electric current may increase the depth of corrugations.

5.8 Conclusions

From the theoretical and experimental investigations reported in this Chapter for roller and ball bearings, the following conclusions are drawn [16] and [20]:

(1) The formation of corrugations occurs on the races of roller/ball bearings in course of time under the influence of electric fields. The corrugations cover almost the entire circumference and act to reduce the actual area of contact.

(2) The width of corrugations on the races is not affected by the frequency of rotation and depends on the load conditions, the pitch and rolling-elements diameters of a bearing, material properties and positions of plane of action of loadings.

(3) The pitch of corrugations on a roller bearing surface depends on kinematics, frequency of rotation, position of plane of action of loading, bearing quality and lubricant characteristics. For a ball bearing pitch depends on circumference of the circle formed by the width of corrugations including the above parameters.

(4) The pitch of corrugations is smaller on the inner race as compared to the outer race, whereas the width of corrugations is higher on the inner race as compared to the outer race.

(5) Under identical conditions, the theoretical ratio of outer to inner race pitch is 1.5, and ratio of inner to outer race width is 1.05 for the ball bearing 6326.

(6) For a ball bearing the ratio of radius of curvature of the corrugation width on the outer to inner race is in general more than unity.

(7) The measured width and pitch of corrugations on a roller/ball bearing races match closely with analytical values.

(8) Without the effect of electric current, corrosion and corrugation pattern as are observed on the bearing surfaces within electric fields are not detected.

(9) The probability of formation of the corrugation pattern on the surface of rolling-elements is remote. However, corrugation may form on the

rolling-elements after prolonged operation of the bearings within electric fields. The pitch and width of corrugations on the rollers/balls may affect the pitch of corrugations already formed on the races.

A bearing using low-resistivity lubricant fails by the mechanism of 'silent' electric discharge accompanied by electrochemical decomposition of the lubricant leading to the corrosion on the bearing surfaces, and by the formation of corrugations on the races by tempering and plastic deformation under localized high temperature and load in the loaded zone of the bearings.

REFERENCES

1. Boyd, J. and Kaufman, H. N. (1959). The causes and the control of electrical currents in bearings. *Lubric. Eng.*, 28–36.
2. Bradford, M. (1984). *Prediction of Bearing Wear due to Shaft Voltage in Electrical Machines.* Era Technology Limited.
3. Chu, P. S. Y. and Cameron, A. (1967). Flow of electrical current through lubricated contacts. *ASLE Trans.*, **10**, 226–234.
4. Kauffman, H. N. and Boyd, J. (1967). The conduction of current through bearings. *ASLE Trans.*, **10**, 226–34.
5. Smith, L. R. (1975). *Bearing Contact Resistance as a Diagnostic Aid.* Shaker Research Corp., N.Y. 12019.
6. Komatsuzaki, S., Vematsu, T. and Nakano, F. (1987). Bearing damage by electrical wear and its effect on deterioration of lubricating greases. *Lubric. Eng.*, **43**(1), 25–30.
7. Furey, J. M. (1961). Metallic contact and friction between sliding surfaces. *ASLE Trans.*, **4**, 1–11.
8. Andreason, S. (1968). Passage of electric current through rolling bearings. *Ball Bearing J.,* **153**, 6–12.
9. Simpson, F. E. and Crump, J. J. (1963). Effect of electrical currents on the life of rolling contact bearings. Lubrication and wear convention, *Inst. Mech. Engrs.*, Paper 27.
10. Morgan, A. W. and Wyllie, D. (1969–1970). A survey of rolling bearing failures. *Proc. Inst. Mech. Engrs.*, **184**, Part 3F, 48–56.
11. Winder, L. R. and Wolfe, O. J. (1968). Valuable results from bearing damage analysis. *Metal Process*, 52–59.
12. Warnock, F. V. and Benham, P. P. (1965). *Mechanism of Solids and Strength of Materials*, Sir Isaac Pitman and Sons, London.
13. Prashad, H., Ghosh, M. and Biswas, S. (1985). Diagnostic monitoring of rolling-element bearings by high frequency resonance technique. *ASLE Trans.*, **28**(4), 439–448.
14. Martin, Ray L. (1970). Detection of ball bearing malfunctions. *J. Instrum. Control Syst.*, 79–82.
15. Prashad, H. (1987). The effect of cage and roller slip on the measured defect frequency response of rolling-element bearings. *ASLE Trans.*, **30**(3) 360–367.

16. Prashad, H. (1988). Investigations of corrugated pattern on the surfaces of roller bearings operated under the influence of electrical fields. ASME/ASLE Tribology Conference, San Antanio Marriott, (October 5–8, 1987), also published in *Lubric. Eng.*, **44**, 710–718.
17. Prashad, H. (1986). Experimental study on influence of electrical fields on behaviour of greases in statically bounded conditions and when used in non-insulated bearings. *BHEL J.*, **7**(3).
18. Prashad, H. (1989). Diagnosis of deterioration of lithium greases used in rolling-element bearings by X-ray diffractrometry. *STLE Trans.*, **32**, 205–214.
19. Haughton P. S. (1976). *Ball and Roller Bearings*, Applied Science Publishers Limited, London.
20. Prashad, H. (1991). Theoretical and experimental investigations on the pitch and width of corrugations on the surfaces of ball bearings. *Wear*, **143**, 1–14.

Nomenclature

C_{ir}, C_{or}	arc length of corrugations on roller track of inner race and outer race, respectively
C_{ib}, C_{ob}	arc length of corrugations on ball track of inner race and outer race, respectively
C_{re}, C_{rb}	arc length of corrugations on surface of roller and ball, respectively
d	diameter of rolling-element
d_1	diameter of rolling-element 1
d_2	diameter of rolling-element 2
D	pitch diameter
E	Young's modulus of elasticity
f_s	shaft rotational frequency
f_b	ball rotational frequency
K	number of rolling-elements in the loaded zone
M_{ir}, M_{or}	number of similar corrugations on an arc length on roller track of inner and outer race, respectively
M_{ib}, M_{ob}	number of similar corrugations on an arc length on ball track of inner race, and outer race, respectively
M_{re}, M_{rb}	number of similar corrugations on an arc length on surface of roller and ball, respectively
N	number of rolling-elements
n	rpm
p	positions of plane of action of loading ($p = 1, 2, 3, \ldots$)
P_1	radial load on bearing at position 1

P_2	radial load on bearing at position 2
P_3	axial load on bearing
P	resultant load on bearing
r	radius of rolling-element
r_{ib}, r_{ob}	radius of curvature of width of corrugation along bearing width on the inner race, and on the outer race, respectively
r_i, r_o	rolling-element track radius of inner race and outer race, respectively
r_t	maximum roughness depth
r_z	average roughness depth
r_a	average roughness
r_{12}	radius of contact area of balls of diameters d_1 and d_2
S_1, S_2	bearing coefficient for roller and ball, respectively
W_{ir}, W_{or}	width of corrugations on roller track of inner race and outer race, respectively
W_{ib}, W_{ob}	width of corrugations on ball track of inner race and outer race, respectively
W_{re}, W_{rb}	width of corrugations on roller and ball, respectively
W_w	width of contact on rollers of diameters d_1 and d_2
Δ_{ir}, Δ_{or}	pitch of corrugations on roller track of inner race and outer race, respectively
Δ_{ib}, Δ_{ob}	pitch of corrugations on ball track of inner race and outer race, respectively
Δ_{re}, Δ_{rb}	pitch of corrugations on surface of roller and ball, respectively
α	contact angle
β	overlapping coefficient of corrugation width on rolling-element due to width of corrugations on races

Chapter 6

ELECTRICAL PARAMETERS OF ROLLING-ELEMENT BEARINGS

6.1 Introduction

In rotating rolling-element bearings, the actual surface area of contact depends on surface roughness, and comprises 'asperity contacts' primarily. The bearing performance is improved if the film thickness is 2–3 times greater than that of the average surface roughness depth. Under the influence of potential drop across a bearing, the minimum film thickness between the race and the rolling-elements offers maximum capacitance and minimum capacitive reactance, depending on the permittivity of the lubricant. The active resistance offered by a bearing is a minimum at a minimum film thickness; however it is primarily governed by resistivity of the lubricant. The electrical interaction between the races and the rolling-elements in the presence of oil film is like a resistor capacitor (RC) circuit and offers an impedance to current flow.

This chapter deals with the studies to analyze theoretically the capacitance and active resistance between the track surfaces of races and the rolling-elements, at the minimum film thickness under different parameters of operation, and to study the equivalent bearing capacitive reactance and resistance in order to analyze the effect of permittivity and resistivity of the used lubricants and the role of equivalent capacitance on the bearing behavior. Besides this, the maximum limit of charge accumulation on the track surface of races was determined at the first threshold voltage, at different operating parameters, which if exceeded produces a momentary flow of current through the bearings, as determined in Chapter 4.

6.2 Electrical Parameters of Roller Bearings

6.2.1 Theoretical Determination of Capacitance and Active Resistance between Races and a Roller (Approximate Method)

6.2.1.1 Capacitance

The width of deformation along the roller length (L) is separated throughout by a minimum film thickness (h_o). This forms a parallel plate capacitor between the inner race and the roller area $W_{ir}L$, in the presence of a lubricant of dielectric constant, ξ, the capacitance of which is given as [1]:

$$C_{ir} = \frac{\xi W_{ir} L}{h_o} \tag{6.1}$$

Similarly, the capacitance between the outer race and a roller is given as:

$$C_{or} = \frac{\xi W_{or} L}{h_o} \tag{6.2}$$

The minimum film thickness of a roller bearing is determined by Grubin's formula given in Chapter 4, and W_{ir} and W_{or} are determined by Eqns. (5.16) and (5.17), respectively.

6.2.1.2 Active Resistance

The width of deformation W_{ir}, in the presence of a lubricant of resistivity ρ and film thickness h_o, forms electrodes of area $W_{ir}L$ on the roller track of inner race and a roller under the influence of the electric fields.

The minimum resistance between them is given as:

$$R_{ir} = \frac{\rho_1 h_o}{W_{ir} L} \tag{6.3}$$

Similarly, the minimum resistance between the outer race and a roller is given by:

$$R_{or} = \frac{\rho_1 h_o}{W_{or} L} \tag{6.4}$$

6.2.2 Theoretical Determination of Capacitance and Active Resistance between Races and a Roller through Bearing Kinematics (Detailed Method)

6.2.2.1 Capacitance

The capacitance of the parallel plate capacitor formed by the roller track of a race and a roller is maximum at the asperity line contact at the minimum film thickness and decreases on both sides of the contact. The film thickness h at an arbitrary point C on a roller at a distance X from the minimum film thickness is given as (Fig. 6.1):

$$h = \mathrm{AC} - \mathrm{AB} \tag{6.5}$$

where

$$\mathrm{AC} = \mathrm{DG} + \mathrm{DE}$$

or

$$\mathrm{AC} = h_o + r(1 - \cos\theta)$$

Since θ is a small angle, $\sin\theta = \theta$, and $\cos\theta = (1 - \theta^2)/2$.
Hence,

$$\mathrm{AC} = h_o + r\left\{1 - \left(1 - \frac{X^2}{2r^2}\right)\right\}$$

or

$$\mathrm{AC} = h_o + \frac{X^2}{2r} \tag{6.6}$$

Since, B is a point on the inner surface of the outer race, so:

$$\mathrm{AB} = \frac{X^2}{2r_o} \tag{6.7}$$

Using Eqns. (6.6) and (6.7), the film thickness from Eqn. (6.5) is determined as:

$$h = \frac{h_o + X^2}{2r} - \frac{X^2}{2r_o}$$

or

$$h = h_o + \delta X^2 \tag{6.8}$$

where

$$\delta = 0.5\left(\frac{1}{r} - \frac{1}{r_o}\right) \tag{6.9}$$

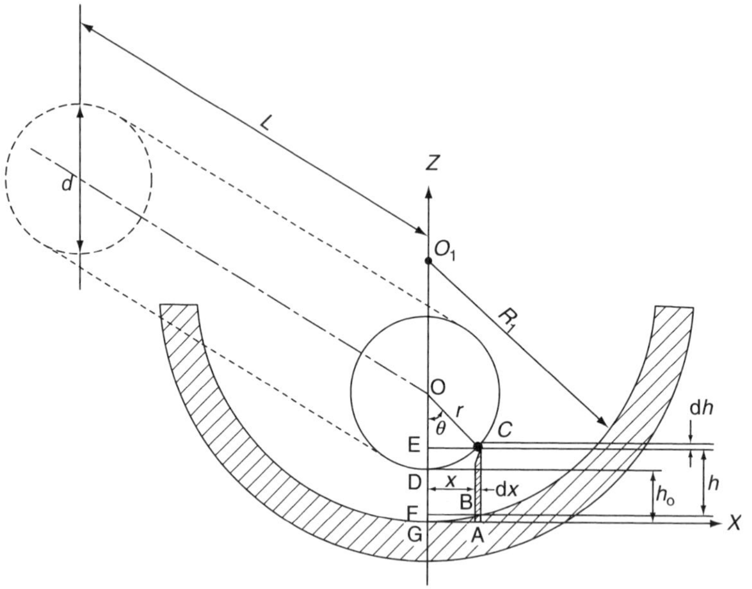

Fig. 6.1. Roller configuration in the outer race of a bearing under rotation

Similarly, from Fig. 6.2, the film thickness h_1 at an arbitrary point C_1 on a roller at a distance X from the minimum film thickness is determined as:

$$h_1 = A_1C_1 + A_1B_1 \tag{6.10}$$

Similar to Eqn. (6.8), the film thickness is determined as:

$$h_1 = h_o + \phi X^2 \tag{6.11}$$

where

$$\phi = 0.5\left(\frac{1}{r} + \frac{1}{r_i}\right) \tag{6.12}$$

From Figs. 6.1 and 6.2, the change in capacitance ΔC between two rectangular plates of width dX, situated in between the distances $X + \mathrm{d}X$ and X along X-axis, having a length L (along the Y-axis), and separated by a film thickness varying between h and $h + \Delta h$, is given by:

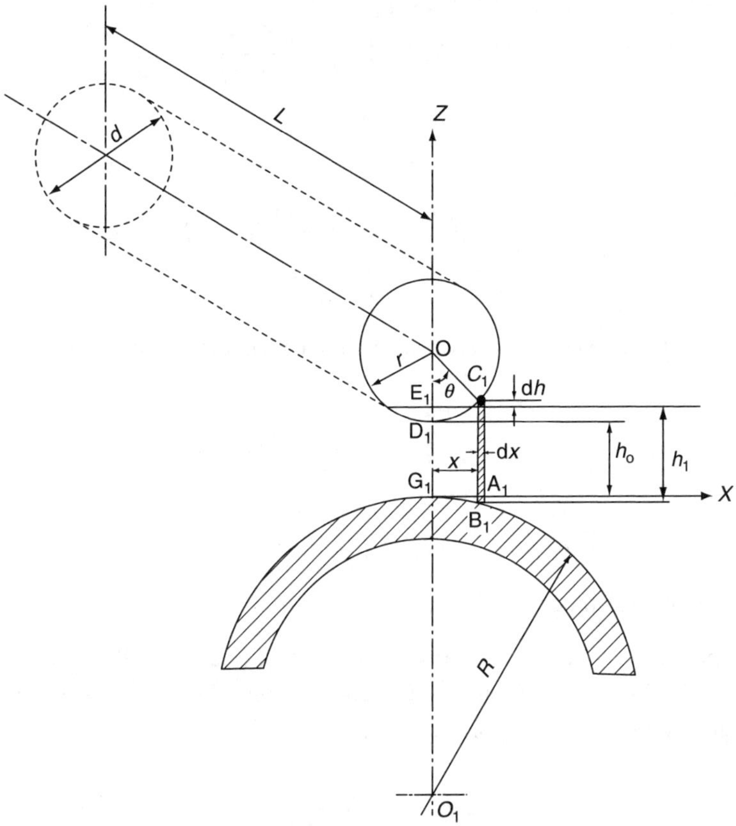

Fig. 6.2. Roller configuration in the inner race of a bearing under rotation

$$\Delta C = \frac{\xi\{(X + dX)L - XL\}}{0.5(2h + \Delta h)} \tag{6.13}$$

The change in film thickness Δh is very small in comparison with dX, and hence can be neglected. Thus the change in capacitance is given as:

$$\Delta C = \frac{\xi dXL}{(h_o + \delta X^2)} \tag{6.14}$$

The maximum change in capacitance is along the width of deformation between the race and a roller. Therefore on integrating between 0 and $W_{or}/2$, and considering the deformation on both sides of the minimum film thickness, the capacitance between the roller track of outer race and a roller is given as [2]:

$$C_{\mathrm{or}} = 2\xi L \int_{0}^{W_{\mathrm{or}}/2} \frac{\mathrm{d}X}{(h_{\mathrm{o}} + \delta X^2)} \tag{6.15}$$

or

$$C_{\mathrm{or}} = \frac{2\xi L \left\{ \int^{\mathrm{o}} \tan^{-1} W_{\mathrm{or}} \frac{(\delta/h_{\mathrm{o}})^{1/2}}{2} \right\}}{(\delta h_{\mathrm{o}})^{1/2}} \tag{6.16}$$

Similarly, the capacitance between the inner race and a roller in the loaded zone is given as:

$$C_{\mathrm{ir}} = \frac{2\xi L \left\{ \tan^{-1} W_{\mathrm{ir}} \frac{(\phi/h_{\mathrm{o}})^{1/2}}{2} \right\}}{(\phi h_{\mathrm{o}})^{1/2}} \tag{6.17}$$

6.2.2.2 Active Resistance

The change in conductance between the outer race and a roller of the element of the area dX L, situated between $X + \mathrm{d}X$ and X along the X–axis, and separated by the average gap between h and $h + \Delta h$ is given as:

$$\Delta F = \frac{\mathrm{d}XL}{\rho_1 (h_{\mathrm{o}} + \delta X^2)} \tag{6.18}$$

The maximum change in conductance/resistance^{-1} is along the width of deformation between the race and a roller. Therefore on integrating [cf. Eqn. (6.15)] between 0 and $W_{\mathrm{or}}/2$, the resistance between the outer race and a roller is determined as:

$$F = \frac{1}{R_{\mathrm{or}}} = \left(\frac{2L}{\rho_1}\right) \int_{0}^{W_{\mathrm{or}}/2} \frac{\mathrm{d}X}{(h_{\mathrm{o}} + \delta X^2)} \tag{6.19}$$

or

$$\frac{1}{R_{\mathrm{or}}} = \frac{2L \left\{ \tan^{-1} W_{\mathrm{or}} \frac{(\delta/h_{\mathrm{o}})^{1/2}}{2} \right\}}{\rho_1 (\delta h_{\mathrm{o}})^{1/2}}$$

and

$$R_{\text{or}} = \frac{\rho_1 (\delta h_{\text{o}})^{1/2}}{2L\left\{\tan^{-1} W_{\text{or}} \frac{(\delta/h_{\text{o}})^{1/2}}{2}\right\}} \tag{6.20}$$

Similarly, the resistance between the inner race and a roller in the loaded zone is given as:

$$R_{\text{ir}} = \frac{\rho_1 (\phi h_{\text{o}})^{1/2}}{2L\left\{\tan^{-1} W_{\text{ir}} \frac{(\phi/h_{\text{o}})^{1/2}}{2}\right\}} \tag{6.21}$$

6.2.3 Determination of Capacitive Reactance and Impedance between Race and a Roller

Capacitive reactance between the outer race and a roller is given as [1]:

$$X_{\text{cor}} = \frac{1}{wC_{\text{or}}} \tag{6.22}$$

and

$$X_{\text{cir}} = \frac{1}{wC_{\text{ir}}} \tag{6.23}$$

The equivalent electric circuit of the outer race and roller is considered as a series RC circuit. The impedance of the equivalent circuit is given as:

$$Z_{\text{or}} = R_{\text{or}} - jX_{\text{cor}} \tag{6.24}$$

Similarly, the impedance between the inner race and a roller is determined as:

$$Z_{\text{ir}} = R_{\text{ir}} - jX_{\text{cir}} \tag{6.25}$$

6.2.4 Equivalent Impedance, Resistance and Capacitive Reactance of a Roller Bearing

The equivalent electric circuit of a bearing is shown in the Fig. 6.3. The equivalent circuits of the inner and outer races with the rollers are parallel RC circuits, whereas both inner and outer race circuits are in series with each other.

The equivalent impedance between outer race and rollers is given as:

$$\frac{1}{Z_{\text{oreq}}} = \frac{1}{Z_{\text{or1}}} + \frac{1}{Z_{\text{or2}}} + \frac{1}{Z_{\text{or3}}} + \ldots + \frac{1}{Z_{\text{ork}}} \tag{6.26}$$

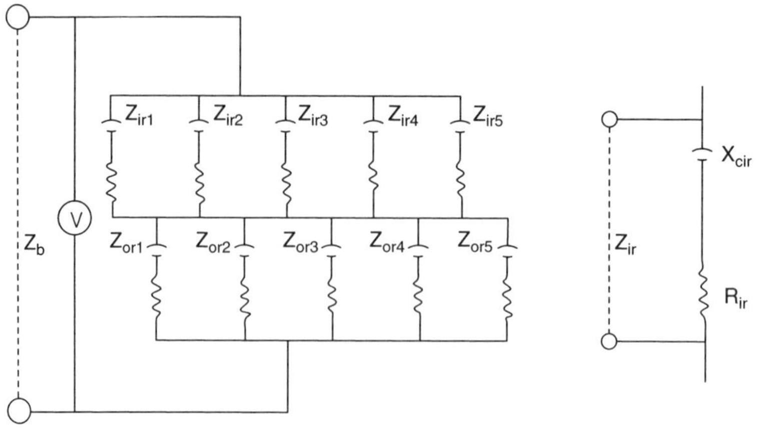

a) Equivalent electric circuit of a bearing

b) Equivalent electric circuit between race and a roller

Fig. 6.3. Equivalent electric circuit of a bearing and between race and a roller

and that between inner race and rollers is determined as:

$$\frac{1}{Z_{\text{ireq}}} = \frac{1}{Z_{\text{ir1}}} + \frac{1}{Z_{\text{ir2}}} + \frac{1}{Z_{\text{ir3}}} + \ldots + \frac{1}{Z_{\text{irk}}} \tag{6.27}$$

Assuming that all the rollers in the loaded zone (K) of a bearing, under the inner and outer races, are equally loaded, impedance between race and an individual roller is equal. Hence,

$$Z_{\text{or1}} = Z_{\text{or2}} = \ldots = Z_{\text{ork}} = Z_{\text{or}}$$

and

$$Z_{\text{ir1}} = Z_{\text{ir2}} = \ldots = Z_{\text{irk}} = Z_{\text{ir}}$$

From Eqns. (6.26) and (6.27), it is evident that

$$\frac{1}{Z_{\text{oreq}}} = \frac{K}{Z_{\text{or}}}$$

or

$$Z_{\text{oreq}} = \frac{Z_{\text{or}}}{K} \tag{6.28}$$

Similarly

$$Z_{\text{ireq}} = \frac{Z_{\text{ir}}}{K} \tag{6.29}$$

and the equivalent bearing impedance is:

$$Z_{\text{b}} = \frac{(Z_{\text{ir}} + Z_{\text{or}})}{K} \tag{6.30}$$

Also, using Eqns. (6.24) and (6.25), the equivalent bearing impedance is given as:

$$Z_{\text{b}} = \frac{\{(R_{\text{ir}} + R_{\text{or}}) - j(X_{\text{cir}} + X_{\text{cor}})\}}{K} \tag{6.31}$$

and the equivalent bearing resistance and capacitive reactance are determined as:

$$R_{\text{b}} = \frac{(R_{\text{ir}} + R_{\text{or}})}{K} \tag{6.32}$$

$$X_{\text{cb}} = \frac{(X_{\text{cir}} + X_{\text{cor}})}{K} \tag{6.33}$$

The ratio of capacitive reactance and resistance is, hence, given as:

$$\frac{X_{\text{cb}}}{R_{\text{b}}} = \frac{(X_{\text{cir}} + X_{\text{cor}})}{(R_{\text{ir}} + R_{\text{or}})} \tag{6.34}$$

6.2.5 Equivalent Bearing Capacitance and Stored Charge

The equivalent bearing capacitance is determined from the equivalent capacitive reactance [Eqn. (6.33)] and is given as:

$$C_{\text{b}} = \frac{K}{W(X_{\text{cir}} + X_{\text{cor}})} \tag{6.35}$$

The equivalent capacitance of a bearing opposes any change in the existing voltage and causes the electric charge to be stored. The stored charge in the bearing is determined as:

$$Q = V \cdot C_{\text{b}} \tag{6.36}$$

A change in Q results in a current flow in the capacitor. It is this current which opposes a voltage change across the capacitor and is determined by [1]:

$$I_b = \frac{C_b \mathrm{d}V}{\mathrm{d}t} \tag{6.37}$$

where $\mathrm{d}V/\mathrm{d}t$ is the rate of voltage change in volts per second.

6.3 Electrical Parameters of Ball Bearings

6.3.1 Capacitance and Active Resistance of Ball Bearings

6.3.1.1 Theoretical Determination of Capacitance

6.3.1.1.1 Capacitance between Outer Race and a Ball

The capacitance of the parallel plate capacitor formed by the ball track of race and a ball is maximum at the point contact at the minimum film thickness, and reduces on both sides of the contact. The film thickness (h) at an arbitrary point C on a ball at a distance X from the minimum film thickness (h_o) between outer race and a ball (Fig. 6.4) is given as:

$$h = h_o + \delta_1 X^2 \tag{6.38}$$

where δ_1 for a ball bearing is the same as δ of a roller bearing given in Eqn. (6.9).

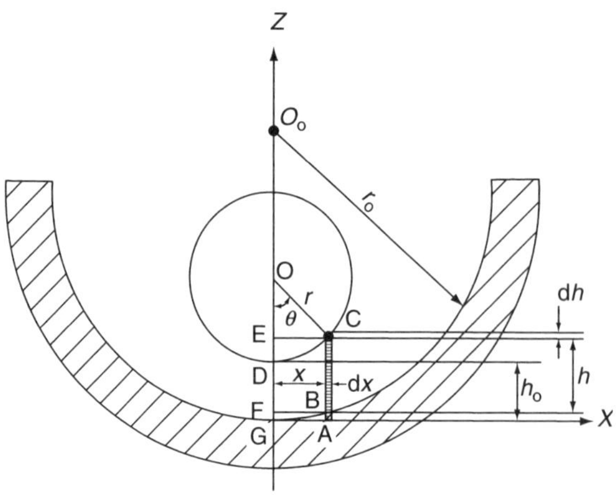

Fig. 6.4. Ball configuration in outer race of a bearing under rotation

The film thickness at an arbitrary point along Y-axis on the ball track surface of the outer race is determined similar to Eqn. (6.38), and is given as:

$$h = h_o + \delta_2 Y^2 \tag{6.39}$$

where δ_2 is derived similar to Eqn. (6.9), and given as:

$$\delta_2 = 0.5\left(\frac{1}{r} - \frac{1}{r_{ob}}\right) \tag{6.40}$$

and r_{ob} is given by Eqn. (5.22) (Chapter 5).

From Eqn. (6.38) and (6.39), it is evident that:

$$Y = \left(\frac{\delta_1}{\delta_2}\right)^{1/2} X = kX \tag{6.41}$$

where

$$k = \left(\frac{\delta_1}{\delta_2}\right)^{1/2}$$

The point contact between the ball track of outer race and a ball is an ellipse with major and minor axis as $2X$ and $2Y$, the area of contact (A) between ball track surface of outer race and a ball is obtained, by using Eqn. (6.41) and given as [3]:

$$A = \mathrm{p}kX^2 \tag{6.42}$$

From Fig. 6.4, the change in capacitance (ΔC), along X-axis between two ellipses having major axis as $X + \mathrm{d}X$ and X, respectively, and separated by a film thickness varying between $h + \Delta h$ and h, is given as:

$$\Delta C = \mathrm{p}k\xi \frac{\{(x + \mathrm{dx})^2 - x^2\}}{\dfrac{(2h + \Delta h)}{2}} \tag{6.43}$$

The change in film thickness (Δh) is very small in comparison to X, and hence can be neglected. Similarly, $\mathrm{d}X^2$ is also neglected. Thus, the change in capacitance using Eqn. (6.38) is given as:

$$\Delta C = \frac{2\mathrm{p}k\xi x dX}{(h_o + \delta_1 X^2)} \tag{6.44}$$

The maximum change in capacitance is along the width of deformation between outer race and a ball, which is determined by Eqn. (5.41). Therefore, on integrating between 0 to $W_{ob}/2$, and considering the deformation on both sides of minimum film thickness, the capacitance between outer race and a ball is given as:

$$C_{ob} = 4\text{p}\xi k \int_0^{W_{ob}/2} X \frac{\mathrm{d}X}{h_o + \delta_1 X^2}$$

or

$$C_{ob} = 2\text{p}\xi k \delta_1^{-1} \log_e \left(\frac{1 + \delta_1 W_{ob}^2}{4h_o} \right) \tag{6.45}$$

6.3.1.1.2 Capacitance between Inner Race and a Ball

The film thickness (h_1) at an arbitrary point on a ball at a distance X from minimum film thickness (h_o) between ball track of inner race and a ball is calculated similar to Eqn. (6.11) and is given as [3]:

$$h_1 = h_o + \phi_1 X^2 \tag{6.46}$$

where ϕ_1 of a ball bearing is the same as ϕ of a roller bearing as given in Eqn. (6.12).

The film thickness (h_1) at an arbitrary point along the Y-axis on the ball track of the inner race is determined similar to Eqn. (6.39), and is given as:

$$h_1 = h_o + \phi_2 Y^2 \tag{6.47}$$

where ϕ_2 is determined similar to Eqn. (6.40), and is given as:

$$\phi_2 = \frac{\left(\frac{1}{r} - \frac{1}{r_{ib}} \right)}{2} \tag{6.48}$$

and r_{ib} is determined by Eqn. (5.21).

From Eqns. (6.46) and (6.47), it is evident that:

$$Y = \left(\frac{\phi_1}{\phi_2} \right)^{1/2}, X = k_1 X \tag{6.49}$$

where

$$k_1 = \left(\frac{\phi_1}{\phi_2} \right)^{1/2}$$

Similar to Eqn. (6.45), the capacitance between ball track of inner race and a ball in the loaded zone is given as:

$$C_{\mathrm{ib}} = 2\mathrm{p}\xi k_1 \phi_1^{-1} \log_e \left(\frac{1 + \phi W_{\mathrm{ib}}^2}{4h_{\mathrm{o}}} \right) \tag{6.50}$$

where W_{ib} is determined by Eqn. (5.40), and $\mathrm{h_o}$ by Grubin's formula given in Chapter 4 [4].

6.3.1.2 Active Resistance

6.3.1.2.1 Active Resistance between Outer Race and a Ball

The change in conductance/(Resistance)$^{-1}$ along X-axis between two ellipses having half major axis as $X + \mathrm{d}X$ and X respectively, and separated by the average gap varying between $h + \Delta h$ and h, is given as:

$$\Delta F = \frac{\mathrm{p}k\{(X + dX)^2 - X^2\}}{0.5\rho_1(2h + \Delta h)} \tag{6.51}$$

The maximum change in conductance is along the width of deformation between ball track of outer race and a ball. Thus on integrating between 0 and $W_{\mathrm{ob}}/2$ and neglecting Δh and $\mathrm{d}X^2$, similar to Eqn. (6.45), the resistance between ball track of outer race and a ball ($R_{\mathrm{ob}} = 1/F$) is determined as:

$$R_{\mathrm{ob}} = \frac{\rho_1 \delta_1}{2\mathrm{p}k} \log_e \left(\frac{1 + \delta_1 W_{\mathrm{ob}}^2}{4h_{\mathrm{o}}} \right) \tag{6.52}$$

6.3.1.2.2 Active Resistance between Inner Race and a Ball

Similar to Eqn. (6.52), the active resistance between ball track of inner race and a ball is given as:

$$R_{\mathrm{ib}} = \frac{\rho_1 \phi_1}{2\mathrm{p}k_1} \log_e \left(\frac{1 + \phi_1 W_{\mathrm{ib}}^2}{4h_{\mathrm{o}}} \right) \tag{6.53}$$

6.3.2 Determination of Capacitive Reactance, Active Resistance and Impedance of Ball Bearings

Similar to the Eqns. (6.22)–(6.34) for roller bearings, the capacitive reactance, active resistance and impedance of ball bearings are derived as per the equivalent electrical circuit shown in Fig. 6.5, and are given in the following equations:

$$X_{\mathrm{cob}} = \frac{1}{wC_{\mathrm{ob}}} \quad \text{and} \quad X_{\mathrm{cib}} = \frac{1}{wC_{\mathrm{ib}}} \tag{6.54}$$

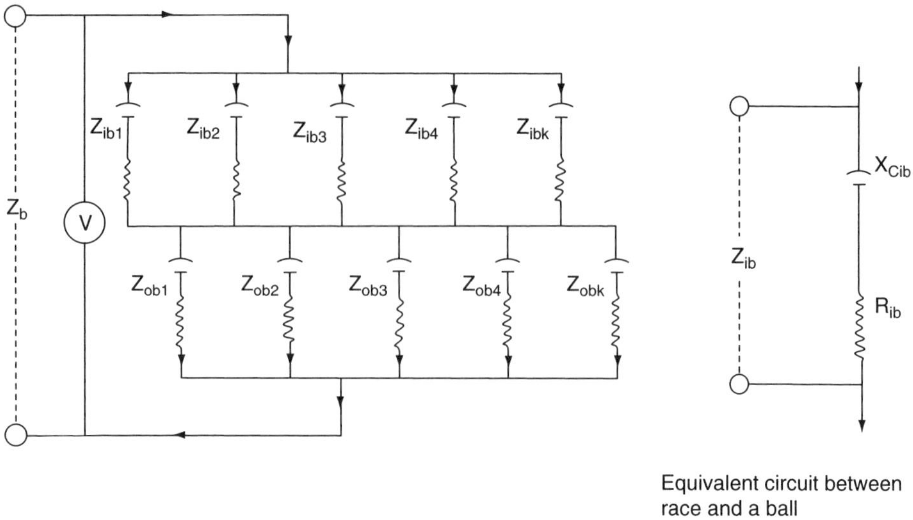

Fig. 6.5. Electrical equivalent bearing circuit

$$Z_{ob} = R_{ob} - jX_{cob} \quad \text{and} \quad Z_{ib} = R_{ib} - jX_{cib} \tag{6.55}$$

$$Z_{obeq} = \frac{Z_{ob}}{K} \quad \text{and} \quad Z_{ibeq} = \frac{Z_{ib}}{K} \tag{6.56}$$

$$Z_{bb} = \frac{[(R_{ib} + R_{ob}) - j(X_{cib} + X_{cob})]}{K} \tag{6.57}$$

and equivalent bearing resistance and capacitive reactance are determined as:

$$R_{bb} = \frac{(R_{ib} + R_{ob})}{K} \quad \text{and} \quad X_{cbb} = \frac{(X_{cib} + X_{cob})}{K} \tag{6.58}$$

The ratio of capacitive reactance to active resistance is given as:

$$\frac{X_{cbb}}{R_{bb}} = \frac{(X_{cib} + X_{cob})}{(R_{ib} + R_{ob})} \tag{6.59}$$

From the above analysis and using Eqns. (6.1)–(6.4), this ratio can be expressed as:

$$\frac{X_{cbb}}{R_{bb}} = \frac{1}{2pf}\xi\rho_1 = \frac{X_{cb}}{R_b} \tag{6.60}$$

6.3.3 Equivalent Bearing Capacitance and Stored Charges

The ball bearing equivalent capacitance, similar to Eqn. (6.35) is determined from the ball bearing capacitive reactance and is given as:

$$C_{bb} = \frac{K}{w(X_{cib} + X_{cob})} \tag{6.61}$$

and charge accumulation is

$$Q = VC_{bb} \tag{6.62}$$

6.4 Data Under Different Operating Parameters

The equivalent capacitance, resistance, capacitive reactance and impedance of roller and ball bearings using lubricant of different resistivities are determined using Eqns. worked out in the Sections 6.2 and 6.3. The data are tabulated in the Tables 6.1–6.4. Figure 6.6 shows the variation of capacitance with speed at different radial loads for the NU 330 bearing, using high-resistivity lubricant. For the evaluation of W_{ir} and W_{or} for a roller bearing, p is taken as unity, since the load applied was unidirectional, and the value of K as 3. For the test ball bearing p is taken as 3, since the three directional loads was applied, and for motor bearing p as 2.The threshold voltages of bearings, determined in the Chapter 4, have been used to determine the charge accumulation on the bearing surfaces.

6.5 Review of Results

6.5.1 Roller Bearings

6.5.1.1 Equivalent Bearing Capacitive Reactance, Resistance and Impedance

For roller bearings using high-resistivity (10^{11} ohm cm) and low-resistivity (10^7 ohm cm) lubricants with a relative permittivity of 2.5, the ratio of capacitive reactance to active resistance is 0.14 and 14×10^2, respectively, and is constant up to the first threshold voltage under the different parameters of operation (Tables 6.2 and 6.4). It is evident from the threshold voltage phenomenon as discussed in Chapter 4 that the bearing with high-resistivity lubricant shows capacitive behavior ($X_{cb}/R_b = 0.14$) and accumulates the electric charge up to the first threshold voltage. On the contrary, a bearing with low-resistivity lubricant behaves like a resistor ($X_{cb}/R_b = 14 \times 10^2$) and passes current without indicating threshold voltage phenomenon. Furthermore, passage of current disintegrates the low-resistivity lubricant and the bearing fails in due course, which suggests that X_{cb}/R_b should be less than unity for capacitive response of a bearing.

Table 6.1. Capacitance, capacitive reactance and resistance between races and a roller for a NU 330 bearing at different operating parameters[a]

Speed (rev min^{-1})	P (× 10) (N)	V_{T1} [6.3] (mV)	W_{ir} (mm) at $p = 1$, $K = 3$	W_{or} (mm) at $p = 1$, K = 3	h_o ($\times 10^{-4}$) (mm)	C_{or} (pϕ)	C_{ir} (pϕ)	X_{cir} ($\times 10^7$) (Ω)	X_{cor} ($\times 10^7$) (Ω)	R_{ir} ($\times 10^7$) (Ω)	R_{or} ($\times 10^7$) (Ω)
450	450	50	0.163	0.195	22.4	86.61	72.42	4.4	3.68	3.53	25.5
450	750	45	0.210	0.252	21.2	118.4	98.67	3.2	2.6	22.4	18.6
450	1000	40	0.243	0.291	20.7	140.03	116.93	2.72	2.3	18.96	15.8
750	450	–	0.163	0.195	32.5	59.76	49.86	6.39	5.33	44.4	37.0
750	750	200	0.210	0.252	30.8	81.44	67.87	4.6	3.83	32.6	27.17
750	1000	150	0.243	0.291	30.1	96.50	80.42	3.96	3.30	27.5	22.9
1000	450	270	0.163	0.195	40.0	48.58	40.58	7.85	6.56	54.5	45.6
1000	750	210	0.210	0.252	38.0	66.06	55.05	5.7	4.8	40.21	33.5
1000	1000	200	0.243	0.291	37.0	78.36	65.39	4.9	4.1	33.8	28.3
1200	450	850	0.163	0.195	45.7	42.5	35.53	8.97	7.5	62.3	52.1
1200	750	–	0.210	0.252	43.4	57.84	48.20	6.60	5.62	45.93	38.27
1200	1000	580	0.243	0.291	42.3	69.58	57.98	5.6	4.6	38.6	32.2

[a] d = 42 mm, $r_i = 96.5$ mm, D = 235 mm, $r_o = 138.5$ mm, Lzz = 45 mm, N = 14, $\delta = 0.020\,mm^{-1}$, $\phi = 0.029\,mm^{-1}$, $\rho_1 = 10^{11}\,\Omega cm$, $\xi = 2.5 \times 8.85419 \times 10^{-12}\,\phi\,m^{-1}$

Table 6.2. Equivalent capacitance, capacitive reactance, resistance, impedance and charge accumulation on NU 330 bearing surfaces at different operating parameters

Speed (rev min^{-1})	P ($\times$ 10) (N)	Z_{ir} ($\times 10^7$) (Ω)	Z_{or} ($\times 10^7$) (Ω)	R_b ($\times 10^7$) (Ω)	X_{cb} ($\times 10^7$) (Ω)	Z_b ($\times 10^7$) (Ω)	I_b ($\times 10^{-10}$) (A)	C_b (pϕ)	Q ($\times 10^{-12}$) (C)	Q (average) ($\times 10^{-12}$) (C)	X_{cb}/R_b	I_b (average) ($\times 10^{-10}$) (A)
450	450	30.82	25.76	18.67	2.69	18.86	2.65	118.46	5.92	6.94	0.14	3.10
450	750	22.63	18.78	13.67	1.93	13.81	3.25	164.7	7.41	6.94	0.14	3.10
450	1000	19.15	15.96	11.59	1.67	11.71	3.41	190.4	7.62	6.94	0.14	3.10
750	450	44.86	37.38	27.13	3.91	27.41	–	81.46	–	21.17	0.14	9.39
750	750	32.92	27.44	19.92	2.81	20.10	9.94	113.1	22.62	21.17	0.14	9.39
750	1000	27.78	23.14	16.80	2.42	16.97	8.4	131.41	19.71	21.17	0.14	9.39
1000	450	55.06	46.03	33.36	4.80	33.70	8.01	66.3	17.90	19.38	0.14	8.68
1000	750	40.61	33.84	24.57	3.50	24.82	8.46	90.85	19.07	19.38	0.14	8.68
1000	1000	34.15	28.59	20.70	3.0	20.90	9.56	106.0	21.20	19.38	0.14	8.68
1200	450	62.94	52.64	38.13	5.49	38.52	22.06	57.9	49.22	51.86	0.14	23.19
1200	750	46.40	38.68	28.06	4.07	28.35	–	78.23	–	51.86	0.14	23.19
1200	1000	39.0	32.53	23.6	3.4	23.84	24.32	94.0	54.5	51.86	0.14	23.19

Table 6.3. Electrical parameters of bearing NU 330 using low-resistivity lubricant

V_{T1} (mV)	h_o ($\times 10^{-4}$) (mm)	C_{ir} (pϕ)	C_{or} (pϕ)	X_{cir} ($\times 10^7$) (Ω)	X_{cor} ($\times 10^7$) (Ω)	R_{ir} ($\times 10^3$) (Ω)	R_{or} ($\times 10^3$) (Ω)
≈ 0.0	28.0	57.98	69.58	5.48	4.57	38.2	31.8
Z_{ir} ($\times 10^7$) (Ω)	Z_{or} ($\times 10^7$) (Ω)	R_b ($\times 10^3$) (Ω)	X_{cb} ($\times 10^7$) (Ω)	Z_b ($\times 10^7$) (Ω)	C_b (pϕ)	X_{cb}/R_b	Q ($\times 10^{12}$) (C)
5.48	4.6	23.33	3.35	3.35	95.0	14×10^2	≈ 0.0

Kinematics: $\rho_2 = 10^7\ \Omega$ cm, $\xi = 2.5 \times 8.85419 \times 10^{-12}\ \phi\,\text{m}^{-1}$ at $P = 4500$ N, $n = 1000$ rev min^{-1}, Wir $= 0.163$ mm, $W_{or} = 0.195\ mm$.

Table 6.4. Capacitance, capacitive reactance, resistance and impedance of a 6326 ball bearing at different operating parameters

Speed (rpm)	1000		1500	
P (N) [12]	11.7×10^3	70×10^3	95×10^3	120×10^3
W_{ib} (av.) [12] (in mm)	1.20	2.40	2.89	2.97
W_{ob} (av.) [12] (in mm)	1.15	2.29	2.77	2.84
$h_o \times 10^{-4}$(mm)	23.02	28.17	27.83	27.31
C_{ib}(pϕ)	15.87	25.74	29.17	29.81
C_{ob}(pϕ)	14.02	24.57	28.49	29.21
$X_{cib} \times 10^7$ (ohms)	20.05	12.36	10.91	10.67
$X_{cob} \times 10^7$ (ohms)	22.71	12.96	11.17	10.89
$R_{ib} \times 10^7$ (ohms)	139.30	85.9	75.8	74.2
$R_{ob} \times 10^7$ (ohms)	157.9	90.1	77.7	75.8
$Z_{ib} \times 10^7$ (ohms)	140.74	86.78	76.58	74.96
$Z_{ob} \times 10^7$ (ohms)	159.52	91.03	78.50	76.58
$R_b \times 10^7$ (ohms)	99.07	58.67	51.17	50.0
$X_{cbb} \times 10^7$ (ohms)	14.25	8.44	7.36	7.19
$Z_{bb} \times 10^7$ (ohms)	100.1	59.27	51.69	50.51
C_{bb}(pϕ)	22.34	37.7	43.2	44.3
X_{cbb}/R_{bb} ($\rho_1 = 10^{11}$ ohm cm)	0.14	0.14	0.14	0.14
X_{cbb}/R_{bb} ($\rho_2 = 10^7$ ohm cm)	14×10^2	14×10^2	14×10^2	14×10^2

Bearing Kinematics and Lubricant Characteristics: $d = 46$ mm, $D = 206$ mm, $r_i = 80$ mm, $r_o = 126$ mm, $N = 9$, $\delta_1 = 0.018$ mm^{-1}, $\delta_2 = 0.0089$ mm^{-1}, $\phi_1 = 0.028$ mm^{-1}, $\phi_2 = 0.007745$ mm^{-1}, $k = 1.42$, $k_1 = 1.9013$, $K = 3$, $\rho_1 = 10^{11}$ ohm cm, $\rho_2 = 10^7$ ohm cm, $\xi = 2.5 \times 8.85419 \times 10^{-12}$ ϕm^{-1}.

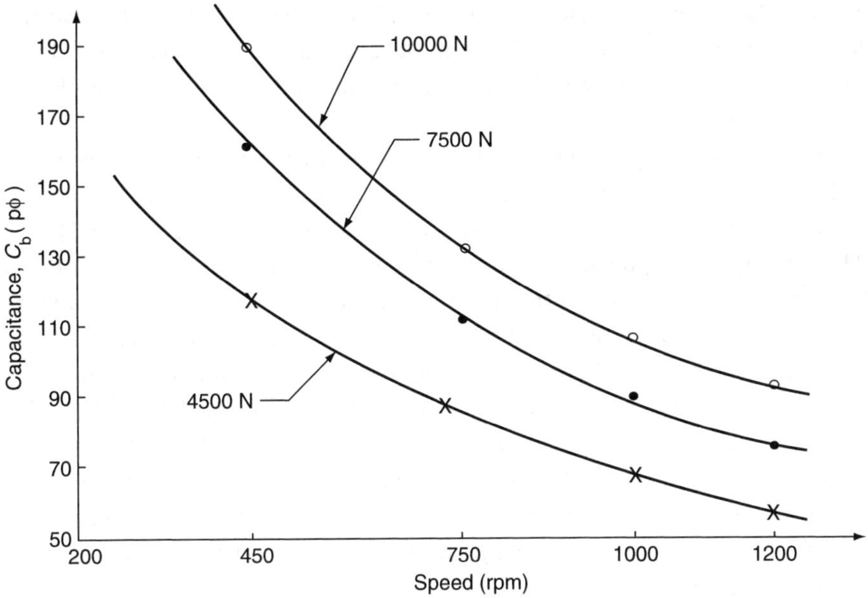

Fig. 6.6. Variation of capacitance with speed at different radial loads in NU 330 type bearing using a high resistivity lubricant

Equivalent bearing resistance, impedance and capacitive reactance at constant load increase with speed (Table 6.2). However, the equivalent resistance rather than the equivalent capacitive reactance up to the first threshold voltage mainly governs bearing impedance. At 10 000 N, Z_b increases from 11.71×10^7 to 23.84×10^7 ohm and R_b increases from 11.59×10^7 to 23.6×10^7 ohm as against X_{cb}, which increases from 1.67×10^7 to 3.4×10^7 ohm as the speed increases from 450 to 1200 rpm (Table 6.2). As the voltage across the bearing exceeds the threshold voltage, the bearing impedance asymptotically drops, and the bearing behaves like a variable resistor with the flow of current increasing with the marginal change in the voltage. The impedance of the NU 330 bearing using high-resistivity lubricant has been measured to be as low as 0.5 ohm at 1 A at different operating parameters [4]. On the contrary, within the first threshold voltage, current flowing through the bearing is as low as 22.06×10^{-10} A and the impedance is as high as 38.52×10^7 ohm at 1200 rpm and 4500 N load (Table 6.2).

6.5.1.2 Equivalent Bearing Capacitance and Charge Accumulation

Under different operating conditions, the capacitance between the outer race and a roller is higher than that between the inner race and a roller. On the contrary, the capacitive reactance and the active resistance between inner race and a roller is more than those between outer race and a roller (Table 6.1). The capacitance and active resistance of the races and rollers depend on the film thickness, width of deformation and are governed by the permittivity and resistivity of the lubricant (Tables 6.1 and 6.3) [5].

The equivalent capacitance of the NU 330 bearing using the high-resistivity lubricant, at a constant load, decreases as the speed increases. However, at the same speed, the capacitance increases with the applied load. At 4500 N and 10 000 N loads, the equivalent capacitance decreases from 118.46 to 57.9 pϕ and 190.4 to 94 pϕ, respectively, as the speed increases from 450 to 1200 rpm (Fig. 6.6).

The charge accumulation on the NU 330 bearings at the first threshold voltage at constant speed and at different loads is more or less constant. The average charge accumulated is 6.94 pC, 21.17 pC, 19.38 pC and 51.68 pC at 450, 750, 1000 and 1200 rpm respectively, at different loads (Table 6.2). NU 2215 bearing shows similar behavior. On the contrary, NU 330 bearing using low-resistivity lubricant does not show any charge accumulation (Table 6.3). This suggests that the voltage applied to the bearing increases the stored charge, provided the dielectric does not get dissociated under voltage.

6.5.2 Ball Bearings

6.5.2.1 Equivalent Capacitive Reactance, Active Resistance and Impedance

Ball bearings behave in the same way as the roller bearings. Similar to a roller bearing, ball bearing impedance is governed by the equivalent resistance rather than the equivalent capacitive reactance. At 1500 rpm Z_b decreases from 59.27×10^7 to 50.51×10^7 ohm and R_b changes from 58.67×10^7 to 50.51×10^7 ohm as against X_{cb}, which decreases from 8.44×10^7 to 7.19×10^7 ohm as the load increases from 70×10^3 to 120×10^3 N (Table 6.4). Within the first threshold voltage, the impedance of 6326 ball bearing remains high, similar to a roller bearing. The impedance of 6326 bearing is determined as 100.1×10^7 ohm at 1000 rpm and 11.7×10^3 N load (Table 6.4).

6.5.2.2 Equivalent Bearing Capacitance

Under different operating conditions, the capacitance between inner race and a ball is higher than outer race and a ball. This is because of higher value of W_{ib} as against W_{ob} (Chapter 5). On the contrary, capacitive reactance and active resistance between outer race and a ball is more than that of inner race and a ball (Table 6.4). As analyzed in the Section 6.3, capacitance and active resistance between races and balls depend on the film thickness, width of deformation, ball radius, track radius,

radius of curvature of width of deformation and are governed by relative permittivity and resistivity of the lubricant.

Equivalent capacitance of 6326 ball bearings using high-resistivity lubricant (10^{11} ohm cm) at the constant speed increases with load. With increase of load from 70×10^3 to 120×10^3 N, capacitance of a 6326 ball bearing increases from 37.7 to 44.3 pϕ at 1500 rpm. This is due to decrease in film thickness and increase in width of contact with load. The comparison of the capacitance data between ball and roller bearing of identical bore indicates that the equivalent capacitance of ball bearing is 2–4 times less than that of a roller bearing under identical operating conditions [3]. This is attributed to the point contact between the ball and the race of a ball bearing as against the line contact between the roller and the race of a roller bearing.

6.6 Conclusions

From the analysis of the electrical parameters of the bearings, the following conclusions are drawn [2–4]:

(1) Under different operating conditions, the capacitance between the outer race and a roller is higher than that between inner race and a roller in a roller bearing. On the contrary, the capacitance between inner race and a ball is higher than that between outer race and a ball in a ball bearing.

(2) The capacitance and active resistance of a rolling-element bearing depend on the film thickness, width of deformation, dimension of a rolling-element, common deformation radius and are governed by the permittivity and resistivity of the lubricant.

(3) The equivalent capacitance of a bearing decreases with increase in speed at constant load but increases with load at constant speed.

(4) The ratios of capacitive reactance to active resistance of a bearing lubricated with high and low-resistivity lubricants having the same relative permittivity are 0.14 and 14×10^2, respectively.

(5) A rolling-element bearing lubricated with high-resistivity as opposed to low-resistivity lubricant, with the same permittivity, behaves like a capacitor up to the first threshold voltage.

(6) Up to the first threshold voltage, the impedance of a bearing using a high-resistivity lubricant is governed by active resistance rather than by capacitive reactance compared with a bearing using low-resistivity lubricant in which it is governed by the capacitive reactance.

(7) Within the first threshold voltage, current flowing through a bearing is negligible (about 10^{-10} A).

(8) For a rolling-element bearing to accumulate charge, the ratio of the capacitive reactance to active resistance should be less than unity.

(9) The charge accumulated on a bearing at the first threshold voltage is more or less independent of load but increases with speed.

(10) The equivalent capacitance of a ball bearing is 2–4 times less than that of a roller bearing under identical operating conditions.

(11) The stored charge depends on the bearing capacitance and the charge increases with the applied voltage provided that the dielectric is not dissociated.

REFERENCES

1. Noble, L. Lockhart, and Ora, E. Rice (1976). *AC Circuit Analysis*, Van Nostrand Reinhold, New York.
2. Prashad, H. (1988). Theoretical analysis of impedance, capacitance and charge accumulation of roller bearings operated under electrical fields. *Wear*, **125**, 223–239. Also, in *BHEL J.*, **9**(2), 21–30.
3. Prashad, H. (1993). Theoretical determination of impedance, resistance, capacitive reactance and capacitance of ball bearings. *BHEL J.*, **14**(2), 40–48.
4. Prashad, H. (1987). Effects of operating parameters on the threshold voltages and impedance response on non-insulated rolling-element bearings under the influence of electrical currents. *Wear*, **117**, 223–240.
5. Prashad, H. (1988). Theoretical evaluation of impedance, capacitance and charge accumulation on roller bearings operated under electrical fields. *Wear*, **125**, 228–239.

Nomenclature

A	area of contact
C_b	capacitance of roller bearing
C_{bb}	capacitance of ball bearing
C_{ir}, C_{or}	capacitance of inner race and a roller, outer race and a roller, respectively
C_{ib}, C_{ob}	capacitance of inner race and a ball, and outer race and a ball, respectively
d	diameter of rolling-element
D	pitch diameter
f	frequency of applied voltage (50 Hz)
F	conductance between track surface of race and roller/ball
h, h_1	oil film thickness

h_o	minimum oil film thickness
I_b	bearing current
K	number of rolling-elements in loaded zone
k, k_1	$(\delta_1/\delta_2)^{1/2}$, $(\phi_1/\phi_2)^{1/2}$
L	roller length
n	rpm
N	number of rolling-elements in a bearing
P	resultant load on bearing
p	plane of action of loading
Q	stored electric charge
r	radius of rolling-element
r_i, r_o	roller/ball track radius of inner race, and outer race, respectively
r_{ib}, r_{ob}	radius of curvature of width of deformation along ball bearing width on inner and outer race, respectively
R_b, R_{bb}	equivalent resistance of roller and ball bearing, respectively
R_{ir}, R_{or}	resistance between inner race and a roller, and outer race and a roller, respectively
R_{ib}, R_{ob}	resistance between inner race and a ball, and outer race and a ball, respectively
V, V_{T1}	voltage, first threshold voltage across bearing
w	2πf
W_{ir}, W_{or}	width of deformation on roller track of inner and outer race, and a roller, respectively
W_{ib}, W_{ob}	width of deformation on ball track of inner and outer race, and a ball, respectively
X	half of a major axis of ellipse at the point contact
X_{cir}, X_{cor}	capacitive reactance between inner and outer race, and a roller, respectively
X_{cib}, X_{cob}	capacitive reactance between inner and outer race, and a ball, respectively
X_{cb}, X_{cbb}	equivalent capacitive reactance of roller and ball bearing, respectively
Y	half of minor axis of ellipse at the point contact
Z_b, Z_{bb}	equivalent impedance of roller and ball bearing, respectively
Z_{ir}, Z_{or}	impedance between inner and outer race, and a roller, respectively
Z_{ib}, Z_{ob}	impedance between inner and outer race, and a ball, respectively
Z_{ireq}, Z_{oreq}	equivalent impedance between inner and outer race, and rollers, respectively
Z_{ibeq}, Z_{obeq}	equivalent impedance between inner and outer race, and balls, respectively

ξ	permittivity of lubricant
ρ_2, ρ_1	resistivity of lubricant
δ, ϕ	outer race and inner race constants of roller bearing $0.5[1/r - 1/r_o]$, and $0.5[1/r + 1/r_i]$, respectively
δ_1, δ_2	outer race constants of ball bearing $0.5[1/r - 1/r_o]$, and $0.5[1/r - 1/r_{ob}]$, respectively
ϕ_1, ϕ_2	inner race constants of ball bearing $0.5[1/r + 1/r_o]$, and $0.5[1/r - 1/r_{ib}]$, respectively

Chapter 7

LIFE ESTIMATION OF ROLLING-ELEMENT BEARINGS DUE TO THE EFFECT OF CURRENT LEAKAGE

7.1 General

It has been well established that in the design of rolling-element bearings, analysis of contact stresses play an important role and the rolling contact failure is intimately connected with the cyclic plasticity [1]. Detailed analysis of the cyclic deformation has been confined to elastic–perfectly plastic stress strain behavior, and has neglected changes in properties with increasing number of cycles. However, recent developments in the related calculations of rolling contact reveal that these features alter the plastic contact deformation and residual stress, and interfere with the initiation of cracks. Cracks grow parallel to the rim at a depth influenced by fracture mechanics considerations, but the contact life of $\sim 10^{10}$ cycles for SAE 52100 (bearing steel) is likely to be dominated by the initiation stage of a crack [2]. However, above a certain value of cyclic stress (the fatigue limit) some crystals on the surface of the specimen develop slip bands as a result of shearing of atomic planes within the crystals. With increase in number of contact cycles, these slip bands broaden and intensify the points where separation occurs within one of the bands and a crack is formed [3]. No such studies have been conducted on the surfaces of rolling-element bearings operated under the influence of electric current.

In this chapter, a study is reported on the theoretical analysis to determine the duration and instant rise in temperature in each shaft rotation in the zone of contacts between rolling-elements and track surface of races of a bearing using different resistivity lubricants and operating under the influence of electric current at different conditions of operation. In addition, the rise of temperature thus calculated was used to determine the semi-range – half of the peak to peak value – of cyclic contact stresses developed on the races, which reduce the fatigue life and lead to the initiation of slip bands and which eventually lead to the formation of corrugations on the rolling tracks of races of a bearing using low-resistivity ($\sim 10^7$ ohm cm) lubricant, and initiation of craters on the track surfaces of a bearing using high-resistivity ($\sim 10^{11}$ ohm cm) lubricant by arcing effect.

7.2 Theoretical Analysis of Life Estimation of Roller Bearings

7.2.1 Duration of Line Contact between Roller Track of Races and a Roller

Assuming that the inner race is rotating and the outer race is stationary, the frequency of rotation of a roller is given in Chapter 5 [Eqn. (5.1)] as:

$$f_r = \frac{Df_s\left[1 - \left(\frac{d}{D}\right)^2\right]}{2d} \tag{7.1}$$

and circumferential speed of a roller is determined as:

$$V_r = \frac{\pi Df_s\left[1 - \left(\frac{d}{D}\right)^2\right]}{2} \tag{7.2}$$

Similarly, the circumferential speed of the roller track of the inner race is calculated as:

$$V_{ir} = \pi(D - d)f_s \tag{7.3}$$

Since the rollers and the inner race rotate in the same direction, the resultant speed at which a line contact on the roller track of the inner race crossed by a roller is given as:

$$V_{irr} = \frac{\pi f_s(3D + d)(D - d)}{2D} \tag{7.4}$$

The resultant speed (since outer race is stationary), at which, a line contact on the roller track of the outer race is crossed by a roller is determined by using Eqn. (7.5).

$$V_{orr} = V_r \tag{7.5}$$

The duration of contact between the roller track of the inner race and a roller to cross the width of contact on the roller track of inner race (W_{ir}) is determined by Eqns. (7.4) and (5.16), and is given as:

$$t_{ir} = \frac{2DW_{ir}}{\pi f_s(3D + d)(D - d)} \tag{7.6}$$

Similarly, from Eqns. (7.5) and (5.17), duration of contact between the roller track of the outer race and a roller to cross the width of contact on roller track of outer race (W_{or}) is determined as:

$$t_{or} = \frac{2DW_{or}}{\pi f_s(D^2 - d^2)} \tag{7.7}$$

7.2.2 Resistance at the Contact Zone

As the film thickness between race and a roller is small, the resistance between the inner race and a roller, and between the outer race and a roller is considered equal ($R_{ir} = R_{or} = R$). Thus, from Eqn. (6.32):

$$R = \frac{KR_b}{2} \tag{7.8}$$

7.2.3 Heat Generated and Instant Temperature Rise in the Contact Zone between Roller Track of Races and a Roller in each Contact by Passage of Electric Current

The passage of a current through the roller track of the inner race and a roller generates heat instantly. The electrical energy q_{ir}, due to current I through a contact resistance R for a duration of t_{ir}, is given as:

$$q_{ir} = I^2 R t_{ir} \tag{7.9}$$

This energy influences the contact zone (of width W_{ir}, length L, and depth H), and increases its temperature after certain number of contact cycles. Hence, by using Eqn. (7.9) and neglecting heat losses by various means, the following is evident:

$$W_{ir} L H T_{ir} c\rho = I^2 R t_{ir} \tag{7.10}$$

So, the instant temperature rise (at $t = 0$) during each contact of a roller with the roller track of inner race is determined by Eqns. (7.6), (7.8) and (7.10), and is given as:

$$T_{\rm ir} = \frac{I^2 R_{\rm b} K D}{\pi f_{\rm s} L H \rho c (3D + d)(D - d)} \quad (7.11)$$

Similarly, instant temperature rise (at $t = 0$) during each contact of a roller with the roller track of outer race is calculated as:

$$T_{\rm or} = \frac{I^2 R_{\rm b} K D}{\pi f_{\rm s} L H \rho c (D^2 - d^2)} \quad (7.12)$$

7.2.4 Instant Temperature Rise in the Contact Zone between Rollers and Roller Track of Races in one Shaft Revolution by Passage of Electric Current

7.2.4.1 Between Roller Track of Inner Race and Rollers

During operation of a bearing, rollers rotate with a frequency higher than that of the inner race because $f_{\rm r}$ has a higher value than $f_{\rm s}$ ($f_{\rm r}/f_{\rm s} = 2.61$ for the bearing NU 326, and 3.7 for the bearing NU 230). Additionally, at any instant, maximum of half of the circumference of an inner race is in the loaded zone [4], and the inner race passes through this zone in $1/2f_{\rm s}$ s in each shaft revolution. Moreover, inner race and rollers travel in the same direction, so time gap between the successive line contacts on the roller track of inner race by the single line position on a roller is $\pi d / V_{\rm irr}$. Thus, the number of such contacts in $1/2f_{\rm s}$ s would be $V_{\rm irr}/2\pi d\, f_{\rm s}$. This is determined using eqn. (7.4), and given as:

$$\beta_{\rm ir} = \frac{(3D + d)(D - d)}{4dD} \quad (7.13)$$

Hence, the instant temperature rise of the roller track of inner race due to a single line position on a roller in each shaft rotation is:

$$T_{\rm irn} = T_{\rm ir} \beta_{\rm ir} \quad (7.14)$$

which is simplified using Eqns. (7.11) and (7.13), and given as:

$$T_{\rm irn} = \frac{I^2 R_{\rm b} K}{4\pi f_{\rm s} d L H \rho c} \quad (7.15)$$

This instant temperature rise would be identical between various line contacts of rollers and the roller track of the inner race. The inner race roller track temperature may stabilize after a certain number of revolutions.

7.2.4.2 Between Roller Track of Outer Race and Rollers

A single line position on a roller to pass over the circumferential length of $\pi(D + d)/2$ (the instant maximum loaded zone of the roller track of the outer race), takes $(D + d)/2f_r d$ s, and in $1/f_r$ s a roller completes one rotation on the roller track of the outer race. So in $1/2f_s$ s (the time taken for the inner race to pass through the loaded zone) a single line position on a roller contacts the roller track of outer race $f_r/2f_s$ times. Thus, the instant temperature rise of the roller track of the outer race caused by the contacts of single line position on a roller in each shaft rotation would be $f_r/2f_s$ multiplied by T_{or}, i.e. $\beta_{or} \times T_{or}$, where $\beta_{or} = f_r/2f_s$. This is simplified using Eqns. (7.1) and (7.12) to give:

$$T_{orn} = \frac{I^2 R_b K}{4\pi f_s LH\rho cd} \tag{7.16}$$

The instant temperature rise of the roller track of the outer race, thus determined, is identical to that of the inner race [Eqn. (7.14)], and would be stabilized in a similar way to that of the inner race track temperature after a few cycles of operation.

7.3 Theoretical Analysis of Life Estimation of Ball Bearings

7.3.1 Duration of Contact between Ball Track of Races and a Ball

Assuming that the inner race is rotating and the outer race is stationary, the frequency of rotation of a ball is given as [Eqn. (5.20]:

$$f_b = \frac{Df_s\left[1 - \left(\frac{d}{D}\right)^2 \cos^2\alpha\right]}{2d} \tag{5.20}$$

and circumferential speed of a ball is determined as:

$$V_b = \frac{\pi Df_s\left[1 - \left(\frac{d}{D}\right)^2 \cos^2\alpha\right]}{2} \tag{7.17}$$

Similarly, circumferential speed of ball track of inner race is calculated as:

$$V_{ib} = \pi(D - d)f_s \tag{7.18}$$

Since balls and inner race rotate in the same direction, the resultant speed, at which, a contact on the ball track of the inner race is crossed by a ball, is given as:

$$V_{ibb} = \frac{\pi f_s(3D^2 - d^2\cos^2\alpha - 2Dd)}{2D} \tag{7.19}$$

and the resultant speed (since outer race is stationary), at which, a contact on the ball track of outer race is crossed by a ball, is determined as:

$$V_{obb} = V_b \tag{7.20}$$

The duration of contact between ball track of inner race and a ball to cross the width W_{ib} is determined by Eqns. (5.21) and (7.19), and is given as:

$$t_{ib} = \frac{2DW_{ib}}{\pi f_s(3D^2 - d^2\cos^2\alpha - 2Dd)} \tag{7.21}$$

Similarly, by Eqns. (5.22) and (7.20), duration of contact between the ball track of outer race and a ball to cross the width W_{ob} is determined as:

$$t_{ob} = \frac{2DW_{ob}}{\pi f_s(D^2 - d^2\cos^2\alpha)} \tag{7.22}$$

7.3.2 Resistance at the Contact Zone

The equivalent circuits of inner and outer races with balls are parallel RC circuits, whereas both inner and outer race circuits are in series to each other. The film thickness between race and a ball being small, the resistance between inner race and a ball, and outer race and a ball is considered equal ($R_{ib} = R_{ob} = R$). Thus, from Eqn. (6.32)

$$R = \frac{KR_b}{2} \tag{7.23}$$

7.3.3 Heat Generated and Instant Temperature Rise in the Contact Zone between Ball Track of Races and a Ball in each Contact by Passage of Electric Current

The passage of a current through the ball track of inner race and a ball generates instant heat. The electrical energy, due to current I through a contact resistance R for a duration of t_{ib}, is given as:

$$q_{ir} = I^2 R t_{ib} \tag{7.24}$$

This energy influences the contact zone i.e. circumference of circle (πW_{ib}) formed by the width of contact/corrugation (W_{ib}) as a diameter, arc length of contact (L_{ib}) on the ball track of inner race, and depth of corrugations (H). This increases the temperature of ball tracks of inner race after a certain number of contact cycles.

The relation, that determines the arc length of contact/corrugation on the ball track on inner race is given as:

$$\frac{L_{ib}}{r_{ib}} = \frac{\alpha 2\pi}{360} \tag{7.25}$$

where

$$r_{ib} = \frac{d(D-d)}{D} \quad \text{Refs. [5, 6]} \tag{5.21}$$

which gives

$$L_{ib} = \frac{\pi \alpha d(D-d)}{180D} \tag{7.26}$$

Hence, by using Eqns. (7.24) and (7.26), and neglecting heat losses by various means, the following is evident:

$$W_{ib} L_{ib} H T_{ib} \rho c = I^2 R t_{ib} \tag{7.27}$$

So, the instant temperature rise (at $t = 0$) during each contact of a ball with the ball track of inner race is determined by Eqns. (7.21), (7.23), (7.26) and (7.27), and is given as:

$$T_{ib} = \frac{5.8 I^2 R_b K D^2}{f_s \alpha H \rho c (D-d)(3D^2 - d^2 \cos^2 \alpha - 2Dd)} \tag{7.28}$$

Similarly, instant temperature rise (at $t = 0$) during each contact of a ball with the ball track of outer race is worked out as:

$$\pi W_{ob} L_{ob} H T_{ob} \alpha \rho c = I^2 R t_{ob} \tag{7.29}$$

where L_{ob} is calculated as:

$$L_{ob} = \frac{r_{ob} \alpha 2\pi}{360} \tag{7.30}$$

and r_{ob} is given as [5, 6]:

$$r_{ob} = \frac{d(D+d)}{(D+2d)} \tag{5.22}$$

Hence, T_{ob} using Eqns. (7.22), (7.29) and (7.30) is determined as:

$$T_{ob} = \frac{5.8 I^2 R_b K D (D+2d)}{f_s H \rho c \alpha d (D+d)(D^2 - d^2 \cos^2 \alpha)} \tag{7.31}$$

7.3.4 Instant Temperature Rise in the Contact Zone between Balls and Ball Track of Races in one Shaft Revolution by Passage of Electric Current

7.3.4.1 Between Ball Track of Inner Race and Balls

During operation of a bearing, balls rotate with a higher frequency than the inner race because of higher value of f_b than f_s($f_b/f_s = 2.61$ for the bearing NU 326, 2.13 for the bearing 6326 and 3.7 for the bearing NU 230). Also, at an instant, a maximum of half of the circumference of an inner race is in the loaded zone, and the inner race passes through this zone in $1/2f_s$ s in each shaft revolution. Moreover, as the inner race and balls travel in the same direction, the time gap between the successive contacts on the ball track of inner race by the single position on a ball would be $\pi d/V_{ibb}$. Thus, the number of such contacts in $1/2f_s$ s would be $V_{ibb}/2\pi d\, f_s$. This is determined using Eqn. (7.19), and is given as:

$$\beta_{ib} = \frac{(3D^2 - d^2\cos^2\alpha - 2Dd)}{4dD} \tag{7.32}$$

Hence, the instant temperature rise of the ball track of inner race due to a single contact position on a ball in each shaft rotation is

$$T_{ibn} = T_{ib}\beta_{ib} \tag{7.33}$$

which is simplified using Eqns. (7.28) and (7.32), and is given as:

$$T_{ibn} = \frac{1.45I^2 R_b KD}{f_s\alpha H\rho c d^2(D-d)} \tag{7.34}$$

This instant temperature rise would be identical between various contact positions of balls and the ball track of inner race. And this would be stabilized inner race ball track temperature after a certain number of revolutions.

7.3.4.2 Between Balls and Ball Track of Outer Race

A single contact position on a ball to pass over the circumferential length of $\pi(D+d)/2$ (instant maximum loaded zone of ball track of outer race), takes $(D+d)/(2f_b d)$ s. And in $1/f_b$ s a ball completes one rotation on ball track of outer race. So in $1/2f_s$ s, the time taken for inner race to pass through the loaded zone, a single position on a ball contacts the ball track of outer race $f_b/2f_s$ times. Thus, the instant temperature rise of the ball track of the outer race due to the contacts of single position on a ball in each shaft rotation would be $f_b/2f_s$ times the T_{ob}, i.e. $\beta_{ob} \times T_{ob}$, where $\beta_{ob} = f_b/2f_s$. This is simplified using Eqns. (5.25) and (7.31), and is given as:

$$T_{\mathrm{obn}} = \frac{1.45I^2 R_{\mathrm{b}} K(D+2d)}{f_{\mathrm{s}} H \alpha \rho c d^2 (D+d)} \tag{7.35}$$

The instant temperature rise of the ball track of the outer race, thus determined, is comparable to that of inner race [Eqn. (7.34)], and would be stabilized similar to the inner race track temperature after a few operating cycles. And the ratio of T_{ibn} to T_{obn} depends on the bearing dimensions and determined as [Eqns. (7.34) and (7.35)]:

$$\frac{T_{\mathrm{ibn}}}{T_{\mathrm{obn}}} = \frac{D(D+d)}{(D+2d)(D-d)} \tag{7.36}$$

7.4 Capacitive Effect of Rolling-Element Bearings

As discussed in Chapter 6, there occurs the accumulation of charges on the bearing elements, when the bearings using high-resistivity lubricant (10^{11} ohm cm) operate under the influence of electric current. A study is carried out on the effects of instant charge leakage from the bearing surfaces, and is used to determine the instantaneous temperature rise and increase in contact stresses on the bearing elements, which reduces the fatigue life and leads to initiate craters on the track surfaces of rolling-elements.

7.4.1 Resistance at the Contact Zone between Race and a Roller

For an instant the resistance under static condition between race and a roller (R_{os}) is determined similar to Eqn. (7.8), and given as:

$$R_{\mathrm{os}} = \frac{KR_s}{2} \tag{7.37}$$

Also, instantaneously the R_{os} can be taken as equivalent to R, the asperity contact resistance between race and a roller during operation of a roller bearing. This is because for a fraction of a second ($\sim 10^{-5}$ s contact duration as shown in Table 7.1) at the asperity contact, race and a roller are stationary during operation of a bearing.

7.4.2 Equivalent Bearing Capacitance and Stored Charge

The equivalent bearing capacitance is determined from equivalent capacitive reactance of a roller bearing and is given in Chapter 6 as:

$$C_{\mathrm{b}} = \frac{K}{W}(X_{\mathrm{cir}} + X_{\mathrm{cor}}) \tag{6.35}$$

Table 7.1. Instantaneous temperature rise and contact duration between races and rollers, stress concentration and fatigue cycles before slip bands initiation on the roller track of races of a NU 326 bearing operated under the influence of an electric current

Analytical values of parameters			
Inner race		Outer race	
W_{ir}(mm) at $p = 2$, $K = 7$	0.120	W_{or}(mm) at $p = 2$, $K = 7$	0.145
t_{ir}(s)	7.8×10^{-6}	t_{or}(s)	25.4×10^{-6}
T_{ir} °C	31.41	T_{or} °C	84.41
T_{irn} °C	109.90	T_{orn} °C	109.90
σ_{irn}(kg mm^{-2})	35.46	σ_{orn}(kg mm^{-2})	35.46
S_a(kg mm^{-2})	17.73	S_a(kg mm^{-2})	17.73

Number of cycles before slip bands initiation after the increase in stabilized bearing temperature = 10^6 cycles.
Hours of operation after the increase in stabilized bearing temperature = 15.15 h.
Total number of hours of operation before the slip bands initiation on the roller track of inner race = 40.15 h.
Dimensions of NU 326 bearing and operating parameters: $d = 38$ mm, $D = 205$ mm, $L = 38$ mm, $N = 14$, $n = 1100$ rpm, $P = 1120$ Kgf, $u = 1.12$ V at 50 A, $H = 0.003$ mm, $R_b = 0.022$ ohm, $\rho_2 = 10^7$ ohm cm.

The equivalent capacitance of a rolling-element bearing opposes any change in the existing voltage and causes an electric charge to be stored. The instant stored charge on the bearing surfaces is determined as (Chapter 6):

$$Q = UC_b \tag{6.36}$$

Also, charges on the inner and outer races (Q_{ir} and Q_{or}) of a bearing are determined assuming that for an instant 'K' rolling-elements are in asperity contact with the races. The charges are given as:

$$Q_{ir} = KUC_{ir} \tag{7.38}$$

and

$$Q_{or} = KUC_{or} \tag{7.39}$$

The equivalent capacitance of inner race and rollers is KC_{ir} and that of outer race and rollers is KC_{or}. Furthermore C_{ir} and C_{or} are in series to each other, and hence related to C_b as [7]

$$C_b = \frac{KC_{ir}C_{or}}{(C_{ir} + C_{or})} \tag{7.40}$$

where C_{ir} and C_{or} are determined by the bearing parameters and given in Chapter 6 [Eqns. (6.1) and (6.2)].

Also, from Eqns. (7.38)–(7.40) it is evident that:

$$\frac{Q_{ir}}{C_{ir}} = \frac{Q_{or}}{C_{or}} = \frac{KQ}{C_b} \tag{7.41}$$

The similar relations hold good for a ball bearing.

7.4.3 Heat Generated and Instant Temperature Rise by Instantaneous Charge Leakage in the Contact Zone between Roller Track of Races and a Roller

The slow passage of charge takes place when the high points on the asperities of the races and rolling-elements come close or when the conducting particles bridge the oil film. These conducting paths are broken, when the asperities become separated by greater thickness of film by the presence of high-resistivity lubricants. Arcing results due to instantaneous charge leakage by vibration effects or by positive and negative cage and roller slip phenomenon occurring under different operating conditions in a bearing [8]. Under these conditions contact resistance between the race and a roller may become instantly minimum, equivalent to that of bearing resistance under static conditions (R_s).

The instantaneous charge leakage through the roller track of inner race and a roller generates instantaneous heat. The electrical energy q_{ir} due to charge leakage Q_{ir} for a duration t_{ir} at the first threshold voltage U through a contact resistance R_{os} between the inner race and a roller of the bearing having capacitance C_{ir}, is given as:

$$q_{ir} = I^2 R_{os} t_{ir} \tag{7.42}$$

or in terms of charge and capacitance, electrical energy can be expressed as [by using Eqns. (7.38) and (7.41)]:

$$q_{ir} = \frac{Q_{ir}^2 t_{ir}}{K^2 C_{ir}^2 R_{os}} \tag{7.43}$$

This energy influences the contact zone (of width W_{ir}, length L_a, and depth H), and increases its temperature after a certain number of contact cycles. Hence, by using Eqn. (7.43) and neglecting heat losses by various other means, the following is evident:

$$W_{ir} L_a H T_{ir}^* \rho c = \frac{Q_{ir}^2 t_{ir}}{K^2 C_{ir}^2 R_{os}} \tag{7.44}$$

So, the instantaneous temperature rise (at $t = 0$) due to sudden charge leakage during each contact of a roller with the roller track of inner race is determined by Eqns. (7.6), (7.37), (7.38), (7.41) and (7.44), and given as:

$$T_{\mathrm{ir}}^{*} = \frac{4Q^2 D}{\pi C_{\mathrm{b}}^2 K f_{\mathrm{s}} L_{\mathrm{a}} R_{\mathrm{s}} H \rho c (3D + d)(D - d)} \tag{7.45}$$

Similarly, instantaneous temperature rise (at $t = 0$) due to charge leakage during each contact of a roller with the roller track of outer race is worked out as:

$$T_{\mathrm{or}}^{*} = \frac{4Q^2 D}{\pi C_{\mathrm{b}}^2 K f_{\mathrm{s}} L_{\mathrm{a}} R_{\mathrm{s}} H \rho c (D^2 - d^2)} \tag{7.46}$$

7.4.4 Instantaneous Temperature Rise due to Charge Leakage in the Contact Zone between Rollers and Roller Track of Races in One Shaft Revolution

7.4.4.1 Between Roller Track of Inner Race and Rollers

The instantaneous temperature rise of the roller track of inner race at a single position due to continuous rolling of rollers in each shaft rotation is determined using Eqns. (7.13), (7.14) and (7.45), and is given as:

$$T_{\mathrm{irn}}^{*} = \frac{Q^2}{\pi C_{\mathrm{b}}^2 K f_{\mathrm{s}} R_{\mathrm{s}} d L_{\mathrm{a}} H \rho c} \tag{7.47}$$

This instantaneous temperature rise would be identical between various asperity contacts of the rollers and the roller track of the inner race. This would cause local damage on the roller track of the inner race after a certain number of revolutions.

7.4.4.2 Between Roller Track of Outer Race and Rollers

The instantaneous temperature rise of an asperity position on the roller track of the outer race due to the contacts of the rollers in each shaft rotation would be $f_{\mathrm{r}}/2f_{\mathrm{s}}$ times the T_{or}, which is simplified using Eqns. (7.3) and (7.46), and given as:

$$T_{\mathrm{orn}}^{*} = \frac{Q^2}{\pi C_{\mathrm{b}}^2 K f_{\mathrm{s}} R_{\mathrm{s}} L_{\mathrm{a}} H d \rho c} \tag{7.48}$$

The instantaneous temperature rise of roller track of outer race, thus determined, is identical to that of the inner race [Eqn. (7.47)], and would cause local damage similar to the inner race after certain hours of operation.

7.5 Thermal Stresses Due to Thermal Transients on Rolling-Element Track Surfaces of Races

During operation of a rolling-element bearing, a transient thermal gradient first occurs, prior to attainment of equilibrium thermal conditions from an initial uniform temperature. If the bearing's inner race temperature is known at any instant, then the transient thermal gradients after various time intervals can be determined. In addition, the thermal stresses can be evaluated for any instant of time. For a very small interval of contact of a roller with roller track of inner race ($t_{\mathrm{ir}} = 7.8 \times 10^{-6}$ s, as shown in Table 7.1), the overall temperature during the contact approaches zero, but for an instant ($t = 0$), on the rolling-element track of races (both for roller and ball bearings), instant tangential stress [9] is:

$$\sigma_{\mathrm{ir}} = \frac{-\Gamma E T_{\mathrm{ir}}}{(1-\mu)}, \quad \sigma_{\mathrm{ib}} = \frac{-\Gamma E T_{\mathrm{ib}}}{(1-\mu)} \tag{7.49}$$

and

$$\sigma_{\mathrm{or}} = \frac{-\Gamma E T_{\mathrm{or}}}{(1-\mu)}, \quad \sigma_{\mathrm{ob}} = \frac{-\Gamma E T_{\mathrm{ob}}}{(1-\mu)} \tag{7.50}$$

The σ_{ir}, σ_{ib} and σ_{or}, σ_{ob} are compressive during heating. The surface material at the contact region during the contact instant expands/tends to grow, but is restricted by adjacent material on the roller/ball track, which is at a lower stabilized temperature. However, compressive stresses at the roller/ball tracks change gradually to tensile stresses at the other surfaces of the tracks, i.e. inner surface of the inner race and outer surface of the outer race.

The instant temperature rise of roller/ball tracks of a bearing using low-resistivity lubricant in each rotation of a shaft ($T_{\mathrm{irn}}/T_{\mathrm{ibn}}$ or $T_{\mathrm{orn}}/T_{\mathrm{obn}}$) stabilizes after a few hours of operation. So, the stresses on the roller/ball tracks are determined as:

$$\sigma_{\mathrm{irn}} = \frac{-\Gamma E(T_{\mathrm{irn}} - T_{\mathrm{a}})}{(1-\mu)} \tag{7.51}$$

and

$$\sigma_{\mathrm{ibn}} = \frac{-\Gamma E(T_{\mathrm{ibn}} - T_{\mathrm{a}})}{(1-\mu)}$$

The operating temperature (T_{o}) of a bearing using high-resistivity lubricant under rolling friction without the effect of the charge leakage affects the contact stresses. And the instantaneous temperature rise of roller tracks due to charge leakage in each rotation of a shaft as per Eqns. (7.47) and (7.48) after a few operating cycles may set residual stresses on the asperity contacts, which are determined as:

$$\sigma^*_{\rm irn} = \frac{-\Gamma E(T^*_{\rm irn} + T_{\rm o})}{(1 - \mu)} \tag{7.52}$$

The similar relation holds good for a ball bearing.

At the stabilized temperature, tangential stresses at the track surfaces of inner race is equal to that of outer race since $T_{\rm irn} = T_{\rm orn}$, and $T_{\rm ibn} = T_{\rm obn}$ [Eqns. (7.15), (7.16), (7.34) and (7.35)].

7.6 Determination of Number of Cycles Before the Initiation of Slip Bands and Craters Formation on Rolling-Element Tracks of Races, and Data Derived

Design fatigue curves for carbon, low alloy series 4XX, high alloy steels and high-tensile steels, for temperatures not exceeding ~370°C [10], have been used for bearing steel SAE 52100, the properties of which are given in Appendix A [11] to determine the number of cycles leading to the initiation of craters and slip bands against the semi-range of stresses ($\sigma_{\rm irn/2}$) [by using Eqns. (7.52) and (7.51)].

The number of cycles, thus determined, indicates in terms of the time leading to crater and slip band initiation on the bearing surfaces due to the instantaneous charge leakage, and shearing of atomic planes within the crystals, respectively, depending on the characteristics and resistivities of the lubricants used.

7.6.1 Experimental Investigations on Passage of Current through Bearings and Theoretical Calculations on Temperature Rise and Initiation of Slip Bands and Craters

The investigations on the passage of current through bearings show that the threshold voltage varies with operating parameters and lubricant characteristics. A low-resistivity lubricant (10^7 ohm cm) having a viscosity 99.92 cSt at 40°C and 9.68 cSt at 100°C, and a relative permittivity of 2.5, begin to disintegrate and conduct electricity at a voltage of 1.21 V and allows a current of 50 A (AC) to pass through the NU 326 roller bearing; tested on roller bearing test rig operating at 1100 rpm under radial loads of 1000 Kgf and 500 Kgf as discussed in Chapter 6.

7.6.2 Bearings Using Low-Resistivity Lubricant

7.6.2.1 Roller Bearing

After 41 h of operation, slip bands were observed on the inner race of the bearing as shown in the Fig. 5.1. Figure 5.3 is an enlarged view of the inner race showing slip bands. However, a developed corrugation pattern was observed after 250 h of operation as shown in Fig. 5.4.

The outer race of the bearing and grease temperature stabilized at temperatures of 35 and 50°C above ambient up to 25 h of periodical testing, although temperature at the contact zone between the roller tracks of the races and rollers would have been higher since heat dissipates through the shaft and the housing, etc. However, a stabilized outer race and grease temperature could be taken as a reasonably accurate indication of a stabilized temperature of the roller tracks.

The maximum depth of slip bands on the roller track of inner race of NU 326 is taken as 0.003 mm, equivalent to the measured depth of corrugations after 250 h of operation for a current 50 A [4]. Calculated values of various parameters derived in Eqns. (7.6), (7.7), (7.15), (7.16) and (7.51) are shown in Table 7.1.

7.6.2.2 Ball Bearing

The ball track condition of inner and outer races of the 6326 ball bearing (lubricated with low-resistivity grease) was monitored after 250 h of operation, and corrugation pattern was observed to have been developed on the bearing races on passing a current of 50 A (AC) as shown in the Figs. 5.6 and 5.7 [5].

The bearing outer race and grease temperatures were stabilized at 31 and 56.4°C above ambient, up to 30 h of periodical testing.

The maximum depth of slip bands on the ball track of inner race of 6326 ball bearing is taken as 0.003 mm, similar to NU 326 roller bearing operated under the influence of electric current of 50 A (AC) [12]. The calculated values of various parameters derived in Eqns. (7.21), (7.22), (7.34), (7.35) and (7.51) are tabulated in Table 7.2.

7.6.3 Bearing using High-Resistivity Lubricant

The calculated values of various parameters derived in the Eqns. (7.45)–(7.50) and (7.52) have been determined and shown in Table 7.3. The charge accumulation on the bearing using high-resistivity lubricant and the threshold voltages are taken from the published work in the references [13 and 14] [Chapter 4 and 6] for the determination of instantaneous temperature rise on the roller tracks of races (Table 7.3). Figure.4.8 shows the craters due to instantaneous charge leakage on roller tracks of the inner race of a bearing using high-resistivity lubricant.

The maximum depth of craters (H) on the roller track of inner race of NU 330 bearing is taken as 0.001 mm [15] for determination of contact temperature due to charge leakage (Table 7.3).

Table 7.2. Instantaneous temperature rise and contact duration between races and balls, stress concentration and fatigue cycles before slip bands initiation on the ball track of races of a 6326 ball bearing operated under the influence of an electric current

Analytical values of parameters			
Inner race		Outer race	
W_{ib}(mm) at p = 3, K = 3	1.30	W_{ob}(mm) at p = 3, K = 3	1.25
t_{ib}(s)	9.61×10^{-5}	t_{ob}(s)	24.3×10^{-5}
T_{ib}°C	45.58	T_{ob}°C	110.49
T_{ibn}°C	128.39	T_{obn}°C	117.92
σ_{ibn}(kg mm^{-2})	44.98	σ_{obn}(kg mm^{-2})	40.49
S_a(kg mm^{-2})	22.49	S_a(kg mm^{-2})	20.25

Number of cycles leading to slip bands initiation after the increase in stabilized bearing temperature $= 6.2 \times 10^4$ cycles.

Total number of hours of operation leading to the slip bands initiation on the ball track of races > 30.0 h.

Dimensions of 6326 ball bearing and operating parameters: $d = 46$ mm, $D = 206$ mm, $\alpha = 15°$, $f_s = 16.67$ Hz, $N = 9$, $n = 1000$ rpm, $P = 1170$ Kgf, $u = 2.1$–3 V at 50 A, $H = 0.003$ mm, $\rho_2 = 10^7$ ohm cm, $R_b = 0.051$ ohm (average).

Table 7.3 Instantaneous temperature rise, contact duration and contact stresses due to charge leakage between races and rollers before initiation of craters on the roller tracks of races of the NU 330 bearing lubricated with a high resistivity lubricant operated under the influence of an electric current

Experimental and analytical values of different parameters				
n (rpm)	450	750	1000	1200
f_s(1/s)	7.5	12.5	16.67	20
$Q \times 10^{-12}$(C)(7.13)	7.62	19.71	21.20	54.5
C_b(pϕ)(7.13)	190.4	131.41	106.0	94.0
$t_{ir} \times 10^{-5}$(s)	3.36	2.02	1.52	1.26
$t_{or} \times 10^{-5}$(s)	0.85	6.51	4.88	4.07
T^*_{or}(°C)	0.084	0.71	0.95	6.62
T^*_{ir}(°C)	0.23	1.92	2.55	17.86
T^*_{irn}(°C)	0.31	2.59	3.46	24.17
T^*_{orn}(°C)	0.31	2.59	3.46	24.17
$\sigma^*_{irn} \times 10^7$(N m^{-2}) and $\sigma^*_{orn} \times 10^7$ (N m^{-2})	4.57	5.58	5.97	15.6
$S_a \times 10^7$ (Nm^{-2})	2.3	2.8	3.0	7.8
C_s(>)	10^{10}	10^{10}	$\geq 10^{10}$	$\geq 10^7$

Dimensions of NU 330 bearing and operating parameters: $d = 42$ mm, $D = 235$ mm, $L = 45$ mm, $N = 14$, $n = 1200$ rpm, $P = 10000$ N, $H = 0.001$ mm, $R_s = 0.5$ ohm, $\rho_1 = 10^{11}$ ohmcm, $\xi = 2.5 \times 8.85419 \times 10^{-12}$ ϕm^{-1}, $T_o = 11$°C, $W_{ir} = 0.243$mm, $W_{or} = 0.291$ mm, $p = 1$, $K = 3$, $L_a = 1$ mm (approximately 2% of L).

7.7 Discussion of Results

7.7.1 Duration of Contact between Track Surface of Races and a Rolling-Element

The duration of contact between roller track of races and a roller depends on the pitch diameter and roller diameter of the bearing. It decreases with rotating speed frequency but increases with width of contact on the roller tracks [Eqns. (7.6), (7.7), (7.21) and (7.22)]. The duration of line contact between the roller track of the outer race and a roller is about 3.2 times higher than the corresponding duration between the roller track of the inner race and a roller. This is because inner race and rollers rotate in the same direction. For the NU 330 bearing operating at 1200 rpm, t_{or} is 4.07×10^{-5} s as against t_{ir} as 1.26×10^{-5} s (Table 7.3). The similar trend is observed for a ball bearing (Table 7.2).

7.7.2 Temperature Rise in the Contact Zone between Track Surface of Races and a Rolling-Element

7.7.2.1 Bearing using Low-Resistivity Lubricant

7.7.2.1.1 Roller Bearing

The temperature rise at a line contact between the roller track of the races and a roller depends on the pitch diameter, the length and diameter of the roller and the number of rollers in the loaded zone of a bearing. The temperature rise decreases with increase in the frequency of rotation but increases with resistance of the bearing and the square of the current passing through the bearing. However, it is independent of the width of contact between the roller tracks and a roller [Eqns. (7.11) and (7.12)].

For the NU 326 bearing T_{or} and T_{ir} are 84.41 and 31.41°C, respectively (Table 7.1). However, without loss of heat the instant temperature rise of both inner and outer race tracks in each shaft rotation is identical and does not depend on the pitch diameter of a bearing [Eqns. (7.15) and (7.16)]. For the NU 326 roller bearing the value of T_{irn} and T_{orn} is determined as 109.9°C under the parameters of operation mentioned in Table 7.1. This instant temperature rise is considered to be stabilized for up to 25 h of periodic operation, based on stabilized temperature of grease and outer race (determined experimentally), and this gives rise to an increase in contact zone temperature of 79.90°C above ambient (30°C being the ambient temperature).

7.7.2.1.2 Ball Bearing

The temperature rise at the point of contact between the ball track of races and a ball depends on the pitch and the ball diameters, the contact angle, and the number of balls in the loaded zone of a bearing [Eqns. (7.28) and (7.31)]. For the 6326 ball bearing, the value of T_{ibn} and T_{obn} are determined as 128.39 and 117.92° C respectively, under the parameters of operation mentioned in Table 7.2. This instant temperature rise is considered as stabilized up to 30 h of periodic operation, based on stabilized temperature of grease and outer race (determined experimentally), which gives rise to increase in contact zone temperature above ambient as 101.39°C of inner race track and 90.92°C of outer race track (27°C being the ambient temperature).

7.7.2.2 Capacitive Effect of Bearings

Under the capacitive effect of a bearing using high-resistivity lubricant (10^{11} ohm cm), the temperature rise at a line contact between roller track of races and a roller depends on the pitch diameter, the roller diameter, the summation of the length of asperity contacts on the circumference of the roller track during contact with a roller (L_a) and the number of rollers in the loaded zone of a bearing. Temperature rise under the influence of electric current decreases with increase in frequency of rotation, but increases with square of the ratio of charge accumulation to the bearing capacitance. However, it is independent of the width of contact between the roller tracks and a roller [Eqns. (7.45) and (7.46)]. Temperature rise between roller track of the outer race and a roller is much higher than the corresponding rise between the roller track of inner race and a roller in each contact. This is attributed to higher value of t_{or} compared with t_{ir} as discussed in the Section 7.8.1. For the NU 330 bearing at 1200 rpm, T_{or} and T_{ir} are 17.86 and 6.62°C, respectively (Table 7.3). However, without loss of heat the instant temperature rise of the inner and outer race tracks in each shaft rotation is identical and does not depend on the pitch diameter of a bearing [Eqns. (7.47) and (7.48)]. For the NU 330 roller bearing the values of T_{irn} and T_{orn} are determined as 24.17°C when the charge accumulation on the bearing surfaces is 54.5×10^{-12} (C) and capacitance is 94 (pϕ) (Table 7.3). With increase in charge accumulation from 7.62×10^{-12} to 54.5×10^{-12} (C), and decrease of capacitance from 190.4 to 94 (pϕ), the T_{irn} and T_{orn} are increased from 0.31 to 24.17 °C (Table 7.3). However, mechanical loading and heating due to friction also contribute to the temperature rise, but their effect is much less than the effect of a current on a bearing [12, 16].

7.7.3 Thermal Stresses and Number of Cycles before Initiation of Craters and Slip Bands on Track Surfaces of Races

7.7.3.1 Bearings using Low-Resistivity Lubricant

7.7.3.1.1 Roller Bearing

The semi-range of contact stresses on the roller track of a NU 326 bearing using low-resistivity lubricant (10^7 ohm cm) at the stabilized temperature of 109.90°C is

17.73 Kg mm^{-2}, which indicates that slip bands will initiate on the roller tracks after 10^6 cycles or 15.15 h of continuous operation [10] as shown in Table 7.1. As per the experimental investigations, outer race temperature of the NU 326 bearing was stabilized for up to 25 h of operation, which indicates that the initiation of slip bands on the bearing would take 40.15 h. This corresponds closely with the first appearance of slip bands on the roller track of the inner race (Fig. 5.1) tested with a current of 50 A (AC) and shown in an enlarged view of the inner race after 41 h of operation (Fig. 5.3).

7.7.3.1.2 Ball Bearing

The semi range of contact stresses on the ball track of 6326 bearing at the stabilized temperature of 128.30°C is 22.49 Kg mm^{-2}, which indicates that slip bands will get initiated on the ball tracks after 6.2×10^4 cycles of continuous operation and after increase in stabilized bearing temperature, as shown in Table 7.2 [10]. As per the experimental investigations outer race temperature of the 6326 bearing was stabilized up to 30 h of operation, which indicates that the bearing would take 6.2×10^4 cycles before the initiation of slip bands after the increase in stabilized bearing temperature. Experimental results reported by Simpson, F.F. and Crump, W.J.J. about the life of rolling contact bearing under electric currents, also match with the results in the above analysis [16].

A roller bearing, using low-resistivity lubricant, takes longer time for the slip bands to get initiated as compared to a ball bearing operating under identical conditions. This is because of less rise in contact temperature and corresponding less value of semi range of contact stresses in the roller bearings than in ball bearings [12] (Table 7.2).

7.7.3.2 Bearings using High-Resistivity Lubricant

Instantaneous thermal stresses due to thermal transients on the roller/ball track of races due to roller/ball contact depend on the instantaneous rise in temperature [Eqns. (7.49) and (7.50)]. The contact thermal stresses with the instantaneous temperature rise are increased and affect the fatigue life [Eqn. (7.52)]. The semi range of contact stresses in each shaft rotation at the asperity contacts on the roller track of NU 330 bearing at the instant temperature rise of 24.17°C is 7.8×10^7 N m^{-2}, which indicates that the craters will initiate on the roller tracks after 10^7 cycles (Table 7.3) or about 150 h of operation at 1200 rpm. This matches closely with the test results. If the semi ranges of contact stresses are less than 3×10^7 N m^{-2}, the craters will initiate after 10^{10} cycles of operation for a bearing with charge accumulation of less than 21.20×10^{-12} (C) and capacitance of more than 106 (pϕ) (Table 7.3).

It may be noted that the life expectancy of a bearing before the craters are initiated will be much higher if a bearing is operated below the threshold voltage. This is because of less rise in contact temperature. In addition, decrease in charge accumulation and increase in bearing capacitance also results in less rise in contact temperature [Eqn. (7.47)]. The craters/slip bands on the rolling-element track of inner race initiate much earlier than on the rolling-element track of the outer race. This is attributed to the rotation of the inner race, and less heat dissipation to its surroundings as compared to the outer race. Furthermore, mechanical loading and thermal stresses due to rolling friction affect the bearing life by increase in subsurface shear stresses and reduction in lambda factor.

7.8 Conclusions

From the analysis and the investigations of this chapter, the following conclusions are drawn [7, 12, 17]:

(1) The duration of contact between the roller/ball track of races and a roller/ball depends on the pitch and the rolling-element diameters of a bearing. It is higher for the roller/ball track of the outer race and a roller/ball than for the roller/ball track of the inner race and a roller/ball. The duration of contact decreases with rotating speed frequency but increases with width of contact.

(2) The temperature rise at each contact between rolling-element track of races and a rolling-element depends on the bearing kinematics, the depth of slip bands/craters, the number of rolling-elements in the loaded zone, and the properties of the bearing material. The temperature rise at each contact decreases with increase in frequency of rotation but increases with resistance of the bearing and with the current.

(3) The temperature rise at the contact between rolling-element track of the outer race and a rolling-element is more than that between the inner race track and a rolling-element.

(4) The instant temperature rise of inner and outer race tracks in each shaft rotation is identical and is a function of the bearing dimensions, the current and resistance of the bearing and operating speed frequency, but does not depend on the pitch diameter of a bearing.

(5) The number of cycles before the initiation of slip bands/craters on the rolling-element tracks depends on the semi range of contact stresses on the surface of the raceways.

(6) A bearing with a higher capacitance and less charge accumulation takes longer time before the craters are initiated on the track surfaces.

(7) Slip bands on the roller tracks of races of the NU 326 bearing using low-resistivity lubricant at a current of 50 A are initiated in less than 41 h of operation, and it takes 15.15 h or 10^6 cycles after the bearing temperature is stabilized. For a 6326 ball bearing slip bands are initiated in less than 6.2×10^4 cycles. At a lower current density more time elapses before slip bands appear on the tracks.

(8) The NU 330 bearing using high-resistivity lubricant with a maximum charge accumulation of 54.5×10^{-12} (C) and minimum capacitance of 94 (pϕ) has the fatigue life of $>10^7$ cycles before the craters are initiated due to charge leakage at the asperity contacts.

This analysis makes it possible to predict the time before the slip bands, craters and cracks are initiated on track surfaces of the bearings lubricated with low-resistivity and high-resistivity lubricants, respectively, under the influence of an electric current.

REFERENCES

1. Bhargava, V., Rubin, C. A. and Hahn, G. T. (1987). The effects of indent residual stresses on mixed mode cyclic crack growth driving forces for bearing inner rings. *ASME Trans.*, **109**, 634–639.
2. Hahn, G. T., Bhargava, V., Rubin, C. A., Chen, Q. and Kim, K. (1987). Analysis of the rolling contact residual stresses and cyclic plastic deformation of SAE 52100 steel ball bearings. *ASME Trans.*, **109**, 618–626.
3. Warnock, F. V. and Benham, P. P. (1965). *Mechanics of Solids and Strength of Materials*, Sir Isaac Pitman and Sons, London.
4. Prashad, H. (1981). Investigations of corrugated pattern on the surfaces of roller bearings operated under the influence of electrical fields. ASLE/ASME Tribology Conference, San Antonio, Texas, 5–8 October.
5. Prashad, H. (1991). Theoretical and experimental investigations on the pitch and width of corrugations on the surfaces of ball bearings. *Wear*, **143**, 1–14.
6. Houghton, P. S. (1976). *Ball and Roller Bearings*, Applied Science Publishers Limited, London.
7. Prashad, H. (1990). Theoretical analysis of the effects of instantaneous charge leakage on roller bearings lubricated with high-resistivity lubricants under the influence of electric current. ASME, *J. Tribol.*, **112**, 37–43.
8. Prashad, H. (1987). The effect of cage and roller slip on the measured defect frequency response of rolling-element bearings. *ASLE Trans.*, **30**(3), 360–367.
9. Harvey, John F. (1963). *Pressure Vessel Design*: *Nuclear and Chemical Application*, D. Van Nastrand Company, Inc., New York.
10. ASME, *Boiler and Pressure Vessel Code*, Section 8, Division 2, United Engineering Centre, 345 east, 47th street, New York 10017.

11. *Metal Handbook, Properties and Selection Iron & Steel*, (1986). ASTM, Vol. 1, Metal Park, Ohio 44073.
12. Prashad, H. (1989). Analysis of the effects of electrical currents on contact temperature, residual stresses, and slip bands initiation on roller tracks of roller bearings. *Wear*, **131**, 1–14.
13. Prashad, H. (1988). Theoretical evaluation of impedance, capacitance, and charge accumulation on roller bearings operated under electrical fields. *Wear*, **125**, 223–239.
14. Prashad, H. (1987). Effects of operating parameters on the threshold voltages and impedance response on non-insulated rolling-element bearings under the action of electrical currents. *Wear*, **117**, 223–240.
15. Rodstein, L. (1974). *Electrical Control Equipments*, Mir Publishers.
16. Simpson, F. E. and Crump J. J. (1963). Effects of electrical currents on the life of rolling contact bearings. Proc. Lubrication and Wear Convention, Bournemouth, 1963, Inst. Mech. Engrs., London, Paper 27.
17. Prashad, H. (1990). Analysis of the effects of electrical current on contact temperature, residual stresses leading to slip bands initiation and formation of corrugations pattern on ball tracks of ball bearings. *BHEL J.*, **11**(1), 39–47.

APPENDIX A

Properties of SAE 52100 (Bearing Steel)

A. Chemical Composition

Carbon	0.98–1.10%
Manganese	0.25–0.45%
Phosphorous, max	0.025%
Sulphur, max	0.025%
Silicon	0.15–0.35%
Chromium	1.30–1.60%
Nickel (max)	0.25%
Copper (max)	0.35%
Molybdenum (max)	0.10%
Maximum operating temperature	150°C

B. Physical Properties

(1)	electrical resistance	3.0×10^{-7} ohm m
(2)	density	7.8×10^{3} Kg m^{-3}
(3)	specific heat	4.5×10^{2} ws kg^{-1} °C^{-1}
(4)	thermal conductivity	45 w m^{-1}°C^{-1}
(5)	Young's modulus of elasticity	210×10^{9} N m^{-2}
(6)	Poisson's ratio	0.29
(7)	coefficient of thermal expansion	15×10^{-6} °C^{-1}
(8)	yield strength	230×10^{6} N m^{-2}

Nomenclature

c	specific heat of bearing material
C_b	equivalent bearing capacitance
C_s	number of contact cycles before initiation of craters on roller tracks
C_{ir}, C_{or}	capacitance between inner race and outer race and a roller, respectively
d	diameter of rolling-element
D	pitch diameter
E	Young's modulus of elasticity
f	frequency of applied voltage (50 Hz)
f_b	ball rotational frequency
f_s	shaft rotational frequency
f_r	roller frequency
H	depth of slip bands and craters on track surfaces of races
I	current passing through bearing
K	number of rolling-elements in the loaded zone
L	length of roller
L_a	summation of the length of asperity contacts on circumference of roller track during contact with a roller
L_{ib}, L_{ob}	arc length of width of contact/corrugation along bearing width on ball track of inner race and outer race, respectively
N	number of rolling-elements in bearing
n	rpm
P	resultant load on bearing
p	positions of plain of action of loading ($p = 1, 2, 3, \ldots$)
Q	charge accumulation on bearing
Q_{ir}, Q_{or}	charge accumulation on inner race and outer race, respectively
q_{ir}	electrical energy due to charge leakage between inner race and a roller

r	radius of rolling-element
r_{ib}, r_{ob}	radius of curvature of width of contact/corrugation in ball bearing along bearing width on inner race and outer race, respectively
R_{ib}, R_{ob}	resistance between ball track of inner race and outer race, and a ball, respectively
R_{ir}, R_{or}	resistance between roller track of inner race and outer race, and a roller, respectively
R_{os}	resistance in static condition between roller track of a race and a roller if $R_{ir} = R_{or}$
R_s	equivalent bearing resistance in static condition
R	resistance between track surface of a race and a rolling-element if $R_{ir} = R_{or}$
R_b	equivalent bearing resistance
S_a	semi-range of contact stresses
T_a	ambient temperature
T_o	rise in temperature of roller tracks under rolling friction without the effect of charge leakage
T_{ib}, T_{ob}	instant temperature rise of ball track of inner race and outer race during each contact with a ball
T_{ibn}, T_{obn}	instant temperature rise of ball track of inner race and outer race in each shaft rotation, respectively
T_{ir}, T_{or}	instantaneous temperature rise of roller track of inner race and outer race during each contact with a roller, respectively
T_{irn}, T_{orn}	instantaneous temperature rise of roller track of inner race and outer race in each shaft rotation, respectively
t_{ib}, t_{ob}	duration of each contact between ball track of inner race and a ball, and outer race and a ball, respectively
t_{ir}, t_{or}	duration of each line contact between roller track of inner race and outer race, and a roller, respectively
u	voltage across bearing
V_b	circumferential speed of ball
V_{ib}	circumferential speed of ball track of inner race
V_{ibb}, V_{obb}	resultant speed of crossing of instant contact between ball track of inner race and a ball, and outer race and a ball, respectively
V_r	circumferential speed of roller
V_{ir}	circumferential speed of roller track of inner race
V_{irr}, V_{orr}	resultant speed of crossing of instant line contact between roller track of inner race and a roller, outer race and a roller, respectively
w	$2\pi f$

W_{ib}, W_{ob}	width of contact/corrugation along ball track of inner race and outer race, respectively
W_{ir}, W_{or}	width of line contact on roller track of inner race and outer race, respectively
X_{cir}, X_{cor}	capacitive reactance between roller track of inner race and outer race, and a roller, respectively
ξ	permittivity (dielectric constant of lubricant)($\phi\,m^{-1}$)
α	contact angle
μ	Poisson's ratio
Γ	coefficient of thermal expansion
ρ	density of bearing material
ρ_2, ρ_1	resistivity of low- and high-resistivity lubricants, respectively
σ_{ib},σ_{ob}	instant tangential stress on ball track of inner race and outer race by each contact of a ball, respectively
σ_{ir}, σ_{or}	instant tangential stress on roller track of inner race and outer race by each contact of a roller, respectively
σ_{ibn}, σ_{obn}	tangential stress on ball track of inner race and outer race, respectively
σ_{irn}, σ_{orn}	tangential stress on roller track of inner race and outer race, respectively
β_{ib}, β_{ob}	number of contacts on ball track of loaded zone of inner race and outer race by a single position on ball in each shaft rotation
β_{ir}, β_{or}	number of line contacts on roller track of loaded zone of inner race and outer race by a single line position on roller in each shaft rotation, respectively

Note: (*) indicates parameters pertaining to capacitive effect of a rolling-element bearing (bearing using high-resistivity lubricant).

Chapter 8

ANALYSIS OF CAPACITIVE EFFECT OF ROLLER BEARINGS ON REPEATED STARTS AND STOPS OF A MACHINE

8.1 Introduction

In this chapter, a study is reported on the capacitive effect of roller bearings on repeated starts and stops of a machine operating under the influence of shaft voltages to determine the increase in charge accumulation with time and the gradual leakage of the accumulated charges on bearing surfaces as the shaft voltage falls as soon as the instant power supply to a machine is switched-off. Under these conditions the variation of shaft revolutions to accumulate charges and discharge of the accumulated charges on the roller tracks of races at various levels of bearing to shaft voltage is analyzed. Also, variation of safe limits of starts and stops with the ratio of bearing to shaft voltage is studied [1].

8.2 Theoretical Analysis to Accumulate and Discharge of the Accumulated Charges

8.2.1 Theoretical Background

Magnetic flux develops in the electric machines due to dissymmetry of the magnetic circuits, which closes in the circumference over the yoke and induces the voltage on the shaft. Shaft voltage and flux can occur in the electric rotating machines due to various reasons which usually results in a localized current at each bearing rather than a potential difference between shaft ends. Furthermore, under the influence of potential drop across a bearing, the varying oil film thickness between races and the rollers form capacitor of varying capacitance depending on the lubricant permittivity, and offer an impedance to current flow.

8.2.2 Time Required to Accumulate Charges on Bearing Surfaces after the Start of a Machine

At the instant when the machine is started, the potential difference (V) across the inner race and rollers as well as rollers and outer race is zero. But this gradually increases and approaches the shaft voltage (E). While the shaft voltage increases, the charge on the bearing (using high-resistivity lubricant 10^{11} ohm cm) comprising of charges on inner and outer races (Q_i and Q_o) builds up. Till then V is changing and transient current is delivered from the shaft voltage.

To maintain continuity throughout the whole circuit, there is a rate of change of flux ($\mathrm{d}Q_b/\mathrm{d}T$) within the dielectric (lubricant), which is given as [1]:

$$I = \frac{\mathrm{d}Q_b}{\mathrm{d}T} \tag{8.1}$$

and, the stored charge on the bearing surfaces is determined as:

$$Q_b = V \times C_b \tag{8.2}$$

and if the current is varying then:

$$I = C_b \times \frac{\mathrm{d}V}{\mathrm{d}T} \tag{8.3}$$

Also, it can be expressed as:

$$I = \frac{(E - V)}{R_b} \tag{8.4}$$

From Eqns. (8.1)–(8.4), it is evident that:

$$V = E - R_b \times C_b \times \frac{\mathrm{d}V}{\mathrm{d}T} \tag{8.5}$$

On integrating Eqn. (8.5) and applying the initial conditions as $T = 0$, $V = 0$, the solution of Eqn. (8.5) is obtained as:

$$V = E \times (1 - e^{-T/C_b R_b}) \tag{8.6}$$

Hence, Eqn. (8.2) becomes:

$$Q_b = E \times C_b(1 - e^{-T/C_b R_b}) \tag{8.7}$$

The time taken (T_{cb}) to develop charge Q_b on bearing surfaces having capacitance C_b and resistance R_b is expressed as [using Eqn. (8.6)]:

$$T_{cb} = -C_b \times R_b \times \log_e (1 - a) \tag{8.8}$$

Similarly time taken to develop charges (Q_{ir} and Q_{or}) on a roller and roller tracks of inner as well as outer races is given as:

$$T_{cir} = -C_{ir} \times R_{ir} \log_e (1 - a) \tag{8.9}$$

$$T_{cor} = -C_{or} \times R_{or} \log_e (1 - a) \tag{8.10}$$

Equivalent capacitance and resistance of inner race (C_i and R_i) as well as outer race (C_o and R_o) with 'K' rollers in the loaded zone are $K\ C_{ir}$, $K\ C_{or}$ and R_{ir}/K, R_{or}/K, respectively. Since rollers within inner and outer races are in parallel, whereas both in inner and outer races are in series to each other for the electric path. So, T_{ci} and T_{co} are equal to that of T_{cir} and T_{cor}, respectively [Eqns. (8.9) and (8.10)].

8.2.3 Current Passing through Bearing

From Eqns. (8.2) and (8.4), shaft voltage is given as:

$$E = \frac{I \times R_b + Q_b}{C_b} \tag{8.11}$$

Using Eqn. (8.1) and applying initial conditions as $T = 0$, $I = E/R_b$, the solution of Eqn. (8.11) is:

$$I = \left(\frac{E}{R_b}\right) \times e^{-T/C_b R_b} \tag{8.12}$$

And potential drop across a bearing is given as:

$$V = I \times R_b = E \times e^{-T/C_b R_b} \tag{8.13}$$

Similarly, potential difference between roller track of inner race as well as outer race and a roller is given by:

$$V = I \times e^{-T/R_{ir} C_{ir}} \tag{8.14}$$

and

$$V = I \times e^{-T/R_{or} C_{or}} \tag{8.15}$$

The above Eqns. (8.14) and (8.15) are also valid for potential difference between roller tracks of inner as well as outer races and rollers since:

$$R_i = \frac{R_{ir}}{K},\ R_o = \frac{R_{or}}{K} \tag{8.16}$$

and

$$C_i = K \times C_{ir},\ C_o = K \times C_{or} \tag{8.17}$$

8.2.4 Time Required to Discharge the Accumulated Charge from Bearing Surfaces During Stop

As soon as the power supply to a machine is switched-off, magnetic flux disappears and the shaft voltage becomes zero instantaneously. The rate of discharge of bearing capacitor (formed during bearing operation) is determined by differentiating Eqn. (8.13), and given as:

$$\frac{-\mathrm{d}V}{\mathrm{d}T} = \left(\frac{E}{C_b \times R_b}\right) e^{-T/C_b R_b} \tag{8.18}$$

At the instant, when machine is stopped ($T = 0$), the rate of change of bearing potential drop is determined as:

$$\frac{\mathrm{d}V}{\mathrm{d}T} = \frac{-E}{C_b \times R_b} \tag{8.19}$$

In time constant ($C_b R_b$), the potential drop of bearing capacitor falls and is determined as [by Eqn. (8.13)]:

$$V = E \times e^{-1} = 0.368E \tag{8.20}$$

The time required (T_{db}) to drop the potential from V to the value of 'a' times the shaft voltage (E) ($V = a{*}E$) is given as [by Eqn. (8.13)]:

$$T_{db} = -R_b \times C_b \log_e a \tag{8.21}$$

Similarly time required to discharge the accumulated charges from roller track of inner as well as outer races and a roller/rollers is given as:

$$T_{dir} = T_{di} = -R_{ir} \times C_{ir} \log_e a \tag{8.22}$$

$$T_{dor} = T_{do} = -R_{or} \times C_{or} \log_e a \tag{8.23}$$

8.2.5 Duration of Contact between Roller Track of Races and a Roller

Assuming that the inner race is rotating and the outer race is stationary, the duration of contact between the roller track of the inner race and a roller crossing the width of contact W_{ir} is determined as [2]:

$$t_{ir} = \frac{2D \times W_{ir}}{\pi \times f_s \times (3D + d) \times (D - d)} \tag{8.24}$$

Similarly,

$$t_{or} = \frac{2D \times W_{or}}{\pi \times f_s \times (D^2 - d^2)} \tag{8.25}$$

Also, the number of contacts between inner race and rollers, and outer race and rollers, in each shaft rotation is determined as β_{ir}, and β_{or}, respectively and given as [2]:

$$\beta_{ir} = \frac{(3D + d) \times (D - d)}{4d \times D} \tag{8.26}$$

and,

$$\beta_{or} = \frac{(D^2 - d^2)}{4Dd} \tag{8.27}$$

8.2.5.1 Number of Shaft Rotation for the Charge Accumulation on Bearing Surfaces after Start of a Machine

The number of contacts between roller track of inner race as well as outer race with a roller (N_{ic}, N_{oc}) for the charge accumulation Q_{ir} and Q_{or}, respectively, after start of a machine can be determined as the ratio of time taken (T_{cir} and T_{cor}) to accumulate charges to the time taken for each contact between roller track of inner race (t_{ir}) and a roller, and similarly between outer race and a roller (t_{or}).

Thus, the number of contacts (N_{ic}, N_{oc}) can be determined as:

$$N_{ic} = \frac{T_{cir}}{t_{ir}}, \quad \text{and} \quad N_{oc} = \frac{T_{cor}}{t_{or}} \tag{8.28}$$

By using Eqns. (8.9), (8.10), (8.24) and (8.25), the N_{ic} and N_{oc} are expressed as:

$$N_{ic} = \frac{-\pi \times f_s \times (3D + d) \times (D - d) \times C_{ir} \times R_{ir} \times \log_e (1 - a)}{2D \times W_{ir}} \tag{8.29}$$

and

$$N_{oc} = \frac{-\pi \times f_s \times (D^2 - d^2) \times C_{or} \times R_{or} \times \log_e(1-a)}{2D \times W_{or}} \tag{8.30}$$

The number of contacts (N_i and N_o) to accumulate charges Q_i and Q_o will be the same as that of N_{ic} and N_{oc}, respectively [Eqns. (8.16) and (8.17)]. The number of shaft rotations to accumulate the charges (Q_{ir} and Q_{or}) is determined as the ratio of number of contacts between a roller and inner race, as well as outer race and a roller to the number of respective contacts in each shaft rotation [Eqns. (8.26) and (8.27)]. Thus N_{icn} and N_{ocn} are expressed as follows using Eqns. (8.26)–(8.30):

$$N_{icn} = \frac{-[2\pi \times f_s \times d \times C_{ir} \times R_{ir} \times \log_e(1-a)]}{W_{ir}} \tag{8.31}$$

$$N_{ocn} = \frac{-[2\pi \times f_s \times d \times C_{or} \times R_{or} \times \log_e(1-a)]}{W_{or}} \tag{8.32}$$

N_{icn} and N_{ocn} will be equal to N_{in} and N_{on}, similar to that of N_{ic} and N_{oc} as equal to N_i and N_o, respectively [Eqns. (8.16) and (8.17)].

8.2.6 Number of Shaft Rotations for the Discharge of the Accumulated Charges after the Stop of a Machine

The number of contacts between roller track of inner as well as outer races with a roller to discharge of the accumulated charges from the bearing surfaces (N_{id} and N_{od}) after the power supply to a machine is switched-off, are determined similar to Eqn. (8.28) and given as:

$$N_{id} = \frac{T_{dir}}{t_{ir}}, \quad \text{and} \quad N_{od} = \frac{T_{dor}}{t_{or}} \tag{8.33}$$

Similar to the logic discussed above, relations for number of contacts between inner race and a roller and outer race and a roller (N_{id}, N_{od}), and N_{idn} and N_{odn} are developed using Eqns. (8.22)–(8.27), and given as:

$$N_{id} = -\left[R_{ir} \times C_{ir} \times \frac{\log_e a}{2D \times W_{ir}}\right] \times \pi \times f_s \times (3D + d) \times (D - d) \tag{8.34}$$

$$N_{od} = -\left[R_{or} \times C_{or} \times \frac{\log_e a}{2D \times W_{or}}\right] \times \pi \times f_s \times (D^2 - d^2) \tag{8.35}$$

$$N_{idn} = -2\pi \times f_s \times d \times R_{ir} \times C_{ir} \times \frac{\log_e -a}{W_{ir}}, \text{ and} \tag{8.36}$$

$$N_{\text{odn}} = \frac{-2\pi \times f_s \times d \times R_{\text{or}} \times C_{\text{or}} \times \log_e a}{W_{\text{or}}} \tag{8.37}$$

The N_{icd} and N_{ocd} are equal to N_{id} and N_{od}, and N_{icdn} and N_{ocdn} are equal to that of N_{idn} and N_{odn}, respectively [Eqns. (8.16) and (8.17)].

8.2.7 Determination of the Ratio of Contacts for Charge Accumulation to Discharge of the Accumulated Charges

The ratio of the number of contacts and shaft revolutions for the charge accumulation to the discharge of the accumulated charges from the inner as well as outer races is determined using Eqns. (8.31)–(8.37), and given as:

$$\frac{N_{\text{icn}}}{N_{\text{idn}}} = \frac{N_{\text{ocn}}}{N_{\text{odn}}} = \frac{N_{\text{in}}}{N_{\text{icdn}}} = \frac{N_{\text{on}}}{N_{\text{ocdn}}} = \frac{\log_e (1-a)}{\log_e a} \tag{8.38}$$

8.2.8 Number of Starts and Stops before the Initiation of Craters on the Bearing Surfaces

The N_{ssi} and N_{sso} can be determined as the ratio of the net time required to initiate craters (C_{si}/f_s and C_{so}/f_s) to the time lapse for charge and discharge of the accumulated charges (T_{Cir} and T_{dir}) in each start and stop of a machine to the number of cycles (C_{sp}) required before the machine come to standstill condition after the power supply to the machine is put off. Thus N_{ssi} and N_{sso} can be determined as

$$N_{\text{ssi}} = -\frac{C_{\text{si}}}{f_s^*}(T_{\text{cir}} + T_{\text{dir}})^* C_{\text{sp}} \tag{8.39}$$

and

$$N_{\text{sso}} = -\frac{C_{\text{so}}}{f_s^*}(T_{\text{cor}} + T_{\text{dor}})^* C_{\text{sp}} \tag{8.40}$$

The number of cycles C_{sp}, before the machine comes to standstill condition, depends on the machine inertia, friction in bearings, etc. Various experimental investigations reveal that, in general C_{sp} vary between 20 and $90f_s$. Using Eqns. (8.9), (8.10), (8.22) and (8.23), the N_{ssi} and N_{sso} are determined as:

$$N_{\text{ssi}} = \frac{C_{\text{si}}}{f_s \times C_{\text{ir}} \times R_{\text{ir}} \times \log_e a(1-a)^* C_{\text{sp}}} \tag{8.41}$$

and

$$N_{\rm sso} = -\frac{C_{\rm so}}{f_{\rm s} \times C_{\rm or} \times R_{\rm or} \times \log_{\rm e} a(1-a) \times C_{\rm sp}} \tag{8.42}$$

and the ratio of $N_{\rm ssi}$ to $N_{\rm sso}$ is given as:

$$\frac{N_{\rm ssi}}{N_{\rm sso}} = \frac{C_{\rm si} \times C_{\rm ir} \times R_{\rm ir}}{C_{\rm so} \times C_{\rm or} \times R_{\rm or}} \tag{8.43}$$

8.3 Theoretical Data

The ratio $N_{\rm icn}/N_{\rm idn}$ is determined for different values of bearing to shaft voltage ($a = V/E$) varying from 0.1 to 0.9, and is shown in Fig. 8.1. The same variation is valid for $N_{\rm ocn}/N_{\rm odn}$, $N_{\rm in}/N_{\rm icdn}$ and $N_{\rm on}/N_{\rm ocdn}$ [Eqn. (8.38)].

The number of starts and stops of a machine before the initiation of craters on the bearing surfaces are determined for $a = 0.1 - 0.9$ using Eqns. (8.41) and (8.42). Figure 8.2 indicates the variation of number of starts and stops ($N_{\rm ssi}$ and $N_{\rm sso}$) at various levels of bearing to shaft voltage (V/E) of the NU 330 bearing operating at 1200 rpm ($f_{\rm s} = 20\,{\rm s}^{-1}$) having $C_{\rm ir}$, $C_{\rm or}$, $R_{\rm ir}$ and $R_{\rm or}$ as 57.98 (pϕ), 69.58 (pϕ), 38.6×10^7 ohm and 32.2×10^7 ohm, respectively [2, 3].

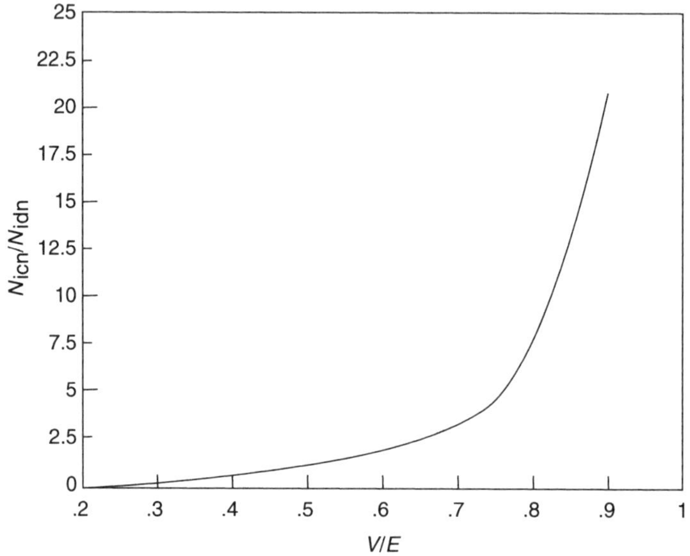

Fig. 8.1. Variation of the ratio of shaft revolutions to accumulate and discharge of accumulated charges ($N_{\rm icn}/N_{\rm idn}$) at various levels of bearing to shaft voltages (V/E) on roller tracks of inner and outer races of a roller bearing operating under the influence of electric current

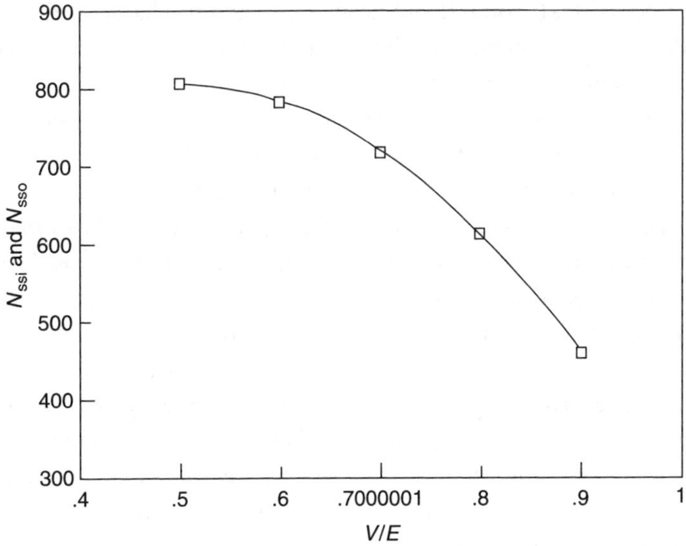

Fig. 8.2. Variation of number of starts and stops of a motor before formation of craters on roller track of inner and outer races (N_{ssi} and N_{sso}) at various levels of bearing to shaft voltage (V/E) of the NU 330 bearing operating under the influence of electric currents

C_{sp} for the machine has been experimentally determined as approximately 20 f_s. C_{si} and C_{so} based on the temperature rise and thermal stress calculations [2, 4] under frequent start and stop regimes are approximately established as 2×10^5 cycles.

8.4 Results and Discussion on Shaft Rotations for Charge Accumulation/ Discharge and the Formation of Craters

8.4.1 Time and Number of Shaft Rotations Required to Accumulate Charges on Bearing Surfaces and Discharge of the Accumulated Charges

The time required for accumulating the charges on the roller tracks of inner race and outer race depends on capacitance (C_i, C_o) and resistance (R_i and R_o). Also, this is a function of natural logarithm of the difference of shaft and bearing voltage to the shaft voltage [Eqns. (8.9) and (8.10)]. Similarly, T_{cb} is a function of R_b, C_b and $\log_e (1 - a)$ [Eqn. (8.8)]. The time required to discharge of the accumulated charges depends on C_i, C_o, R_i and R_o, and is a function of natural logarithm of the ratio of bearing to the shaft voltage [Eqns. (8.22) and (8.23)]. Also, T_{db} is a function of R_b, C_b and $\log_e a$ [Eqn. (8.21)].

The number of shaft rotations to accumulate and discharge of the accumulated charges depends on the diameter of rolling-element, the shaft rotational frequency, the width of contact between the roller track of the inner and outer races and a rolling-element, and the natural logarithm of the shaft to bearing voltage, besides C_{ir}, C_{or}, R_{ir} and R_{or} [Eqns. (8.28)–(8.37)]. However, the ratio of shaft revolutions to accumulate and discharge of the accumulated charges from inner race and outer race with a roller/rollers is independent of all the parameters and depends only on the ratio of difference of natural logarithm of shaft and bearing voltage to the voltage across bearing, to the ratio of the natural logarithm of shaft and the bearing voltage [Eqn. (8.38)].

The ratio of shaft rotations to accumulate and discharge of the accumulated charges (N_{icn}/N_{idn}) increases with V/E. As the V/E increases from 0.1 to 0.9, the (N_{icn}/N_{idn}) increases from 0.05 to 20.91 (Fig. 8.1). It is evident that for the successive higher value of 'a', the ratio of N_{icn}/N_{idn} is higher than for the immediate lower value of V/E (Fig. 8.1). The ratio of shaft rotations to accumulate/discharge of the accumulated charges by a single roller and by all the rollers on a bearing is identical [Eqn. (8.38)]. Also, complete charge accumulation take the same number of shaft revolutions equivalent to that of the charge accumulation by a single rolling-element and roller track of races [Eqns. (8.16), (8.17), (8.31) and (8.32)].

8.4.2 Number of Start and Stop Cycles before the Initiation of Craters on Roller Track of Races due to Discharge of the Accumulated Charges

The number of starts and stops before the initiation of craters on the roller track of inner and outer races depend on the number of cycles required to initiate craters, the shaft rotational frequency, respective capacitance and resistance of the races and V/E [Eqns. (8.41) and (8.42)]. However, the ratio of N_{ssi} and N_{sso} is independent to shaft voltage and shaft rotational frequency [Eqn. (8.43)].

The number of starts and stops to initiate craters on the roller tracks decreases as the ratio of bearing to shaft voltage increases. For the NU 330 bearing number of starts and stops to initiate craters on the roller track of inner and outer races decreases from 803.63 to 463.5 as the bearing to shaft voltage 'a' increases from 0.5 to 0.9 (Fig. 8.2).

8.5 Conclusions

The following conclusions are drawn based on the investigation given in this chapter [1, 2]:

(1) The time required for accumulating and discharge of the accumulated charges on the roller track of races depends on the capacitance and

resistance between the rollers and roller track of races, and depends on the natural logarithm of the ratio of bearing to shaft voltage.

(2) The number of shaft rotations to accumulate and discharge of the accumulated charges on roller track of races depends on the shaft rotational frequency, the diameter of rolling-element and the width of contact between roller track of races and rolling-element besides capacitance and resistance of races, and the ratio of bearing-to-shaft voltage.

(3) The number of rotations required accumulating and discharge of the charges between a roller and the roller track of races of a bearing is the same as that of all rollers and roller track of races.

(4) As the ratio of bearing to shaft voltage increases, the ratio of shaft rotation to accumulate and discharge of the accumulated charges increases.

(5) With increase of bearing-to-shaft voltage, the number of starts and stops to initiate craters on the roller track of races decreases.

This analysis besides establishing the effect of capacitive response of the bearings is useful for transient performance analysis under the effect of shaft voltages.

REFERENCES

1. Prashad, H. (1992). Theoretical of analysis of capacitive effect of roller bearings on repeated starts and stops of a machine operating under the influence of shaft voltages. ASME, *J. Tribol.*, **114**, 818–822.
2. Prashad, H. (1990). Theoretical analysis of the effects of instantaneous charge leakage of roller tracks of roller bearings lubricated with high-resistivity lubricants under the influence of electric current. ASME, *J. Tribol.*, **112**, 37–43.
3. Prashad, H. (1988). Theoretical evaluation of impedance, capacitance and charge accumulation of roller bearings operated under electrical fields. *Wear*, **125**, 223–239.
4. Prashad, H. (1989). Analysis of the effects of an electric current on contact temperature, contact stresses and slip bands initiation on roller tracks of roller bearings. *Wear*, **131,** 1–14.

Nomenclature

a	ratio of potential difference across bearing to shaft voltage ($a = V/E < 1$)
C_b	equivalent bearing capacitance
C_{si}, C_{so}	number of cycles before initiation of craters on roller track of inner race and outer race, respectively
C_{sp}	number of cycles before the machine comes to stand still condition after the power supply is put off

C_{ir}, C_{or}	Capacitance between inner race and a roller, and outer race and a roller, respectively
C_i, C_o	equivalent capacitance between inner race and rollers, and outer race and rollers, respectively
d	diameter of roller
D	pitch diameter
E	shaft voltage
f_s	shaft rotational frequency
I	current passing through bearing
K	number of rollers in the loaded zone
N_{ic}, N_{oc}	number of contacts between inner race and a roller, and outer race and a roller, to accumulate charges (Q_{ir}) and (Q_{or}), respectively
N_{icn}, N_{ocn}	number of shaft rotations to accumulate charges (Q_{ir}) and (Q_{or}) respectively
N_i, N_o	number of contacts between roller track of inner race and rollers, and outer race and rollers, to accumulate charges (Q_i) and (Q_o) respectively
N_{in}, N_{on}	number of shaft rotations to accumulate charges (Q_i) and (Q_o), respectively
N_{id}, N_{od}	number of contacts between inner race and a roller, and outer race and a roller to discharge of the accumulated charges (Q_{ir}) and (Q_{or}), respectively
N_{icd}, N_{ocd}	number of contacts between roller track of inner race and rollers, and outer race and rollers to discharge of the accumulated charges (Q_i) and (Q_o), respectively
N_{idn}, N_{odn}	number of shaft rotations to discharge of the accumulated charges (Q_{ir}) and (Q_{or}), respectively
N_{icdn}, N_{ocdn}	number of shaft revolutions to discharge the accumulated charge (Q_i) and (Q_o), respectively
N_{ssi}, N_{sso}	number of starts and stops before the formation of craters on the roller track of inner race, and outer race, respectively
Q_b	electric charge accumulation on bearing
Q_{ir}, Q_{or}	charge accumulation between roller track of inner race and a roller, and outer race and a roller, respectively
Q_i, Q_o	charge accumulated between inner race and rollers, and outer race and rollers, respectively
R_b	equivalent bearing resistance
R_{ir}, R_{or}	resistance between roller track of inner race and a roller, and outer race and a roller, respectively

R_i, R_o	resistance between roller tracks of inner race and rollers, and outer race and rollers, respectively
t_{ir}, t_{or}	duration of each line contact between roller track of inner race and a roller, and outer race and a roller, respectively
T	time
T_{cb}	time required to accumulate charge (Q_b) on bearing surfaces
T_{cir}, T_{cor}	time taken to accumulate charges (Q_{ir}) and (Q_{or}), respectively
T_{ci}, T_{co}	time taken to accumulate charges (Q_i) and (Q_o), respectively
T_{dir}, T_{dor}	time required to discharge accumulated charges from roller track of inner race and a roller, and outer race and a roller, respectively
T_{db}	time required to discharge of accumulated charges from bearing surfaces
T_{di}, T_{do}	time required to discharge accumulated charges from roller track of inner race and rollers, and outer race and rollers, respectively
V	potential drop across bearing
W_{ir}, W_{or}	width of line contact on roller track of inner race, and outer race, respectively
β_{ir}, β_{or}	number of line contacts at one single position on roller track of loaded zone of inner race and outer race, respectively, by a single line position on rollers in each shaft rotation

Chapter 9

EFFECT OF CURRENT LEAKAGE ON ELECTRO-ADHESION FORCES IN ROLLING FRICTION AND MAGNETIC FLUX DENSITY DISTRIBUTION ON TRACK SURFACES OF ROLLING-ELEMENT BEARINGS

9.1 General

This chapter deals with the investigations carried out on the various rolling-element bearings after being operated under the influence of electric fields, and pure rolling friction on the roller bearing test machine. The significant magnetic flux density was detected on surfaces of the bearings lubricated with low-resistivity lubricant under the influence of electric fields. No such phenomenon was observed either on bearings using high or low-resistivity lubricant under pure rolling friction or on bearings lubricated with high-resistivity lubricant under the influence of electric current. New bearing surfaces do not show significant magnetic flux density but it has been detected after long operation on different motor bearings, lubricated with low-resistivity greases. The electro-adhesion forces in the bearings using low-resistivity lubricant increase under the influence of electric fields in contrast to those with high-resistivity lubricants. Under the pure rolling friction, resistivity of lubricants do not affect the electro-adhesion forces. The investigations reported in this chapter along with the study of damaged/corrugated surfaces, deterioration of the used lubricants (Chapter 3), and flux density distribution in the bearing surfaces, the leakage of current leading to failure of the non-insulated motor bearings can be established.

The residual magnetic flux density on the track surfaces of a rolling-element bearing produces forces of attraction, hysteresis loss and eddy current, and can lead to premature bearing failure. This chapter brings out a theoretical model to determine the magnetic flux density developed on the inner and outer surfaces of inner race and outer race, and on the surface of rolling-elements, of a rolling bearing operating under the influence of electric current. The flux density, analytically determined, is found to agree well with the measured flux density developed on the surfaces of races and rolling-elements of the bearings tested in a test rig under the influence of electric current. Also, the magnetic flux density on the surfaces of damaged bearings of motors and alternators has been measured, and the theoretical model has been used to determine the amount of current flow through the damaged bearings. The value of current flow through the bearings, thus established, has been found to be close to that evaluated by the measurement of shaft voltage and bearing resistance.

Furthermore, analysis given in this chapter besides having a potential to ascertain the cause of failure by passage (leakage) of current, can establish the amount of the flow of the leakage current through the bearings by determining the magnetic flux density on the surfaces of rolling-element bearings. Also, the current flow, thus established, along with the measurement of the shaft voltage, leads to establish the bearing impedance.

9.2 Introduction

Normally rolling-element bearings are not expected to carry electric current, yet, there are instances in which they are to carry current. The current may flow in the bearings owing to several reasons:

(1) A bearing may carry current as a necessary part of an electric circuit,
(2) Current may be self-induced as a result of the design characteristics of the machine, and
(3) Current may result from an electrostatic phenomenon.

When the rolling-element bearings of the motors are damaged in service, the question often arises as to whether the damage is due to leakage electric current. Although, considerable investigations and analysis have been carried out dealing with bearing current (Chapters 2–8) still the diagnosis of the cause of failed bearings – using lubricants having different characteristics – due to electric current, has not been well understood. Hence, a study was undertaken to understand the effect of electric fields on the roller bearings, lubricated with greases of different resistivities, by diagnosis of magnetic flux density on the bearing surfaces. Also, to diagnose the change in the electro-adhesion forces in the bearings operated in pure rolling friction without electric fields and under the influence of electric fields by analysis of the used lubricants.

Shaft voltages exist in electric machines as a result of asymmetry of faults, winding faults, unbalanced supplies, electrostatic effects, air-gap fields, magnetized shaft or other machine members, asymmetries of the magnetic fields, etc. The latter are caused by rotor eccentricity, poor alignment, manufacturing tolerances, uneven gaps, segmental lamination punching, variation in permeability and various other unforeseen reasons. In general, magnetic flux develops in electric machines due to asymmetry of the magnetic circuit, which closes in the circumference over the yoke and induces voltage on the shaft. This results in a localized current at each bearing rather than a potential difference between shaft ends. A current path, however, along shaft, bearings and frame results in a potential difference between the shaft ends as discussed in Chapter 5 [1–3].

This phenomenon can be elaborated as follows. When asymmetry of the magnetic circuits exists due to various reasons, as explained above, including that of rotor and stator sagging, it produces variable magnetic flux. Since shaft continuously rotates even under these conditions (under the influence of magnetic flux in the magnetic field), this leads to induce voltage on the shaft. Since variable magnetic asymmetry is localized because of localized uneven gaps in segmental punching and tolerance, the voltage generated is localized and variable. The localized variable shaft voltage induces voltage on the rolling-elements and bearing outer race by mutual induction, and localized loop of current between shaft/inner race and rolling-elements, and rolling-elements and outer race appears depending on localized bearing impedance. This is a complex phenomenon. But, in principle, it happens as explained. The combination of various defects including axial shaft flux due to residual magnetization, rotor eccentricity and asymmetrical rotor winding increase shaft voltage, and a current path, along shaft bearings and frame, results in a potential difference between shaft ends.

At a certain threshold voltage depending on the resistivity of the lubricant and operating conditions, electrical breakdown occurs and current flows through the bearing [4]. Thus, circular current in the inner race leaks through the rolling-elements to the outer race by following a path of least resistance, and establishes the field strength leading to development of residual magnetic flux on the track surface of races, rolling-elements and, inner and outer surface of races in due course [5].

The investigations given in this chapter were undertaken to determine and compare experimentally and theoretically the developed magnetic flux density distribution on the track surface, inner and outer surface of races, and rolling-elements under the influence of electric current. Also, current flow through the damaged bearings of motors and alternators has been established by the developed theoretical model, by measurement of residual flux density on the track surfaces, inner/outer surfaces of races and rolling-elements. The level of current, thus determined, is compared with the derived value of current obtained by the measurement of shaft-voltage and bearing resistance.

This chapter does not deal with hydrodynamic bearings used in turbogenerators. It deals with rolling-element bearings used in motors/alternators. Frequent failure (30% of total failure) has been reported because of this phenomenon.

9.3 Theoretical Model and Approach for Determination of Field Strength

In a rolling-element bearing, the current enters the inner surface of the inner race, through its bore in a distributed form, and flows around and outward until it concentrates at one or several rolling contacts. As the rolling-elements orbit, the current carrying contact travels. Currents then flow through each of the conducting rolling-elements, between two diametrically opposite contact areas. As the rolling-elements rotate with respect to the contact areas, these currents gradually sweep out 360 degrees around a major circle. In the outer race, the same situation as in the inner race occurs, but from the contact surface outward rather than inward. If there were only one conducting rolling-element at a time, then two arc currents would travel in the races towards its contact areas – one clockwise and one counter clockwise. Depending on the resistance, these two currents may or may not be equal in magnitude. If more than one rolling-element conducts, then of course, the situation is more complex. In the theoretical model, current flow in one direction, both in inner and outer race, is assumed.

9.4 Field Strength on Track Surface of Races and Rolling-Elements

9.4.1 Field Strength on the Track Surface of the Inner Race of a Rolling-Element Bearing due to the Flow of Circular Current in the Outer Race and Rolling-Elements

Under the effect of shaft voltage, circular current flows through the inner race before it leaks through rolling-elements to outer race and establishes the field strength in the bearing [4–6]. The field strength on the track surface of inner race depends on the flow of a circular current in the arc of the outer race before it leaks to the ground through the path of least resistance and flow of current through the rolling-elements before it passes to the outer race. The field strength follows Fleming's left hand rule. Field lines surrounding an arc of a race due to flow of current are shown in Fig. 9.1.

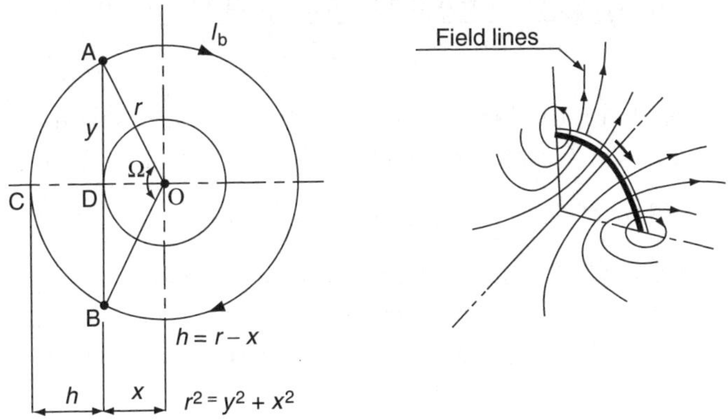

Fig. 9.1. Field strength at locations O and D and field lines due to circular current I_b

9.4.1.1 Due to Current Flow in the Arc of the Outer Race

The developed field strength on the track surface of an inner race, H_{iro}, due to the flow of circular current, I_b, in the outer race, is determined by using R_{ir} for X and $(R_{or}^2 - R_{ir}^2)^{0.5}$ for Y in Eqn. (ix) of the Appendix, and as shown in Fig. 9.1, is given as:

$$H_{iro} = \frac{2\pi I_b (R_{or}^2 - R_{ir}^2)}{R_{or}^3} \tag{9.1}$$

9.4.1.2 Due to Flow of Circular Current in Rolling-Elements

The circular current flows in a few rolling-elements before it leaks from the inner race to the outer race by following a path of least resistance. Since there are a number of rolling-elements in a bearing, the system of rolling-elements is treated as the ring of radius R equal to that of the bearing pitch radius. The field strength on the track surface of inner race due to the current flow in rolling-elements is determined similar to Eqn. (9.1) and is given as:

$$H_{irr} = \frac{2\pi I_b (R^2 - R_{ir}^2)}{R^3} \tag{9.2}$$

9.4.1.3 Equivalent Field Strength on the Track Surface of the Inner Race

The equivalent field strength (H_{ir}) on the track surface of the inner race due to the direction of the flow of circular current in the outer race and rolling-elements is the summation of H_{iro} and H_{irr}, since on rotation, rolling-elements change the polarity and outer race is stationary. So, the equivalent field strength will be either addition or difference of H_{iro} and H_{irr}, and is determined, using Eqns. (9.1) and (9.2), as:

$$H_{\rm ir} = \frac{2\pi I_{\rm b}[R^2 R_{\rm or}^2 (R \pm R_{\rm or}) - R_{\rm ir}^2 (R^3 \pm R_{\rm or}^3)]}{R_{\rm or}^3 R^3} \tag{9.3}$$

9.4.2 Field Strength on the Inner Surface of the Inner Race due to Flow of Circular Current in the Outer Race and Rolling-Elements

9.4.2.1 Due to Flow of Circular Current in Outer Race

Similar to Section 9.4.1, the developed field strength on the inner surface of the inner race, $H_{\rm irio}$, due to flow of circular current, $I_{\rm b}$, in the outer race determined using $R_{\rm iri}$ for X and $(R_{\rm or}^2 - R_{\rm iri}^2)^{0.5}$ for Y in Eqn. (ix) of the Appendix, and as shown in Fig. 9.1, is given as:

$$H_{\rm irio} = \frac{2\pi I_{\rm b}(R_{\rm or}^2 - R_{\rm iri}^2)}{R_{\rm or}^3} \tag{9.4}$$

9.4.2.2 Due to Flow of Circular Current in Rolling-Elements

The field strength on the inner surface of the inner race due to the current flow in rolling-elements, $H_{\rm irir}$, is determined similar to Eqn. (9.2) by replacing $R_{\rm ir}^2$ by $R_{\rm iri}^2$, and is given as:

$$H_{\rm irir} = \frac{2\pi I_{\rm b}(R^2 - R_{\rm iri}^2)}{R^3} \tag{9.5}$$

9.4.2.3 Equivalent Field Strength on the Inner Surface of the Inner Race

The equivalent field strength, $H_{\rm iri}$, on the inner surface of the inner race is determined similar to Eqn. (9.3) by summation of $H_{\rm irio}$ and $H_{\rm irir}$, and is given as:

$$H_{\rm iri} = \frac{2\pi I_{\rm b}[R^2 R_{\rm or}^2 (R \pm R_{\rm or}) - R_{\rm iri}^2 (R^3 \pm R_{\rm or}^3)]}{R_{\rm or}^3 R^3} \tag{9.6}$$

9.4.3 Field Strength on the Track Surface of the Outer Race due to the Flow of Circular Current in the Inner Race and Rolling-Elements

9.4.3.1 Due to the Flow of Circular Current in the Inner Race

The field strength on the track surface of the outer race, $H_{\rm ori}$, due to the flow of current in the inner race, is evaluated using $R_{\rm or}$ for X and $R_{\rm or}\,(R_{\rm or}^2 - R_{\rm ir}^2)^{0.5}/R_{\rm ir}$ for Y (as determined in Fig. 9.2) in Eqn. (ix) of the Appendix, and is determined as:

$$H_{\rm ori} = \frac{2\pi I_{\rm b} R_{\rm ir}(R_{\rm or}^2 - R_{\rm ir}^2)}{R_{\rm or}^4} \tag{9.7}$$

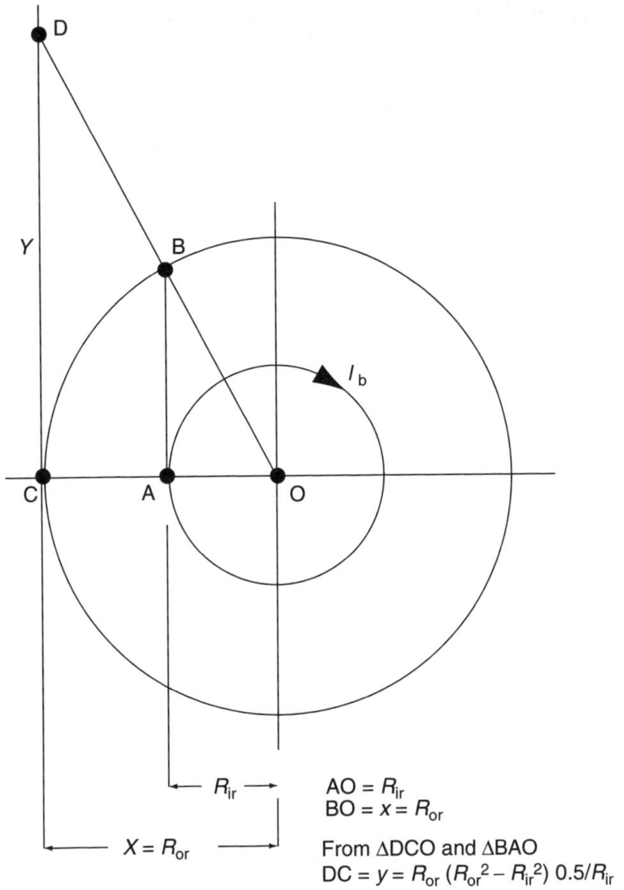

Fig. 9.2. Field strength at location C due to circular current I_b

9.4.3.2 Due to the Flow of Circular Current in the Rolling-Elements

The flow of current through the rolling-elements creates the field strength both in the inner race and outer race. Similarly to Eqn. (9.2) for the field strength on the bearing inner race due to the flow of current in the rolling-elements, the field strength on the track surface of outer race is determined by replacing R_{ir} with R in Eqn. (9.7), and is given as:

$$H_{orr} = \frac{2\pi I_b R[R_{or}^2 - R^2]}{R_{or}^4} \tag{9.8}$$

9.4.3.3 Equivalent Field Strength on the Track Surface of Outer Race of a Bearing

The equivalent field strength on the track surface of the outer race due to the flow of circular current in the inner race and rolling-elements is determined as the summation of H_{ori} and H_{orr} using Eqns. (9.7) and (9.8), since on rotation rolling-elements change the polarity with respect to rotating inner race. So, the equivalent field strength will be either difference or addition of H_{ori} and H_{orr}, and is given as:

$$H_{\text{or}} = \frac{2\pi I_{\text{b}}[R_{\text{or}}^2(R_{\text{ir}} \pm R) - (R_{\text{ir}}^3 \pm R^3)]}{R_{\text{or}}^4} \tag{9.9}$$

9.4.4 Field Strength on the Outer Surface of the Outer Race due to the Flow of Circular Current in the Inner Race and Rolling-Elements

9.4.4.1 Due to Flow of Circular Current in the Inner Race

Similarly to Section 9.4.3, the developed field strength on the outer surface of outer race, H_{oroi}, due to flow of circular current I_{b} in the inner race is determined by using R_{oro} for X and $R_{\text{oro}}(R_{\text{oro}}^2 - R_{\text{ir}}^2)^{0.5}/R_{\text{ir}}$ for Y (as determined by Fig. 9.2) in Eqn. (ix) of the Appendix, and is determined as:

$$H_{\text{oroi}} = \frac{2\pi I_{\text{b}} R_{\text{ir}}(R_{\text{oro}}^2 - R_{\text{ir}}^2)}{R_{\text{oro}}^4} \tag{9.10}$$

9.4.4.2 Due to Flow of Circular Current in Rolling-Elements

The field strength on the outer surface of the outer race, H_{oror}, due to flow of current in rolling-elements is determined by replacing R_{ir} by R in Eqn. (9.10), and is given as:

$$H_{\text{oror}} = \frac{2\pi I_{\text{b}} R(R_{\text{oro}}^2 - R^2)}{R_{\text{oro}}^4} \tag{9.11}$$

9.4.4.3 Equivalent Field Strength on the outer Surface of the Outer Race

The equivalent field strength, H_{oro}, on the outer surface of the outer race is the summation of H_{oroi} and H_{oror}, and is determined similar to Eqn. (9.9), and is given as:

$$H_{\text{oro}} = \frac{2\pi I_{\text{b}}[R_{\text{oro}}^2(R_{\text{ir}} \pm R) - (R_{\text{ir}}^3 \pm R^3)]}{R_{\text{oro}}^4} \tag{9.12}$$

9.4.5 Field Strength on the Rolling-Elements due to the Flow of Circular Current in the Inner and the Outer Races of the Bearing

9.4.5.1 Due to Flow of Circular Current in the Outer Race

The field strength on the rolling-elements due to flow of current in the outer race is determined using R in place of R_{ir} in Eqn. (9.1) and is given as:

$$H_{ror} = \frac{2\pi I_b(R_{or}^2 - R^2)}{R_{or}^3} \tag{9.13}$$

9.4.5.2 Due to Flow of Circular Current in the Inner Race

The field strength on the surface of rolling-elements due to the flow of current in the inner race is determined using R in place of R_{or} in Eqn. (9.7) and is given as:

$$H_{rir} = \frac{2\pi I_b R_{ir}(R^2 - R_{ir}^2)}{R^4} \tag{9.14}$$

9.4.5.3 Equivalent Field Strength on Rolling-Elements

On rotation, rolling-elements change the polarity, and outer race being stationary, the equivalent field strength on the surface of rolling-elements depends on the direction of current flow in the races. The equivalent field strength will be either difference or addition of H_{rir} and H_{ror}. So, by using Eqns. (9.13) and (9.14), it is given as:

$$H_r = \frac{2\pi I_b[R^2(R_{ir}R_{or}^3 \pm R^4) - R_{or}^2(R_{or}R_{ir}^3 \pm R^4)]}{R_{or}^3 R^4} \tag{9.15}$$

9.5 Magnetic Flux Density on Bearing Surfaces

9.5.1 On the Track Surface of the Inner Race of the Bearing

The magnetic flux density on the track surface of the bearing's inner race, B_{ir}, due to field strength, H_{ir}, in an oil medium of relative permeability, U_r, with respect to free space, is given by:

$$B_{ir} = UU_r H_{ir} = 4\pi \times 10^{-7} U_r H_{ir} \text{ (in tesla)} \tag{9.16}$$

On using Eqn. (9.3) for equivalent field strength on the track surface of the inner race (H_{ir}), the flux density on the track surface of the inner race (in gauss), B_{ir}, is determined as:

$$B_{ir} = \frac{78.96 \times 10^{-3} U_r I_b [R^2 R_{or}^2 (R \pm R_{or}) - R_{ir}^2 (R^3 \pm R_{or}^3)]}{R^3 R_{or}^3} \tag{9.17}$$

9.5.2 On the Inner Surface of the Inner Race

Using Eqn. (9.6), flux density, B_{iri}, on the inner surface of the inner race is determined by the relation given in Eqn. (9.16), which yields:

$$B_{iri} = \frac{78.96 \times 10^{-3} U_r I_b [R^2 R_{or}^2 (R \pm R_{or}) - R_{iri}^2 (R^3 \pm R_{or}^3)]}{R_{or}^3 R^3} \tag{9.18}$$

9.5.3 On the Track Surface of the Outer Race

Using Eqns. (9.9) and (9.16), the flux density on the track surface of the bearing's outer race (in gauss) is determined as:

$$B_{or} = \frac{78.96 \times 10^{-3} U_r I_b [R_{or}^2 (R_{ir} \pm R) - (R_{ir}^3 \pm R^3)]}{R_{or}^4} \tag{9.19}$$

9.5.4 On the Outer Surface of the Outer Race

Similarly, using Eqns. (9.12) and (9.16), flux density on the outer surface of the outer race is determined as:

$$B_{oro} = \frac{78.96 \times 10^{-3} U_r I_b [R_{oro}^2 (R_{ir} \pm R) - (R_{ir}^3 \pm R^3)]}{R_{oro}^4} \tag{9.20}$$

9.5.5 On the Surface of Rolling-Elements

Using Eqns. (9.15) and (9.16), the minimum/maximum residual magnetic flux density on a few rolling-elements (in gauss) is determined as:

$$B_r = \frac{78.96 \times 10^{-3} U_r I_b [R^2 (R_{ir} R_{or}^3 \pm R^4) - R_{or}^2 (R_{or} R_{ir}^3 \pm R^4)]}{R_{or}^3 R^4} \tag{9.21}$$

It may be noted that the theoretical and average measured flux densities match closely. According to Table 9.1, the theoretical and the average measured flux densities differ by less than a factor of 1.3 in all cases.

Table 9.1. Comparison of theoretical and measured values of magnetic flux density on the surfaces of races and rolling-elements of various bearings

Bearing type	Bearing parameters and operating conditions	Location	Flux density (gauss) Theoretical	Flux density (gauss) Measured (average)	Flow of current through bearing
NU 326	$R_{ir} = 83.5$ mm	B_{ir}	30.04–4.06	34.50	50 A (measured)
	$R_{or} = 121.5$ mm	B_{iri}	45.65–0.18	–	
	$R = 102.5$ mm				
	$R_{iri} = 65.0$ mm	B_{or}	19.725–3.93	15–25	
	$R_{oro} = 140.0$ mm	B_{oro}	21.10–1.10	–	
	After testing NU 326 bearing for 250 h on exposure of 50 A (AC) at 1.12–2.30 V on bearing test machine at 1100 rev min^{-1} under 5000 N horizontal, 10 000 N radial loads	B_r	19.94–1.99	14–18 (on a few rolling-elements)	
NU 228	$R_{ir} = 79$ mm	B_{ir}	9.51–2.5	12.2	
	$R_{or} = 93$ mm				
	$R = 86$ mm	B_{iri}	16.95–1.45	–	
	$R_{iri} = 70$ mm				
	$R_{oro} = 125$ mm	B_{or}	7.90–2.18	9.0	25 A (measured)
	After about 6000 h of motor operating at 2880 rpm, bearing damage detected. Shaft voltage and resistance measured approximately 2.5 V and 0.1 ohm, respectively	B_{oro}	11.72–0.27	–	
		B_r	6.35–0.22	5–7 (on a few rolling-elements)	32.11 (calculated)

(Continues)

Table 9.1. (*Continued*)

Bearing type	Bearing parameters and operating conditions	Location	Flux density (gauss)		Flow of current through bearing
			Theoretical	Measured (average)	
NU 311	$R_{ir} = 35$ mm	B_{ir}	15.64–2.29	14	12 A (measured)
	$R_{or} = 52$ mm				
	$R = 43.5$ mm	B_{iri}	26.20–0.05	–	
	$R_{iri} = 27.5$ mm				
	$R_{oro} = 60$ mm	B_{or}	11.27–2.13	10	
	A number of bearings of alternator	B_{oro}	11.51–0.65	–	
	operating at 750 rpm damaged at different intervals varying between 500 h and 2500 h of operation. Shaft voltage and resistance measured approximately 6 V and 0.5 ohm, respectively.	B_r	11.64–0.70	10–12 (on a few rolling-elements)	10.65 A (calculated)
NU 230	$R_{ir} = 91$ mm	B_{ir}	15–2.93	15	32.6 A (calculated)
	$R_{or} = 75$ mm				
	$R = 105$ mm	B_{iri}	25.11–0.98	~20	
	$R_{iri} = 119$ mm				
	$R_{oro} = 135$ mm	B_{or}	10.44–2.61	–	
	A number of bearings of 1700 kW	B_{oro}	12.71–1.30	–	
	motors operating at 2880 rpm damaged at different intervals of operation under the influence of different levels of shaft voltages.	B_r	9.78–0.65	–	

9.6 Electro-Adhesion Forces

9.6.1 Role and Assessment of Electro-Adhesion Forces in Rolling Friction

The genesis of intermolecular forces during rolling fiction involves electrostatic attraction or repulsion between electrons and nuclei, and interaction of electro-dynamic, magnetic and exchange forces between the atoms [7]. The higher temperature rise of bearings operated under electric field (45°C temperature rise of the outer race of NU 326 bearing as against 25°C under pure rolling friction under identical operating conditions) suggests the reduction in the fatigue life, and increase in resistance to the rolling due to the forces of electro-adhesion (Chapter 3). However, the function of lubricant is to reduce the adhesion forces across an interface [8].

The mechanism of adhesion, friction and wear on the bearing surfaces in presence of lubricating film is quite complex. The process of adhesion involves the formation of a junction between the asperities contact, which may finally lead to elastic and plastic deformation under load [2, 9]. The energies of atomic nature are exchanged at the asperities, which may be affected by cage and roller slip due to close interaction of rolling-elements with the races [10].

It is rather difficult to estimate the electro-adhesion forces in the rolling friction. But, these can be assessed with a reasonable accuracy by SRV analysis; the change in coefficient of friction, the profile depth and the ball scar diameter of the used greases recovered from the active zone of the bearings. This is because the activities of zinc additive i.e. zinc dithiosphosphate or zinc dialkyldithiophosphate (ZDTPs) used as multifunctional additives in the grease, protects the rubbing metal surfaces under pure rolling friction and contributes to friction and wear reduction, and depends, partly, on the amount of additive on these surfaces. Physisorption and chemisorption processes precede the chemical reactions with metal; therefore, it is probable that load-carrying capacity is related to these processes. Correlation between ZDTPs adsorption data and wear is shown and also discussed that ZDTPs are reversibly physisorbed on iron at 25°C, but at 50°C undergo chemisorption reactions [11]. On the other hand decomposition of ZDTPs in the lithium base greases under the influence of electric fields leads to the formation of lithium zinc silicate ($Li_{3.6}Zn_{0.2}SiO_2$) in presencc of high relative percentage of free lithium and silica impurity in the grease under high temperature in the asperity contacts along with the formation of gama lithium iron oxide (γ-$LiFeO_2$). Besides this the original structure of lithium stearate changes to lithium palmitate as discussed in Chapter 3. On the contrary, these changes are not detected under pure rolling friction [12].

The above changes in the lubricating medium of the bearings operated under electric fields are reflected in SRV analysis (Table 9.2), and are related to electro-adhesion forces in the rolling friction. Also, these are contributed by the medium-metal interaction, rate of chemical reaction, affinity of lubricating medium components for the metal, availability of free metal and temperature rise. If the grease, recovered from such bearings is put in a new bearing, the bearing may have premature failure due to higher temperature rise even under pure rolling friction.

9.6.2 Role of Lubricant in Rolling Friction

The bearing using high-resistivity (10^{11} ohm cm) and low-conductivity lubricant accumulate electric charges on the surfaces up to the threshold limit at which discharge leaks at the asperity contacts. On the contrary, a bearing using low-resistivity (10^{7} ohm cm) and high-conductivity lubricant does not accumulate electric charges, but the 'silent' discharge occurs between the interacting surfaces through the lubricant, which gradually gets decomposed [12]. The type of lubricant affects a bearing in a characteristic manner and develops distinguished failure pattern on the surfaces (Figs. 5.1–5.5).

A bearing is influenced by the adsorption potential and chemical activity of the rubbing surfaces, the boundary layer properties of the lubricant, the temperature gradient and the electric field strength in the crevice-shaped space between rollers and races. The phenomenon in the rolling friction is similar to the adhesion on separation like gas discharge luminescence, emission of high energetic electrons and X-ray radiation [7].

9.7 Experimental Facilities and Investigations

9.7.1 Bearing Test Machine

The bearing test machine developed in-house, shown in Fig. 3.2 of Chapter 3 and in [2, 3, 4], was used to evaluate the performance of various sizes of rolling-element bearings at different operating parameters, and to programme the investigations in the tribological areas.

For evaluating the bearing performance under shaft current, a silver-lined slip-ring assembly was used. The current is allowed through the shaft, a test bearing and the housing to complete the electric circuit; the support bearings were fully insulated to avoid the passage of current. For applying different levels of potential across the test bearing, a variable transformer and dimmerstat were used.

9.7.2 Schmierstoff- / Lubricant- / Material (SRV)Test System

The SRV test system has been used to evaluate the characteristics of the fresh and used greases. After sampling the small quantities of the grease from bearing's active zone (less than 0.1 g), the lubricity has been examined by determining the friction coefficient and the wear, under 100 N load and 1000 μm amplitude and 50 Hz frequency by running ball on flat surface for 2 h test duration at 50°C on SRV. Table 9.2 shows analysis of the fresh and deteriorated grease samples and Figs. 3.23 and 3.24 show typical SRV plot of the deteriorated grease samples after 41 h and 250 h of operation under the influence of electric current, respectively.

9.8 Test Conditions and Procedure

9.8.1 Non-Insulated Bearings under Shaft Current

The bearing NU 326, lubricated with low-resistivity grease (10^5 ohm m) was tested at 1100 rpm under 10^4 N and 5×10 N radial loads (90° to each other) for a duration of 250 h (at 1.12–2.3 V) by passing 50 A (AC). Current of 50 A was controlled throughout the test. Corrugations and corrosions were detected on the track surface of races and rolling-elements of the bearing as discussed in Chapters 3 and 5 [2, 3]. The tests were repeated on new bearings without the passage of electric current.

9.8.2 Study of Magnetic Flux Density on Bearing Surfaces

The surface of bearings, tested under the influence of electric current and without the passage of current, and used damaged bearings of motors and alternators have been examined. The magnetic flux density distribution at various angular locations on the lower and upper width sides on inner and outer surfaces of the races have been studied with Hall Probe and Magnetic Field Flux Monitor (manufactured by Bell Inc.). Also, the same was studied on track surfaces, the inner surface of the inner race and on the outer surface of the outer race of the various bearings. The technical particulars of the Hall Probe and Magnetic Field Flux Monitor are given in Table 9.3.

Figure 9.3 shows the probe location for the measurement of flux density immersing out from the track surface, at various angular locations. Figure 9.4 shows the residual flux density distribution on the upper and lower width sides of the inner race of NU 326 bearing after passing 50 A for 250 h. Figure 9.5 shows the same on inner race of the damaged bearing, NU 228, used in a motor for about 6000 h. Figures 9.6–9.9 show the residual flux density distribution on different surfaces of NU 230 damaged motor bearings. The residual flux density distribution on different damaged alternator bearings of type NU 311 was also studied.

Table 9.2. SRV analysis of greases

S. No	Test data	Fresh grease before filling in the bearings		Grease from active zone of bearings after 250 h of operation under current		Grease from active zone of bearings after 250 h of operation under pure rolling friction
		Low resistivity	High resistivity	Low resistivity	High resistivity	Low resistivity
	Coefficient of friction					
	μ (min)	0.105	0.125	0.132	0.122	0.10
	μ (max)	0.1125	0.132	0.40	0.145	0.120
	Profile depth point (μm)	0.5	0.90	1.6	0.90	0.50
	Ball scar diameter (mm)	0.52	0.48	0.66	0.48	0.48
	Maximum. ratio of coefficient of friction as compared to fresh grease	–	–	3.81	1.16	1.14
	Ratio of profile depth as compare to fresh grease	–	–	3.2	1.0	1.0
	Ratio of ball scar diameter as compared to fresh grease	–	–	1.27	1.0	1.0

Table 9.3. Technical particulars of hall probe and magnetic field flux monitor

1. Field range:	0–10 gauss, 0–100 gauss different ranges with 10 X probe
2. Linearity % of reading:	0.1% to 10 KG
3. Instrument accuracy:	± 0.05% of reading plus ± 0.02% of FS plus probe error
4. Stability:	−20–70°C, time ± 0.1% FS maximum

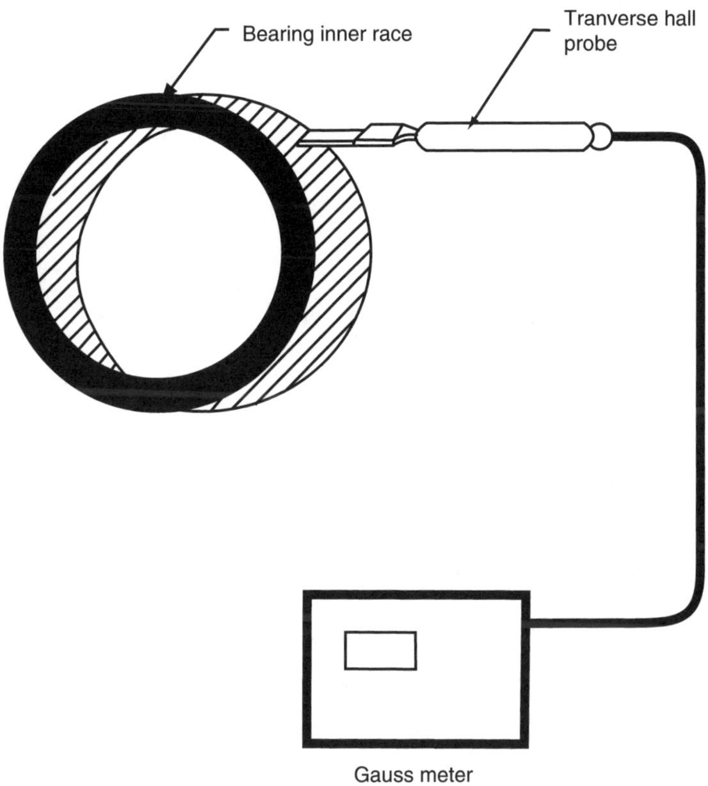

Fig. 9.3. Probe location on bearing track surface for measurement of flux density

9.9 Theoretical and Experimental Data on Flux Density

9.9.1 Experimental

9.9.1.1 Bearing Lubricated with Low-Resistivity Lubricant

Significant magnetic flux density distribution as shown in Fig. 9.4, along with corrosion and corrugations were detected on the track surfaces of races and rolling-elements of the NU 326 roller bearing after operation under electric current [3, 4]. On the contrary, these changes were not detected on bearings after operation without the influence of electric current. A maximum residual flux density of 2.5 gauss on NU 326 bearing surfaces was found without the influence of electric current, which was more or less the same as on the new bearings type NU 326, NU 230, NU 311, and NU 228.

Significant changes were found in SRV analysis of the friction and wear characteristics of the used greases (Table 9.2). On the contrary, these changes were not detected after operation with pure rolling friction. Under the influence of electric fields corrosion and corrugations were detected on the 6326 ball bearing surfaces, but the significant flux density distribution was not detected.

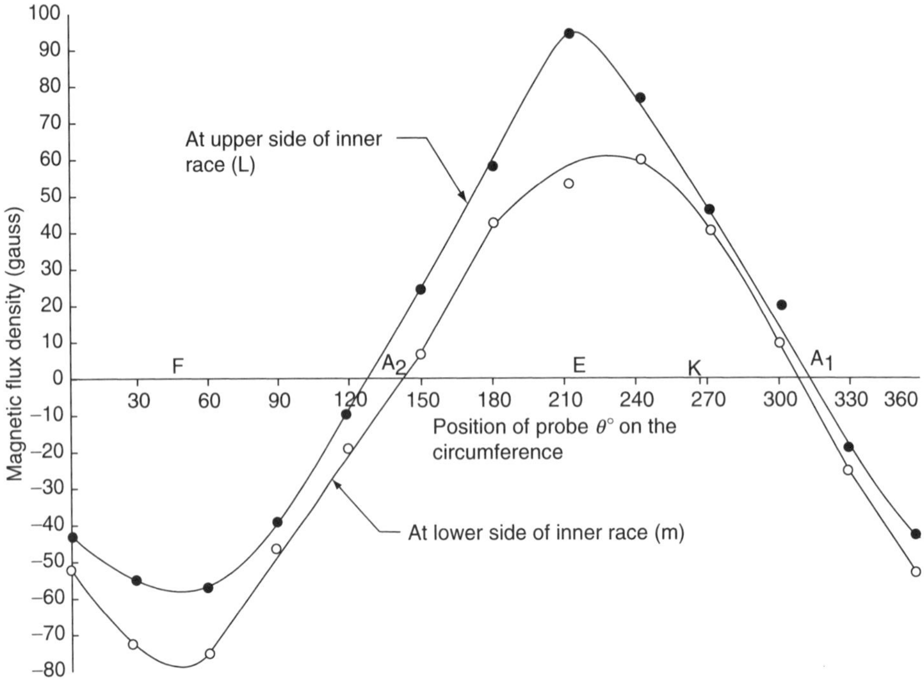

Fig. 9.4. Magnetic flux density distribution around the track surface of inner race of NU 326 bearing after passing 50 A (AC) at 1.2–2.3 V for 250 h

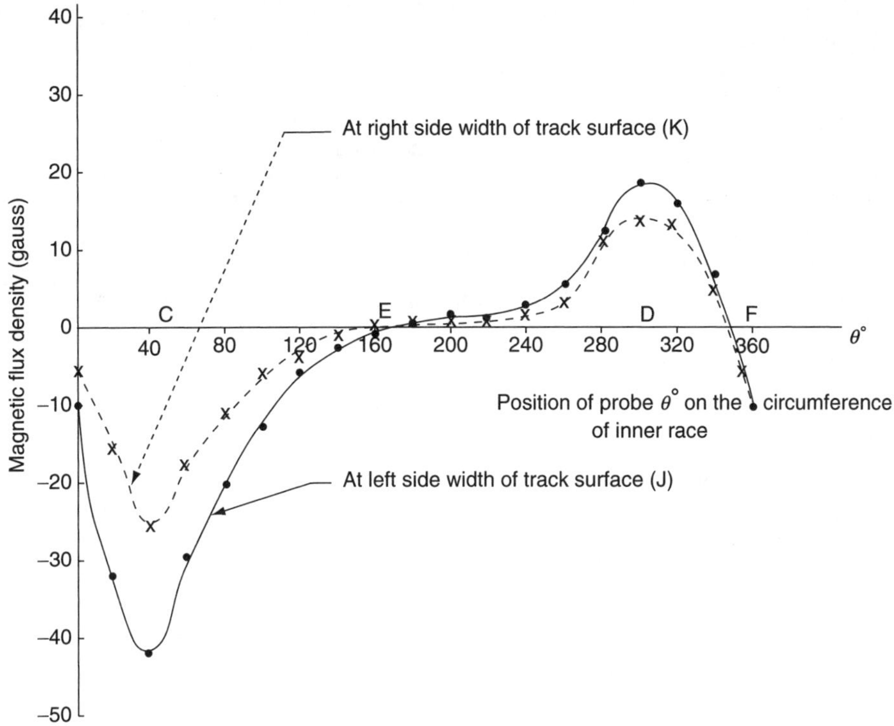

Fig 9.5. Magnetic flux density distribution on width of inner race track surface of NU 228 damaged motor bearing

9.9.1.2 Bearings Lubricated With High-Resistivity Lubricant

Under the influence of electric fields and pure rolling friction, the significant flux density (2.2 gauss) on the surface of roller bearing NU 330 and NU 2215 was not detected, and also significant changes in SRV analysis of the used grease (Table 9.2). The bearing NU 2215 was damaged under electric fields, but no corrugation pattern and corrosion were observed (Fig. 4.8). Also, no damage was found showing corrugations on the NU 330 motor bearing surfaces.

9.9.2 Theoretical

The theoretical values of magnetic flux density were determined on the inner and outer surface of inner and outer races, and rolling-elements of different bearings (NU 326, NU 228, NU 230 and NU 311) using Eqns. (9.17)–(9.21), and are shown in Table 9.1. The residual flux density, thus determined, on the NU 326 bearing was compared with the measured values of the tested bearing under the influence of a

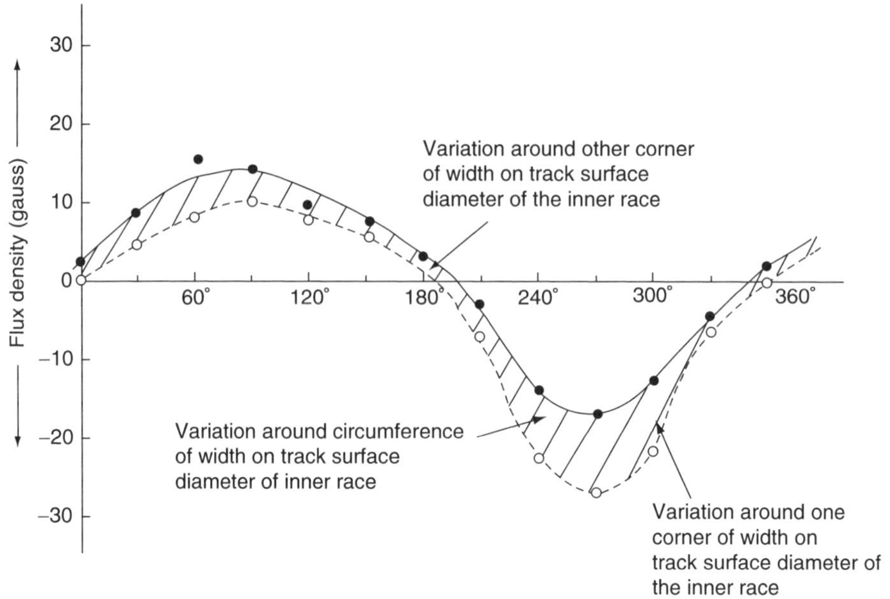

Fig. 9.6. Variation of flux density along the circumference of the width of track, surface diameter of the inner race of NU 230 bearing

50 A current (Fig. 9.4). Similarly, the flow of current through the damaged bearings type NU 228 and NU 230 of the motors was established by measuring the flux density on the bearing surfaces using Fig. 9.5 and Figs. 9.6–9.9, respectively. The flow of current through NU 311 alternator bearing was also established. The same was compared with the measured values of shaft voltage and bearing resistance.

9.10 Discussion on Investigations

9.10.1 Magnetic Flux Density Distribution on Races and Rolling-Elements of Test Bearing

The magnetic flux density on the inner race of the tested NU 326 roller bearing under the influence of electric current varies between 95 and −80 gauss (Fig. 9.4). The two positions ('A_2' and 'A_1') of zero flux density at 180° apart (at 130° and 310° angular locations) indicate that the inner race has become a two-pole magnet with 'E' and 'F' (90° away from 'A_1' and 'A_2') as the points of maximum flux density (Fig. 9.4).

The current from the inner to the outer race tends to flow through the asperity contacts through a path of least resistance towards 'A_2' (Fig. 9.4). The current

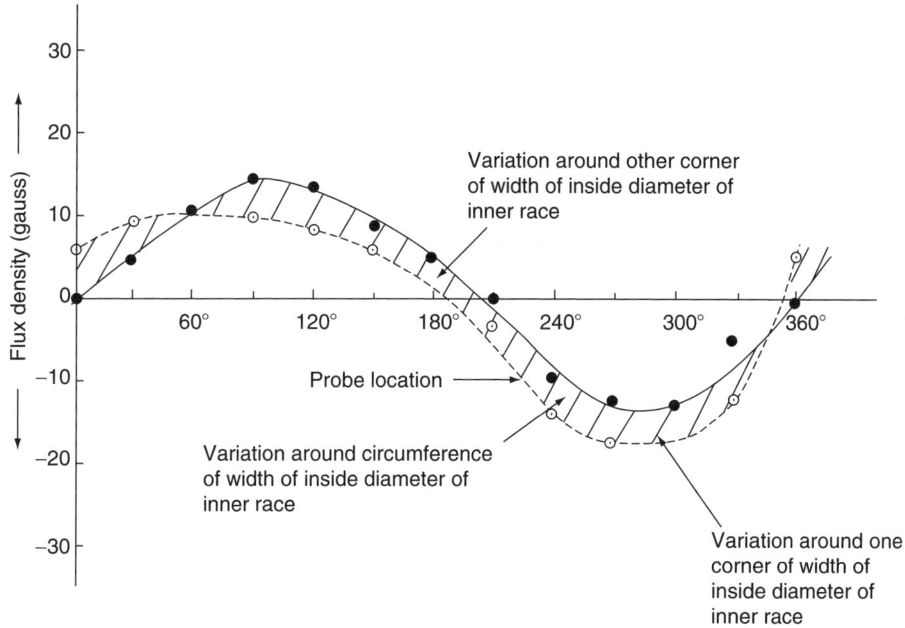

Fig. 9.7. Variation of flux density along the circumference of the width of inside diameter of the inner race of NU 230 bearing

entering the outer race at the farthest point from 'A_2' (at 180° opposite) tends to flow in the clockwise and anti-clockwise directions to reach 'A_2'. The ring current thus flows in the outer race and creates the residual alternating magnetic flux density distribution on the inner race and rolling-elements. The difference in the direction of current flow, at points 90° away from 'A_2', creates equal and opposite residual flux densities on the inner race (the maximum being 95 gauss at 'E' and –80 gauss at 'F', both 180° apart). The maximum difference of 35 gauss at 'E' and –25 gauss at F between the curves L and m, across the width of inner race, is attributed to the difference in the film thickness that could have occurred due to misalignment, and corrosion on the surfaces due to electrochemical decomposition of the lubricant [7].

The average residual flux density has been measured on the track surface of inner race of NU 326 bearing as 34.5 gauss (Fig. 9.4).The flux density on the track surface of outer race varies between 15 and 25 gauss and that on a few rolling-elements varies between 14 and 18, and most of the rolling-elements have a flux density between 2 and 3 gauss. The average to minimum flux density determined by the developed relations varies between 30.04 and 4.06, 19.725 and 3.93 and 19.94 and 1.99 gauss on the track surfaces of the inner race, outer race and on rolling-elements

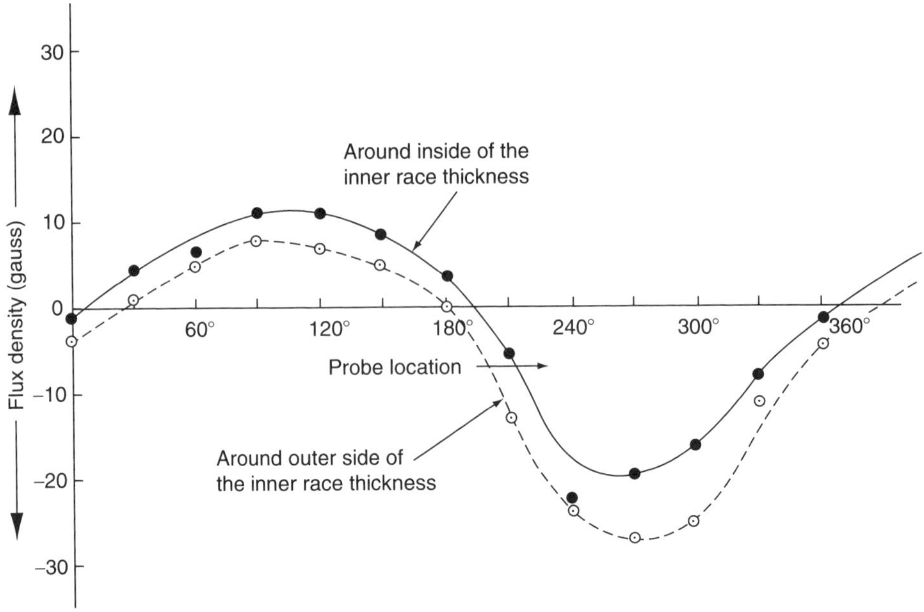

Fig. 9.8. Variation of flux density along the circumference of the thickness of the inner race of NU 230 bearing (side A)

respectively, under the influence of 50 A electric current, as shown in Table 9.1. Also, theoretical values of the similar flux density on the inner surface of the inner race, B_{iri}, and outer surface of the outer race, B_{oro}, were determined to vary between 45.65 and 0.18, 21.10 and 1.10 gauss, respectively, and were found to match the measured values (Table 9.1).

9.10.2 Residual Magnetic Flux Density Distribution on Races and Rolling-Elements of Damaged Bearings of Motor and Alternator

9.10.2.1 Damaged NU 228 Bearing of a Motor

The residual flux density distribution on the track surface of the inner race of the damaged NU 228 motor bearing varies between +20 and −42 gauss as shown in Fig. 9.5. The average residual flux density on the track surface of the inner and outer races were detected as 12.2 and 9 gauss, respectively, and on a few rolling-elements as 5–7 gauss (maximum). However, most of the rolling-elements generally have residual flux density of 1–2 gauss. From the measured residual flux density data and using Eqns. (9.17)–(9.21), the current flow through the motor bearing was estimated as 32.11 A. This matches approximately with a current flow

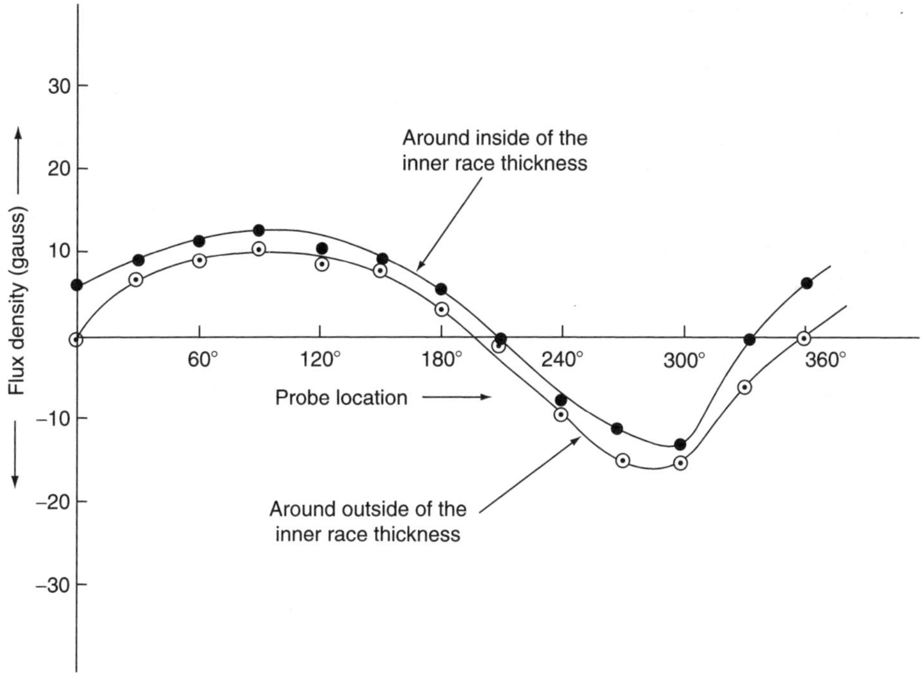

Fig. 9.9. Variation of flux density along the circumference of the thickness of the inner race of NU 230 bearing (side B)

of 25 A determined by the measurement of resistance and the shaft voltage (Table 9.1). The analytically determined flux density on the track surface of the inner race, outer race and on a few rolling-elements varies between 9.51 and 2.5, 7.90 and 2.18 and 6.35 and 0.22 gauss, respectively, using 25 A flow of current through the bearing. The residual flux density on the inner surface of inner race and outer surface of the outer race has been determined to vary in the range 16.96–1.45, 11.72–0.27 gauss respectively. These values match closely with that of the average measured values (Table 9.1).

The two positions C and D (260 degree apart) as the points of maximum residual flux density, with the positions E and F (180° apart) as the locations of zero flux density indicate that the inner ring has become a two-pole magnet (Fig. 9.5). Also, before it is grounded, the circular current flows in the outer race in such a way that it creates the maximum residual flux at locations C and D with a zero flux at locations E and F in the inner race. The maximum difference in flux density of 16 gauss, between curves J and K across the width of the inner race, can be attributed to misalignment and difference in film thickness (Fig. 9.5).

9.10.2.2 Damaged NU 230 Bearing of a Motor

The average residual flux density on the track surface of the inner race, B_{ir} and inner surface of the inner race, B_{iri} of the damaged NU 230 bearing is measured as 15 and 20 gauss, respectively, as shown in Figs. 9.6 and 9.7. This matches closely with the calculated average to minimum values of 15–2.93, and 25.11–0.98 gauss, for B_{ir} and B_{iri} respectively, with a flow of current as 32.6 A (Table 9.1). Residual flux density variation on the track surface and outer surface of the outer race (B_{or} and B_{oro}), and on rolling-elements, B_r, is analytically determined as 10.44–2.61, 12.71–1.30 and 9.78–0.65 gauss, respectively, which matches approximately with the measured values.

Figures 9.8 and 9.9, showing the variation of flux density along the circumference of the thickness of the inner race on both sides indicate a pattern similar to that of the variation on the track surface and inner surface of the inner race (Figs. 9.6 and 9.7).

9.10.2.3 Magnetic Flux Density on Damaged NU 311 Alternator Bearings

The residual magnetic flux density has been measured on various damaged bearings of type NU 311. The average residual flux density on the track surface of the inner race, outer race and rolling-elements was measured as 14, 10, 10–12 gauss, respectively, which matches the theoretically determined average to minimum flux density variation of 15.64–2.29, 11.27–2.13 and 11.64–0.70 gauss respectively, by taking the flow of current as 12 A determined through the measured shaft voltage of 6 V under the bearing resistance of 0.05 ohm, as shown in Table 9.1. Also, residual flux density variation of average to minimum on the inner surface of the inner race, B_{iri}, and the outer surface of the outer race, B_{oro}, has been analytically determined as 26.20–0.05 and 11.51–0.65 gauss, respectively.

In general, it may be noted that the residual flux density on the inner surface of the inner race and the outer surface of the outer race is more than that of the residual flux density on the respective track surfaces of races of the bearings (Table 9.1).This is because of the effect of flow of current and the positional location of the surface with respect to the bearing center.

9.10.2.4 Electro-Adhesion Forces in Bearings

Under the influence of electric fields, the increase in ratio of maximum coefficient of friction, profile depth and ball scar diameter of the used and fresh greases in SRV analysis are 3.81, 3.2 and 1.27 in the bearings using low-resistivity greases as against 1.16, 1.0 and 1.0, respectively using high-resistivity greases (Table 9.2, Figs. 3.23 and 3.24). This may be correlated with decomposition of the grease and the presence of lithium iron oxide and lithium zinc silicate in the medium [12]. This suggests that the electro-adhesion forces increase in the bearings using low-resistivity grease as compared to high-resistivity grease under pure rolling friction (Table 9.2). The deterioration of the low-resistivity grease gradually corrodes bearing surfaces, which finally leads to increased wear and failure of bearing [2, 12].

9.10.2.5 Effect of lubricants on Bearings and Flux Density

Under the influence of the electric current, bearing lubricated with low-resistivity lubricant allows the passage of current, which corrodes the bearing surfaces and flux density is developed on the track surfaces. On the contrary under the influence of electric fields, a bearing lubricated with high-resistivity grease accumulates electric charges on the surfaces, and does not allow passage of current, thereby significant magnetic flux density is not developed. However, when the accumulated charges exceed the threshold value, the sudden discharge takes place at the asperity contacts, and this damages the bearing surfaces [4].

The current starts flowing when the high points on the asperities of interacting surfaces come close or when the conducting particles bridge the oil film. These conducting paths are broken as the asperities become separated either by the higher film thickness, or by the vibration effects, or a combination of both, the arcing results, which damages the surfaces by the arc welding and high temperature effect (Chapters 4, 5 and 11).

9.10.3 Magnetic Flux Density – An Overall View

The developed magnetic flux density on the races and rolling-elements is governed by the field intensity due to the flow of electric current and the permeability of the medium. The flow of current in a bearing depends on the shaft voltage, lubricant characteristics and operating parameters [2, 3]. These are the variable parameters during the course of operation, and accordingly the field intensity becomes variable. Furthermore, the leakage/flow of current from inner race to rolling-elements and to the outer race follows the path of least resistance, which itself is a variable function and is governed by different parameters including that of asperity contacts and temperature-rise besides rolling-element frequency, cage and roller slip and the machine characteristics [4]. To develop a theoretical model for determination of field intensity due to the variable circular current is a complex investigation. However, the developed model is an initial approach work in this direction, which facilitates to determine field intensity depending on the radius of track surfaces, including that of the pitch radius, and the flow of current through the bearing. The flux density varies on the races and the rolling-elements. This is because of relative dimensions of races and rolling-elements and unsteady flow of current under the dynamic conditions of operation of a rolling-element bearing.

9.10.4 Effects of Residual Magnetic Flux on Bearings

The residual magnetic flux density on the track surfaces of a rolling-element bearing can lead to premature failure through the mechanical forces or heat generated. The flux passing through a bearing produces forces of attraction, hysteresis loss and eddy current [13]. The parasitic energy losses are supplied from the mechanical shaft power, and, therefore, increase the torque required to turn the shaft.

The bearings made of hard steel have relatively large magnetic hysteresis loss. The loss can be due to variation in magnitude of flux. During rotation of a bearing, load is distributed between a limited numbers of rolling-elements. This may lead to further instantaneous changes in fluxes and, hence, cyclic change in attractive forces on bearing elements. This may cause premature failure of bearings due to non-uniform wear of bearing surfaces [13].

9.11 Conclusions

Based on the above study and investigations, following conclusions and recommendations are drawn [5, 14, 15, 16]:

(1) Under the effect of current, significant flux density is not developed on the surfaces of ball bearing using low-resistivity grease.
(2) Under the influence of electric current, a bearing using high-resistivity grease does not develop significant magnetic flux density distribution on its surfaces.
(3) Under the influence of electric fields, there occur changes in electro-adhesion forces when bearings are lubricated with low-resistivity grease than with grease of high-resistivity. Whereas, pure rolling friction does not affect electro-adhesion forces considerably.
(4) Detection of corrugations on bearing surfaces does not necessarily indicate damage by electric current.
(5) By the study of magnetic flux density distribution along with the study of damaged and corrugated bearings surfaces, and also by the analysis of deterioration of greases used in the bearings, a diagnosis of leakage current through a non-insulated bearing of an electric motor can be established.
(6) Under the influence of an electric current, a rolling-element bearing using low-resistivity lubricant develops field intensity, which leads to magnetic flux density distribution on the surfaces of races and rolling-elements.

(7) The flow of current in inner race and rolling-elements affects the flux density distribution on outer race track and outer surfaces, and the flow of current in outer race and rolling-elements affects the flux density distribution on inner race track and inner surfaces, and that of flux density on rolling-elements is affected by current flow in the races.

(8) The developed flux density on the different surface of races and rolling-elements also depends on track radii, respective race radii, pitch radius and relative permeability of the lubricant.

(9) In general, the developed flux density is higher on the track surface of the inner race as compared with that of the outer race and rolling-elements. Also, residual flux density on the inner surface of inner race and outer surface of the outer race is more than that of the respective track surfaces of the races.

(10) Using the developed analytical model, and by experimental determination of the flux density on the surfaces of the damaged bearing, the level of flow of electric current through the bearing can be ascertained without the measurement of shaft voltage and bearing impedance.

(11) The magnetic flux density analytically determined on races and rolling-elements, by the developed model, match closely with the experimentally evaluated average residual flux density on the bearing elements.

Study of the magnetic flux density under the influence of variable current density and mutual interaction of flux densities on the races and rolling-elements of a bearing under different modes of operation, is a complex analytical model, and needs further investigation.

REFERENCES

1. Boyd, J. and Kaufman, H. N. (1959). The causes and the control of electrical currents in bearings. *Lubric. Eng.*, **15**(1), 28–35.
2. Prashad, H. (1988). Investigations of corrugated pattern on the surface of roller bearings operated under the influence of electrical fields. *Lubric. Eng.*, **44**(8), 710–718.
3. Prashad, H. (1994). Investigations of damaged rolling-element bearings and deterioration of lubricants under the influence of electric fields. *Wear*, **176**, 151–161.
4. Prashad, H. (1987). Effects of operating parameters on the threshold voltages and impedance response of non-insulated rolling-element bearings under the action of electrical current. *Wear*, **117**(2), 223–240.
5. Prashad, H. (1988). The effects of current leakage on electroadhesion forces in rolling friction and magnetic flux density distribution on the surface of rolling-element bearings. ASME, *J. Tribol.*, **110,** 448–455.

6. Starling G. S. (1960). *Electricity and Magnetism*, Longmans, Green and Co. London.
7. Postnikov, S. N. (1978). *Electrophysical and Electrochemical phenomena in Friction, Curing and Lubrication*, VNR.
8. Buckley, H. and Donald (1981). *Surface Effect in Adhesion, Friction, Wear and Lubrication*, Tribology Series 5, Elsevier Scientific Publishing Company, Oxford.
9. Brainin, E. I., Zbarskii, L. I., Ostankovich, E. V. et al (1983). Investigations of damage to roller bearings of electrical machines under the action of current. *Electrotacknika*, **54**(1), 43–47.
10. Prashad, H. (1987). The effect of cage and roller slip on defect frequency response of rolling-element bearings, Presented in ASME/ASLE, Tribology Conference in Pittsburgh, Pennsylvania, October, 20–22, 1986. Also published in *ASLE Trans.*, **30**(3), 360–368.
11. Plaza, S. (1987). The adsorption of zinc dibutyldithiophosphates on iron and iron oxide powders. *ASLE Trans.*, **30**(2), 233–240.
12. Prashad, H. and Murthy, T. S. R. (1988). Behaviour of greases in in statically bounded conditions and when used in non-insulated anti-friction bearings under the influence of electrical fields. *Lubric. Eng.*, **44**(3), 239–246.
13. Tasker, J. L. and Graham, R. S. (1985). Effects of Magnetic Flux on Rolling-Element Bearings. IEE Conference, September 17–19, Publication 254, pp 152–156.
14. Prashad, H. (1996). Magnetic flux density distribution on the track surface of rolling-element bearings – An experimental and theoretical investigation. *Tribol. Trans.*, **39**(2), 386–391.
15. Prashad. H. (1999). Determination of magnetic flux density on the surfaces of rolling-element bearings as an indication of the current that has passed through them – An investigation. *Tribol. Int.*, **32**, 455–467.
16. Prashad, H. (2000). Determination of magnetic flux density on the surfaces of rolling-element bearings – An investigation. *BHEL J.*, **21**(2), 49–66.

APPENDIX

Calculation of Field Strength at the Centre Due to Circular Current

If a current of strength I_b flows in the race of a rolling-element bearing (in a circular path), it can be treated as a circular magnetic shell of strength ρ. And strength ρ can be taken equivalent to I_b[6].

To find a magnetic potential at a point O on the axis at the bearing centre due to flow of current in the race, the solid angle, Ω, subtended by the circle is determined (Fig. 9.1). To do this, a sphere with centre O is drawn such that circle of radius r lies on the sphere. The solid angle (Ω) subtended by the slice ACB as shown in Fig. 9.1, is given as:

$$\Omega = \frac{\text{Area of slice } ACB}{r^2} \tag{i}$$

The area (A) of slice of a sphere lying between two parallel planes is equal to the area of circumscribing cylinder between the plane, whose axis is perpendicular to these planes. So

$$A = 2\pi rh \tag{ii}$$

where

$$h = r - X \tag{iii}$$

and

$$\Omega = \frac{2\pi rh}{r^2} = 2\pi\left(1 - \frac{X}{r}\right) \tag{iv}$$

But, magnetic potential at the bearing centre is given as:

$$V = \rho\Omega = I_b\Omega = 2\pi I_b\left(1 - \frac{X}{r}\right) \tag{v}$$

Since from Fig. 9.1

$$r^2 = X^2 + Y^2, \tag{vi}$$

Magnetic potential at the bearing centre is given as (by Eqns. (v) and (vi)

$$V = 2\pi I_{\mathrm{b}}\left[1 - X\left(X^2 + Y^2\right)^{-1/2}\right] \tag{vii}$$

By symmetry, it is evident that the magnetic field strength due to the circular current in the inner as well as outer race is directed along the axis, and its value is therefore given as [6]

$$H = \frac{-\mathrm{d}V}{\mathrm{d}X} \tag{viii}$$

Using Eqn. (vii), the field strength is expressed as:

$$H = 2\pi Y^2 I_b \left(X^2 + Y^2\right)^{-3/2} \tag{ix}$$

For a point at the centre of bearing, when $X = O$, the field strength is given as:

$$H_o = \frac{2\pi I_b}{Y} \tag{x}$$

Nomenclature

A	area of element
B_{ir}	magnetic flux density on track surface of inner race
B_{iri}	magnetic flux density on inner surface of inner race
B_{or}	magnetic flux density on track surface of outer race
B_{oro}	magnetic flux density on outer surface of outer race
B_r	magnetic flux density on rolling-elements
h	height of element
H	field strength
H_o	field strength at the centre
H_{iro}	field strength on track surface of inner race due to the flow of circular current in outer race
H_{irio}	field strength on the inner surface of inner race due to flow of circular current in outer race
H_{irr}	field strength on track surface of inner race due to flow of circular current in rolling-elements
H_{irir}	field strength on the inner surface of inner race due to flow of circular current in rolling-elements
H_{ir}	equivalent field strength on track surface of inner race
H_{iri}	equivalent field strength on inner surface of inner race
H_{ror}	field strength on rolling-elements due to flow of circular current in outer race
H_{rir}	field strength on rolling-elements due to flow of circular current in inner race
H_r	equivalent field strength on rolling-elements
H_{ori}	field strength on track surface of outer race due to flow of circular current in inner race

H_{oroi}	field strength on outer surface of outer race due to flow of current in inner race
H_{orr}	field strength on track surface of outer race due to flow of circular current in rolling-elements
H_{or}	equivalent field strength on track surface of outer race
H_{oror}	field strength on outer surface of outer race due to flow of circular current in rolling-elements
H_{oro}	equivalent field strength on outer surface of outer race
I_b	bearing current
R_{ir}	track radius of inner race
R_{iri}	inside radius of inner race
R_{or}	track radius of outer race
R_{oro}	outside radius of outer race
R	pitch radius of bearing
r	radius of circle
v	magnetic potential
X	distance from bearing centre
Y	length of the element
Ω	solid angle
ρ	magnetic strength
U_r	relative permeability oil with respect to free space (= 1)
U_o	permeability of free space (4×10^{-7} henry m^{-1})

Chapter 10

TIME SPAN FOR DEVELOPMENT OF FLUTES AFTER THE APPEARANCE OF SLIP BANDS ON THE TRACK SURFACE OF ROLLING-ELEMENT BEARINGS UNDER THE INFLUENCE OF ELECTRIC CURRENT

10.1 General

A theoretical approach is developed in this chapter to determine the time span for development of flutes after the appearance of slip bands on the track surface of rolling-element bearings operating under the influence of electric current. A theoretical model has been developed using the continuum theory of Griffith to determine the energy per unit area required for the development of corrugations after the appearance of slip bands on the track surfaces. Also, an expression is deduced for the net energy input per unit area on the track surfaces by the amount of electric current passed through the bearing at the measured shaft voltage in a given span of operation. By the formulation of pitch of corrugations derived using bearing dimensional and operational parameters along with developed model and expressions, the time span for the development of corrugations has been determined. A similar time span has also been ascertained experimentally. Theoretical and experimental values of time span, thus determined, have been compared and found to match closely.

The analysis given in this chapter can also be used for the determination of the shaft voltage or bearing current if the time for the formation of corrugation is known.

10.2 Introduction

Chapters 2–9 have dealt with the origin of the phenomenon of shaft voltage in electric machines and the flow of current through rolling-element bearings and rotating contacts. Flow of current [1–3] depends on the bearing impedance, which is a function of the lubricant characteristics, resistivity, oil film thickness and voltage across a bearing [4]. In general, the effect of electric currents on contact temperature, slip bands initiation and the life of rolling contact bearings has been discussed [1, 5]. In addition, investigations have been carried out on corrugation patterns and the theoretical evaluation of impedance, capacitance and instant charge accumulation on the surfaces of rolling-element bearings operated under the influence of electric current [6, 7]. The deterioration of lubricant in rolling-element bearings has been studied and found to be accelerated when the bearings are exposed to electric current [8]. It has been established that cracks grow parallel to the rim at a depth influenced by fracture mechanics considerations, but 10^{10} contact cycles life of SAE 52100 (bearing steel) is likely to be dominated by the initiation stage of a crack [9]. However, above a certain level of cyclic stress (the fatigue limit) some crystals on the surface of the specimen develop slip bands as a result of shearing of atomic planes within the crystals. With increase in the number of contact cycles these slip bands broaden and intensify the points where separation occurs within one of the bands and a crack is formed [10]. No study has been reported regarding appearance of crack of certain width i.e. corrugation on the track surface of a roller bearing after initiation of slip bands for a bearing operated under the influence of electric current. This chapter deals with this study and investigations [11].

10.3 Background of Electric Current Damage and the Formation of Slip Bands/Corrugations

Electric current damage of the bearings is of two types. In the first type, when the low-resistivity lubricant ($< 10^5$ ohm m) is used in the roller bearings, the 'silent' discharge occurs through the bearing elements under the influence of electric current. This decomposes the used lubricant and corrodes the bearing surfaces, which lowers the fatigue life of the bearings [12]. The original structure of the lubricant also changes [13]. Passage of current causes surface heating, which leads to low-temperature tempering. This accelerates the formation of slip bands and corrugations on the track surface of races. Magnetic flux density is also developed on the track surfaces [14].

In the second type, when the high-resistivity lubricant ($> 10^9$ ohm m) is used in the roller bearings, an accumulation of charges occur on the track surfaces until it reaches the threshold critical value when the breakdown takes place. The surface damage is caused by arcing effect. This is accompanied by mass transfer and elevated local temperatures on the asperities of the contact surfaces [4, 7].

10.3.1 Process of Formation of Slip Bands

At each revolution of the shaft, part of the circumference of the inner race passes through a zone of maximum radial force, and Hertzian pressure between the rolling-elements and raceway (at the line contact) lead to a maximum shear stress. The maximum shear stress is taken as the criterion for yielding, and this occurs in the sub-surface at a depth approximately equal to half the radius of the contact surface [6, 8]. It is generally at this point the failure of the material, leading to slip bands formation will initiate. The formation of slip bands is accelerated by the passage of current, corrosion and oxidation of surfaces, lubricant characteristics and quality of a bearing [6].

The number of cycles has been determined in terms of time after the stabilization of the temperature rise of the roller tracks, when some crystals on the roller track of the inner or the outer race develop slip bands as a result of shearing of atomic plane within the crystals [5].

10.3.2 Process of Formation of Flutes/Corrugations on the Track Surfaces of Races and Rolling-Elements

The slip bands are formed in the sub-surface prior to the appearance of flutes/corrugations on the track surfaces. The followings may take place before the initiation of corrugations:

(1) Generation of sub-surface dislocation;
(2) Pile up of dislocation;
(3) Coalescence of voids leading to slip bands formation;
(4) Generation of persistent slip bands [PSB] (these are not due to piled up dislocations at a tilt wall and are not preceded by coalesced voids);
(5) Cracks form along PSB and initiated from the tip of the slip bands, and finally, and
(6) Appearance, formation and propagation of flute/corrugation of the track surface.

Since cracks form along PSB, and flutings/corrugations form along PSB as established experimentally [5], it is assumed that the cracks are the precursor of corrugations/flutings.

After the crack/slip band under the track sub-surface is initiated, the process that governs the propagation leading to formation of corrugations/flutes is considered by the continuum theory of Griffith [15].

10.4 Theoretical Analysis for Energy Requirement for the Appearance of Corrugations

10.4.1 Energy Required for the Formation of Corrugations on Track Surfaces of Bearings

For a flute/corrugation of width W to appear on track surface under a stress ∂, the surface energy required for the opening of a tip of the slip band into diameter/width W of the surface flute is given as:

$$U = 2\pi q_{\text{cr1}}\left(\frac{W}{2}\right)^2 \tag{10.1}$$

Some of the energy input in the material will be absorbed in plastic flow, which can deform the material plastically. The distortion in front of the tip of the corrugation on the track surface to be formed from the existing slip band, is of the order of ∂/E. It extends over dimension of the order of $W/2$ from both sides of the slip band. Thus an opening of flute/corrugation in the distorted material on the track surface leads to an energy change, which is given as [15]:

$$U = 2\pi q_{\text{cr1}}\left[\frac{W}{2}\right]^2 - \left[\frac{\partial^2}{2E}\right]\frac{4}{3}\pi\left[\frac{W}{2}\right]^3 \tag{10.2}$$

When the energy input in the bearing material by the applied stress exceeds the surface energy of the bearing material, the tip of the corrugation surface area will increase. Thus, for the minimum energy required for opening of corrugation tip on the track surface, it is evident that;

$$\frac{dU}{dW} = 0 \tag{10.3}$$

which gives

$$\partial_{\text{corr}} = 2\left[\frac{Eq_{\text{cr1}}}{W}\right]^{1/2} \tag{10.4}$$

And energy per unit area for the appearance of opening of a tip of the flute of diameter/width of corrugation, W, on the track surface is determined as:

$$q_{\mathrm{cr1}} = \frac{\partial^2_{\mathrm{corr}} W}{4E} \tag{10.5}$$

The gradual opening of a slip band after distortion of a single crack tip from the location just below the track surface gets extended over the complete width, B, of the track surface. The number of such openings one after the other covering the entire width of the track surface for a single corrugation is the ratio of width of the track surface to that of diameter/width of the tip of the corrugation and is given as:

$$\beta = \frac{B}{W} \tag{10.6}$$

The energy per unit area required for appearance of a single corrugation on the track surface after the formation of slip band is the product of the energy per unit area required for the opening of the tip of corrugation [q_{cr1}] and the number of such tips of diameter W in the width of the track surface B, and is given as [using Eqns. (10.5) and (10.6)]:

$$q_{\mathrm{cr}} = q_{\mathrm{cr1}}\beta = \frac{\partial^2_{\mathrm{corr}} B}{4E} \tag{10.7}$$

The number of corrugations on the track surface of the inner race is the ratio of the circumference of the track surface of the inner race and the pitch of the corrugation. This can be determined as [6]:

$$N_{\mathrm{ir}} = \frac{\pi D_{\mathrm{ir}}}{\Delta_{\mathrm{ir}}} \tag{10.8}$$

Similarly

$$N_{\mathrm{or}} = \frac{\pi D_{\mathrm{or}}}{\Delta_{\mathrm{or}}} \tag{10.9}$$

and

$$N_{\mathrm{r}} = \frac{\pi d}{\Delta_{\mathrm{r}}} \tag{10.10}$$

The pitch of corrugations on inner race, outer race and rolling-elements are determined by bearing dimensional and operational parameters [6, 8].

The energy per unit area required for the appearance of corrugation on the track surface of races and rolling-elements is determined as:

$$q_{\mathrm{ir}} = q_{\mathrm{cr}} N_{\mathrm{ir}} = \frac{\partial^2_{\mathrm{corr}} B D_{\mathrm{ir}}}{4E\Delta_{\mathrm{ir}}} \tag{10.11}$$

$$q_{\mathrm{or}} = q_{\mathrm{cr}}N_{\mathrm{or}} = \frac{\partial^2_{\mathrm{corr}} BD_{\mathrm{or}}}{4E\Delta_{\mathrm{or}}} \tag{10.12}$$

$$q_{\mathrm{r}} = q_{\mathrm{cr}}N_{\mathrm{r}}N = \frac{\partial^2_{\mathrm{corr}} LDN}{4E\Delta_{\mathrm{r}}} \tag{10.13}$$

The overall net energy per unit area ($\mathrm{J\,m^{-2}}$) required for the appearance of corrugations on the track surface of a bearing is given as:

$$q_{\mathrm{b}} = q_{\mathrm{ir}} + q_{\mathrm{or}} + q_{\mathrm{r}}$$

or

$$q_{\mathrm{b}} = \frac{\partial^2_{\mathrm{corr}}[BD_{\mathrm{ir}}/\Delta_{\mathrm{ir}} + BD_{\mathrm{or}}/\Delta_{\mathrm{or}} + LdN/\Delta_{\mathrm{r}}]}{4E} \tag{10.14}$$

10.4.2 Determination of Electrical Energy for the Formation of Flutes on the Track Surface of Bearings

The net electrical energy input to the bearing by passing current I at V volts for a duration of t seconds is given by:

$$Q = VIt \text{ [J]} \tag{10.15}$$

The area of the track surface of the inner race, outer race and rolling-elements is determined as:

$$A = \pi(BD_{\mathrm{ir}} + BD_{\mathrm{or}} + LdN) \tag{10.16}$$

The net electrical energy per unit area on the track surface of a bearing is determined as:

$$q = \frac{Q}{A} = \frac{VIt}{\pi[BD_{\mathrm{ir}} + BD_{\mathrm{or}} + LdN]} \tag{10.17}$$

Considering the heat loss by various means, the electrical energy per unit area on the track surfaces responsible for the formation of corrugations is given as (Clause 10.5.3 – clause on heat transfer in bearings):

$$q_{\mathrm{c}} = \frac{VIt}{2\pi[BD_{\mathrm{ir}} + BD_{\mathrm{or}} + LdN]} \tag{10.18}$$

10.5 Experimental Investigations and Other Related Aspects

10.5.1 Experimental Investigations on the Passage of a Current through Bearings

The experimental investigations of the passage of a current through NU 326 bearings and periodical damage of bearings are given in Chapter 3–5. Figures 5.1–5.5 show the bearing damage by slip bands initiation and formation of corrugations on bearing surfaces. The dimensional and operating parameters of the NU 326 bearing are given in Table 10.1. The measured values of the pitch and width of corrugations on the track surface of the races of the NU 326 bearing after 250 h of operation is also given in Table 10.1 [6].

The outer race of the bearing and grease temperatures stabilized at temperatures of 35 °C and 50 °C above ambient for up to 25 h of periodic testing, although the temperature at the contact zone between roller tracks of the races and rollers would have been higher since heat dissipates through the shaft and housing, etc. However, a stabilized outer race and grease temperature could be taken as a reasonably accurate indication of a stabilized temperature of the roller track [5].

The maximum depth of slip bands on the track surface of the inner race of the NU 326 bearing is taken as 0.003 mm, equivalent to the measured depth of corrugations after 250 h of operation [6].

10.5.2 Duration/Number of Cycles Before Initiation of Slip Bands on Roller Track of Races

Instant thermal stresses due to thermal transients on the roller tracks of races caused by roller contact under the influence of electric current depend on the instant rise in temperature. As the instant rise in temperature stabilizes the contact thermal stresses increase and affect the fatigue life [5]. Experimental investigations indicated as per the test conditions given in Clause 10.5.1 that the first appearance of slip bands on the roller track of inner race tested with current of 50 A (AC) has taken about 41 h [6], or 10^6 cycles after stabilization of bearing temperature. Also, the life expectancy of a bearing before slip bands are initiated will be higher with a low value of current density. This is because of lesser rise in the contact temperature with decrease in current intensity.

On comparing the electric energy per unit area (q) and total energy per unit area for the appearance of corrugation (q_b) on the track surfaces of the bearing as given in the Eqns. (10.14) and (10.18), respectively, the duration (the additional time) the bearing would have taken after the slip bands formation before the appearance of flutes/corrugations on the track surface of races and rolling-element is determined as:

$$t = \frac{(\pi\partial_{\text{corr}})^2(BD_{\text{ir}} + BD_{\text{or}} + LdN)(BD_{\text{ir}}\Delta_{\text{ir}}^{-1} + BD_{\text{or}}\Delta_{\text{or}}^{-1} + LdN\Delta_{\text{r}}^{-1})}{2VIE} \tag{10.19}$$

Also, shaft voltage or current passed through the bearing can be determined provided the duration (t) of the appearance of corrugation is assessed/established.

10.5.3 Heat Transfer in Bearings

The temperature level at which a rolling-element bearing operates is a function of operating parameters, lubricant characteristics, housing design and operating environment. Three fundamental modes of heat transfer exists; conduction of heat within solid structure, convention of heat from solid to fluids in motion, and radiation. Because of discontinuities of the structure, which comprises a rolling-bearing assembly, classical methods of heat transfer cannot be applied to obtain a solution describing the system temperature. However, the finite difference method can be used to obtain mathematical solution with approximation [16]. For different design it is found that between 40 and 50 per cent of heat is approximately conducted away and lost by various means from the source of generation with the environment via the shaft, housing and bearing. The heat generated in the bearing by rolling friction is very small as compared to that of passage of current [5]. Hence it has not been considered for analysis. So for determining the duration of the appearance of corrugations on track surface of races, half of the electrical energy supplied per unit area has been considered for analysis as shown in Eqn. (10.18).

10.6 Comparison of Theoretical and Experimental Data

From the test data given in Chapter 3 and dimensional parameters of the NU 326 bearing as given in Table 10.1, the energy per unit area required for the appearance of corrugations on the track surface of the inner and outer races as well as rolling-elements, has been determined using Eqn. (10.14). The energy input per unit area on the track surfaces responsible for the formation of corrugations by passage of electric current through the bearing has been determined by Eqn. (10.18). The time required for appearance of flutes/corrugations on the track surface of NU 326 bearing has been determined by Eqn. (10.19) using the data given in Table 10.1. The duration, thus determined, theoretically, has been compared with the duration determined by the investigations carried out on the bearing test rig by the passage of electric current through the bearings.

Table 10.1.

Dimensional and operating parameters of the NU 326 bearing	
d	38 mm
D	205 mm
D_{ir}	167 mm
D_{or}	243 mm
L/B	$\sim$1
L	38 mm
N	14
n	1100 rev. min^{-1}
P	112×10^2 N
U	1.12 V
I	50 A
Experimental and theoretical data	
W_{ir}	0.120 mm
W_{or}	0.145 mm
H	0.003 mm
E	210×10^9 N m^{-2}
Δ_{ir}	0.381 mm
Δ_{or}	0.392 mm
t	209 h (experimental)
t	190.29 h (theoretical)
∂_{corr}	700×10^6 N m^{-2}

10.7 Discussion on Energy and Time Required for the Appearance of Corrugations

10.7.1 Energy for opening of the Tip of Corrugation on the Track Surface

The energy per unit area required for the appearance of the opening of a tip of flute/corrugation on the track surface of races and rolling-elements of a bearing is directly proportional to the square of the stress limit (∂_{corr}) from the existing slip band, and width of the tip of a flute, and is inversely proportion to Young's modulus of elasticity of the bearing material [Eqn. (10.5)]. Furthermore, for the full developed corrugation pattern on the bearing track surfaces after initiation of slip bands, energy per unit area is a function of ∂^2_{corr}, E, width of the track surface, diameter of the track surface and pitch of the corrugations. The less the pitch of the corrugations and the greater the number of rolling-elements in a bearing, more energy is needed per unit area for the appearance of corrugations on the bearing surfaces [Eqns. (10.11)–(10.14)].

10.7.2 Electrical Energy

The electrical energy supplied to the bearing depends on the shaft voltage, bearing current and the time for bearing operation under the influence of electric current. The track surfaces of the inner and outer races and rolling-elements of a bearing are affected by the bearing current. The net electrical energy per unit area on the track surfaces is determined by Eqn. (10.18) and depends on width, number of rolling-elements, diameter of rolling-element and diameter of track surfaces of a bearing.

It may be pointed out that the momentary energy density delivered to the surfaces is not uniform since the current is conducted across narrow paths at asperity contacts. However, this energy will uniformly heat the entire surface by conduction in due course. Thus energy delivered over time average to a uniform total and the flutes will correspondingly develop fairly uniformly on track surfaces. Furthermore, as soon as the lubricant deterioration is initiated and corrosion appears on the track surfaces, the rolling-element frequency changes. Rolling-elements do not follow standard pattern and do not travel uniformly over the non-uniformly corroded rings. This is the reason that the fluting phenomenon creates periodic damage of any pitch in due course [5, 8].

10.7.3 Time Required for the Appearance of Corrugations on the Track Surfaces

Theoretically time, t, required for appearance of corrugations on the track surfaces after the slip bands formation depends on the pitch of corrugations on the track surfaces, width and diameters of the inner race and outer race tracks, diameter and length of the rolling-elements, shaft voltage, current passed through the bearing and the stress limit for appearance of the opening of the tip of the flute on the track surface [Eqn. (10.19)]. The value of t, thus established theoretically (190.29 h) using Eqn. (10.19), match closely with that of experimental value (209 h) as shown in Table 10.1. Also, if the time and shaft voltage for the appearance of corrugations is known, then the amount of current flow through the bearing can be established theoretically by using Eqn. (10.19).

10.8 Conclusions

Based on the above investigations and analysis, the following conclusions are drawn [11]:

(1) The energy per unit area for the appearance of an opening of a tip of the flute on the track surface of a rolling-element bearing depends on material properties, level of contact stresses and width of the corrugations.

(2) The energy per unit area for full development of corrugation pattern on the track surfaces after initiation of slip bands depends on material properties, level of contact stresses, diameter and width of track surfaces, length and diameter of rolling-elements, and pitch of corrugations.

(3) The less the pitch of corrugations and greater the number of rolling-elements in a bearing, the more energy per unit area is needed for the appearance of corrugations after the formation of slip bands.

(4) The diameter and width of track surfaces and diameter and length of rolling-elements affect the electric energy distribution per unit area of a rolling-element bearing.

(5) The time required for the appearance of corrugations after the slip bands formation depend on the dimensions of the track surfaces, pitch of corrugations, number of rolling-elements, shaft voltage, current and properties of the bearing material.

This analysis make it possible to predict the time, shaft voltage or bearing current for the appearance corrugation pattern after the slip bands are formed on the track surface of bearings lubricated with low-resistivity lubricant under the influence of an electric current.

REFERENCES

1. Kaufman, H. N. and Boyd, J. (1959). The conduction of current in bearings. *ASLE Trans.*, **2**(1), 67.
2. Anderson, S. (1968). Passage of electric current through rolling bearings. *Ball Bearing J.*, **153**, 6–12.
3. Simpson, F. E. and Crump, J. J. (1963). Effects of electric currents on the life of rolling contact bearings. Proc. Lubrication and Wear Convention, Bournemouth, 1963, Inst. Mech. Engrs., London, Paper 27.
4. Prashad, H. (1987). Effects of operating parameters on the threshold voltage and impedance response on non-insulated rolling-element bearings under the action of electric current. *Wear*, **117**, 223–240.

5. Prashad, H. (1989). Analysis of the effects of an electric current on contact temperature, contact stresses and slip band initiation on the roller tracks of roller bearings. *Wear*, **131**, 1–14.
6. Prashad, H. (1988). Investigations on corrugated pattern on the surfaces of roller bearings operated under the influence of electrical fields. *Lubric. Eng.*, **44**(8), 710–718.
7. Prashad, H. (1990). Theoretical analysis of the effects of instantaneous charge leakage on roller track of roller bearings lubricated with high-resistivity lubricants. ASME Transaction. *J. Tribol.*, **112**, 37–43.
8. Prashad, H. (1994). Investigations of damaged rolling-element bearings and deterioration of lubricants under the influence of electric current. *Wear*, **176**, 151–161.
9. Hahn, G. T., Bhargava, V., Rubin, C. A. et al. (1987). Analysis of the rolling contact residual stresses and cyclic plastic deformation of SAE 52100 steel ball bearings. *Trans. ASME*, **109**, 618–626.
10. Wornock, F. V. and Benham P. P. (1965). *Mechanisms of Solids and Strength of Materials*, Sir Isaac Pitman and Sons, London.
11. Prashad, H. (1998). Determination of time span for the appearance of flutes on the track surface of rolling-element bearings under the influence of electric current. *Tribol. Trans.*, **41**(1), 103–109.
12. Prashad, H. and Murthy, T. S. R. (1988). Behaviour of greases in statically-bounded conditions and when used in non-insulated anti-friction bearings under the influence of electrical fields. *Lubric. Eng.*, **44**(3), 239–246.
13. Prashad, H. (1989). Diagnosis of deterioration of lithium base greases used in rolling-element bearings by X-ray diffractrometry. *Tribol. Trans.*, **32**(2), 205–214.
14. Prashad, H. (1996). Magnetic flux density distribution on the track surface of rolling-element bearings – An experimental and theoretical investigation. *Tribol. Trans.*, **39**(2), 386–393.
15. Czichos, Host (1957). *Tribology Series 1*, Elsevier Sequoia, UK.
16. Harris, A. T. (1966). *Rolling Bearing Analysis*, Wiley, New York.

Nomenclature

A	area of track surface of inner race, outer race and rolling-elements of bearing
B	width of track surface
d	diameter of rolling-element
D	pitch diameter
D_{ir}	outer diameter of inner race
D_{or}	inner diameter of outer race
E	Young's modulus of elasticity
H	depth of slip bands on roller track of races
I	current passing through bearing

L	length of rolling-element
n	rev. $\min^{-1}$
N	number of rolling-elements in bearing
N_{ir}, N_{or}, N_r	number of corrugations on inner race, outer race and rolling-elements, respectively
P	resultant load on bearing
q	electrical energy distribution per unit area on the track surface
q_{cr}	energy per unit area for appearance of a single flute from existing slip band on track surface
q_{cr1}	energy per unit area for opening a tip of the slip band into diameter/width W of flute on the track surface
q_{ir}, q_{or}, q_r	energy per unit area required for the appearance of corrugations on the track surface of the inner race, outer race and rolling-elements, respectively
q_b	energy per unit area for appearance of corrugations on track surface of bearing
q_c	electrical energy per unit area responsible for appearance of corrugations on the track surface of bearing
Q	net electrical energy
t	time required for appearance of corrugations after slip band formation on track surfaces
U	net energy required for conversion of slip bands into the surface flutes/corrugations
V	voltage across bearing
W	width of corrugation/flute
W_{ir}, W_{or}	width of corrugation on the inner race and outer race, respectively
β	L/W
Δ_{ir}, Δ_{or}	pitch of corrugation on inner race, outer race and rolling-elements, respectively
∂	stress
∂_{corr}	stress for flute appearance from opening of the tip of flute from an existing slip band on the track surface

Chapter 11

APPEARANCE OF CRATERS ON TRACK SURFACE OF ROLLING-ELEMENT BEARINGS BY SPARK EROSION

11.1 General

A theoretical investigation has been carried out to determine the time span for the appearance of craters due to spark erosion on the track surfaces of rolling-element bearings operating under the influence of shaft voltages. An analytical model has been developed using the continuum theory of Griffith to determine energy per unit area required for the development of craters on the track surfaces. Also, an expression is deduced for the net energy transmitted per unit area on the track surfaces by the leakage of stored charge energy resulting in the appearance of craters under the effect of shaft voltage and bearing capacitance depending on operational and dimensional parameters of a bearing in a given span of operation. By the formulation of capacitive stored energy in a bearing using high-resitivity lubricant along with the developed model and expressions using the ratio of cycles to accumulate and discharge of the accumulated charge energy at the various levels of bearing to shaft voltage, the time span/cycles for the appearance of craters of specified dimensions by spark erosion has been determined. Matching time span/cycles has been ascertained by the investigation of a case history of the diagnosis of bearings that failed by crater formation [1]. The analysis can also be used for determination of shaft voltage if the time span/cycles for the formation of known size craters and bearing capacitance are established.

11.2 Introduction

Previous chapters have dealt with the origin of the phenomenon of shaft voltage in electrical machines and the flow of current through rolling-element bearings. The flow of current depends on the bearing impedance, resistivity of lubricant, oil film thickness and voltage across a bearing [2]. In general, the effects of electric currents on contact temperature, slip bands initiation and the life of rolling contact bearings have been analyzed [3]. Investigations have been carried out on corrugation patterns and the theoretical evaluation of impedance, capacitance and instant charge accumulation on the surfaces of rolling-element bearings operated under the influence of electric current [4, 5]. Threshold voltage has been established exceeding which the flow of current through a bearing increases instantaneously. Also, the investigations have been reported for determination of time span for the appearance of flutes/corrugations on the track surface of rolling-element bearings using low-resistivity lubricant under the influence of electric current [6].

In this chapter, a theoretical model has been developed using the continuum theory of Griffith to determine energy per unit area, and time span/cycles required for the appearance of craters of specified dimensions by spark erosion on the track surfaces of bearings using high-resistivity lubricants.

11.3 Background and Principle Involved in the Formation of Craters

Damage of the bearings due to electric current is of two types. In the first type, when the low-resistivity lubricant ($\phi < 10^5$ ohm m) is used in the rolling-element bearings, the 'silent' discharge passes through the bearing elements under the influence of electric current. This decomposes the used lubricant and corrodes the bearing surfaces, which lowers the fatigue life of the bearings. The original structure of the lubricant also changes [7]. Passage of current causes surface heating, which leads to low-temperature tempering. This accelerates the formation of slip bands and corrugations on the track surfaces of races. Magnetic flux density is also developed on the track surfaces [3, 8].

In the second type, when the high-resistivity lubricant ($\phi > 10^9$ ohm m) having viscosities 127.33 cSt at 40 °C and 10.43 cSt at 100 °C, and relative permittivity of 2.5, is used, the bearing acts like a capacitor. An accumulation of charges occur on the track surfaces, until it reaches the threshold critical value depending on the shaft voltages, bearing dimensional and operational parameters, exceeding which the discharge takes place continuously between races and rolling-elements and leads to bearing failure in due course similar to that of the first type. Under these conditions bearing cease to behave like a capacitor and charge accumulation does not take place on the track surfaces. However, below the threshold critical voltage, the bearing acts like a capacitor and charge gets accumulated on the track surfaces. But, even under these conditions, due to various reasons, operational instabilities cause track surface damage by spark erosion and craters of different sizes are formed. In general, the behavior of oil film below the threshold voltage is ohmic, and it becomes non-ohmic/conductive above the threshold voltage [2, 5].

11.4 Theoretical Analysis for Energy Requirement Leading to the Formation of Craters

11.4.1 Energy Required for Formation of Craters on Track Surfaces

For a crater of diameter W_c to appear on the track surface under a stress ∂, the surface energy required for the opening of a tip of the crater from the subsurface is given as [9]:

$$U = 2\pi q_{cr}\left(\frac{W}{2}\right)^2 \tag{11.1}$$

Some of the energy input in the material is absorbed in plastic flow, which can deform the material plastically. The distortion to appear at the crater location on the track surface from the existing stress location at the subsurface is of the order of ∂/E. The distortion extends in different sides over a dimension of the order of the diameter of the crater. Thus an opening of a crater in the thermally distorted material on the track surface leads to an energy change, which is given as [9]:

$$U = 2\pi q_{cr}\left(\frac{W_c}{2}\right)^2 - \left(\frac{\partial^2}{2E}\right)\frac{4\pi}{3}\left(\frac{W_c}{2}\right)^3 \tag{11.2}$$

In general, this equation shows how much energy is required to initiate a crater of diameter W_c in a material with an existing stress of ∂. When the energy input in the bearing material by the applied stress exceeds the surface energy of the bearing material at the localized area, the crater formation propagates. Thus, for the minimum energy required for an appearance of a crater tip on the track surface, it is evident that

$$\frac{\mathrm{d}U}{\mathrm{d}W_c} = 0 \tag{11.3}$$

which gives

$$\partial_c = 2\left(\frac{Eq_{cr}}{W_c}\right)^{0.5} \tag{11.4}$$

This shows what the critical ∂_c must be for a crater of size W_c to grow spontaneously. Furthermore this equation is then re-arranged in terms of ∂_c and W_c [as shown in Eqn. (11.5)] to determine how much energy per unit area, q_{cr}, is needed to form any crater on the track surfaces in absence of an existing stress, a quantity needed later to determine how much energy should be accumulated. From Eqn. (11.4) it is evident that

$$q_{cr} = \frac{\partial_c^2 W_c}{4E} \quad (11.5)$$

The critical stress determined in Eqn. (11.4) can be used to produce spontaneous propagation of existing/initiated craters. In addition, the significance of Eqn. (11.4) is that Eqn. (11.2) can be re-written in the following form by re-arrangement of mathematical terms as shown, for further analysis:

$$U = 2\pi q_{cr}\left(\frac{W_c}{2}\right)^2\left[1 - \frac{2}{3}(\partial^*)^2\right] \quad (11.6)$$

where $\partial^* = \partial/\partial_c$ is just the existing stress normalized to the critical value appropriate to the size of crater chosen, W_c. So, in the first case when $\partial^* = 1$, it still costs energy to create the crater but it costs no more to cause it to grow. Thus, the net energy, U_1, to create the crater is given as

$$U_1 = \left(\frac{2}{3}\right)\pi q_{cr}\left(\frac{W_c}{2}\right)^2 \quad (11.7)$$

In the second considered case when $\partial^* = 0$, the net energy U_2 to create the crater is given as

$$U_1 = 2\pi q_{cr}\left(\frac{W_c}{2}\right)^2 \quad (11.8)$$

From Eqns. (11.7) and (11.8), it is evident that the energy required in the second case, when the existing stress is zero, is three times more than that of the first case when existing and critical stress levels are equal, i.e. ($U_2 = 3U_1$). Furthermore, Eqn. (11.6) shows that the total energy U corresponds to an area twice that of the crater itself, since it is basic to the Griffith's model that two surfaces of area A are created by the cratering process.

For M craters of diameter W_c to appear on the track surfaces simultaneously, the net energy required will be M times that of U, U_1, U_2 under different conditions as discussed in Eqns. (11.6)–(11.8).

11.4.2 Electrical Model and Equivalent Capacitance of Rolling-Element Bearings

Electrically, a bearing is modeled as a parallel Combination of K identical inner-ring contacts in series with a parallel combination of K identical outer-ring contacts, where K is the number of rolling-elements within the loaded zone. Because of the lubricant film present, each contact is taken to be capacitor and resistor combination, so that the net equivalent bearing electrical component reduces to a combination of R_b and C_b [5].

The equivalent capacitance of, C_b, a roller bearing opposes any change in the existing voltage and causes an electrical charge to be stored. The instant stored charge on the bearing surfaces is determined as [10]

$$Q_c = VC_b \tag{11.9}$$

Also, charges on the inner and outer races (Q_{ir} and Q_{or}) of a bearing are determined assuming that for an instant 'K' rollers are in loaded zone with the races. The charges are given as

$$Q_{ir} = KVC_{ir}, \quad \text{and} \quad Q_{or} = KVC_{or} \tag{11.10}$$

The equivalent capacitance of inner race and rollers being in parallel is KC_{ir}, and that of the outer race and rollers is KC_{or}. Furthermore C_{ir} and C_{or} are in series to each other, and related to C_b [5], and given in Chapter 6 as

$$C_b = \frac{KC_{ir}C_{or}}{C_{ir} + C_{or}} \tag{11.11}$$

where C_{ir} and C_{or} are determined by the bearing parameters [5], and given in Chapter 6 as

$$C_{or} = \frac{2\xi L \tan^{-1}\left[W_{ir}(\beta/h_o)^{1/2}/2\right]}{(\beta h_o)^{1/2}} \tag{11.12}$$

and

$$C_{or} = \frac{2\xi L \tan^{-1}\left[W_{or}(\delta/h_o)^{1/2}/2\right]}{(\delta h_o)^{1/2}} \tag{11.13}$$

A change in charge results in a current flow in the capacitor. It is this current that opposes a voltage change across the bearing capacitor and is determined by

$$I = C_b \frac{dV}{dt} \tag{11.14}$$

where dV/dt is the rate of voltage change in volts per second.

11.4.3 Voltage–Current Characteristics

Voltage at which the spark discharge commences is called ignition or the threshold voltage. It has been established that the voltage drop along the spark decreases as the spark current increases, which indicates that the gap resistance drops quicker than the current rises. The resistance of the spark gap and voltage drop along the spark for a short duration (10^{-4} to 10^{-5} s) may be treated as constant and time invariant quantities. For an instant, the resistance drop at the asperity contact is taken as equivalent to that of bearing resistance under static conditions [10], and given in Chapter 7.

11.4.4 Width of Contact on Track Surfaces Between Races and Rolling-Elements

For each shaft revolution, part of the circumference of the inner race of a bearing passes through a zone of maximum radial force. The Hertzian pressure between rolling-elements and raceways leads to a maximum shear stress. The latter is taken as criterion for yielding of roller tracks. The deformation is made by 'K' rolling-elements in the loaded zone on the track surface of races. The width of contact by the resultant load 'P' as a result of 'p' positions of plane of action of loading by 'K' rollers on the roller track of the inner race is determined in Chapters 5 and 6, and given as [4, 5]

$$W_{ir} = 2.15\left(\frac{Pd(D-d)}{pKELD}\right)^{1/2} \tag{11.15}$$

Similarly, width of contact on the roller track of the outer race is determined as

$$W_{or} = 2.15\left(\frac{Pd(D+d)}{pKELD}\right)^{1/2} \tag{11.16}$$

The initiation of crater formation due to spark erosion occurs in the width of contact W_{ir} or W_{or} of a rolling-element bearing.

11.4.5 Spark Erosion due to Instantaneous Leakage of Stored Charge Energy Between Roller Track of Races and Rolling-Elements

When a bearing is operating below the threshold voltage, the leakage of accumulated charge energy on the track surfaces takes place when the high points on the asperities of track surfaces of races and rolling-elements come close, or when the conducting particles bridge the oil film by vibration effects, or by positive and negative cage and roller slip phenomenon occurring under different operating conditions, or when a roller passes from a loaded zone to the unloaded zone [11]. Under these conditions the instantaneous passage of charge energy between rollers and races through oil film causes spark erosion during bearing operation in microsecond (10^{-6} s). This leads the contact resistance between races and rolling-elements to become instantaneously minimum. Thus, repeatedly instantaneous leakage of accumulated charge energy on the track surfaces at the asperity contacts between rollers and races generates an instantaneous heat, which lead to initiate craters in due course after a certain number of shaft revolutions. In such conditions, when conducting paths between rollers and races are broken, the asperities become separated by higher thickness of film, and charges re-start accumulating on the track surfaces by the induction effect. Since the film thickness between inner race and rolling-elements is less than that of the outer race and rolling-elements, the probability of the crater formation is more on inner-race track surface. Figure 11.1 shows the initiation of craters on the inner-race track surface of the damaged bearing of a motor.

11.4.6 Cycles Required to Accumulate and Discharge of Accumulated Charges on Bearing Track Surfaces

The rate at which the charge builds up on the surfaces of a bearing capacitor when connected to a DC source, e, is assumed to vary exponentially with time t, with time constant $R_b C_b$, while similarly, if the source is replaced by a zero resistance, the capacitor discharges with the same time constant. Under this assumption, the times (or equivalently the number of shaft rotations) required for the potential difference across the bearing capacitor either to build up from 0 to V or to decay can be determined.

At the instant when machine is started, the potential difference (V) across the inner race and rolling-elements as well as the rolling-elements and outer race is zero. But this gradually increases by the phenomenon of electrical induction and approaches the shaft voltage (e). While the shaft voltage increases, the charge on the bearing using high-resitivity lubricant, comprising charges on the inner and outer races (Q_{ir} and Q_{or}), builds up. Till then bearing voltage, V, is changing, transient current is delivered from the shaft voltage. To maintain continuity throughout the circuit, there is rate of change of flux within the dielectric (lubricant).

The ratio of shaft revolutions to accumulate and discharge of the accumulated charges on/from the inner race, outer race and a roller/rollers is independent of various parameters and depends only on the ratio of difference of the natural logarithm of the shaft and bearing voltage to the shaft voltage, to the ratio of the natural logarithm of the shaft to the bearing voltage [12], and is given in Chapter 8 as

$$\frac{N_{icn}}{N_{idn}} = \frac{\mathrm{Log}_e(1 - a)}{\mathrm{Log}_e\, a} \tag{11.17}$$

The ratio of shaft revolutions to accumulate and discharge the accumulated charges (N_{icn}/N_{idn}) increases with the ratio of bearing to shaft voltage (V/e). As the ratio V/e increases from 0.1 to 0.9, the ratio N_{icn}/N_{idn} increases from 0.05 to 20.91 [12]. It is evident that for the successive higher values of bearing to shaft voltage, a, the ratio of N_{icn}/N_{idn} is higher than for the immediate lower value of a (V/e). Furthermore, it takes a longer time to accumulate the charges as compared to that of the dissipation of the accumulated charges as $N_{icn} > N_{idn}$ [12]. The ratio of shaft rotations to accumulate and discharge the accumulated charges by a single rolling-element or by all the rolling-elements and races of a bearing is identical. Also, complete charge accumulation takes the same number of shaft revolutions equivalent to that of the charge accumulation by a single rolling-element and roller track of races [12] as explained in Chapter 8.

The number of shaft revolutions, N_{idn}, to discharge of the accumulated charges is based on the time for voltage to decay. The physical significance of N_{idn} indicates that while the bearing inner race, shaft and rolling elements are in rotation, the fraction of a second (10^{-5} s) when dissipation of charge energy between races and rollers takes place, shaft simultaneously continues to rotate. Similarly, for the accumulation of charges on the track surfaces depending on the time constant, the shaft makes revolutions (N_{icn}). The charge accumulation and dissipation is a continuous process during bearing operation under the influence of shaft voltages. Furthermore, result from many experiments indicate that a rolling bearing Hertzian contact behaves like a leaky capacitor, with a capacitance given basically by the EHD lubricant film thickness and a resistance, determined by random asperity contacts through this film, assisted by vibrations, depending on the source micro-topography. The bearing contact resistance, in parallel with the capacitor, varies with high frequency as the asperities move through the Hertzian zone. As a result of this high frequency resistance variation, the potential drop across the contact is never equal to the equilibrium value, which means that the rolling-elements are continuously charging and discharging, with a time constant also continuously varying, however, with more time for charging than time for discharging. This is considered by taking the ratio of N_{icn}/N_{idn} for analysis as explained in Chapter 8 [12].

11.4.7 Determination of Electrical Energy Required for Formation of Craters on the Track Surfaces of Bearings

The net stored energy of charge between track surfaces of races and rolling-elements is given by [13]

$$Q = 0.5 C_b V^2 (\mathrm{J}) \tag{11.18}$$

The area of a crater of diameter W_c appearing on the track surface of a bearing is determined as

$$A = \pi \left(\frac{W_c}{2} \right)^2 \tag{11.19}$$

By the Griffith model two surfaces of area A are created by the cratering process. Therefore the net area is considered as $2A$ for further analysis as discussed in Section 11.4.1. It is evident that the number of times accumulation and dissipation of charge energy in each cycle between races and rollers is a very complex random phenomenon. Hence, it is assumed that the net stored energy is dissipated in each cycle between the track surfaces. Thus electric energy transmitted per unit area per cycle on the track surfaces is determined as

$$q = \frac{Q}{2A} = \frac{C_b V^2}{\pi W_c^2} \tag{11.20}$$

The minimum energy dissipated per unit area in N_c cycles responsible for the formation of a crater on the track surfaces is determined as

$$q_n = \frac{C_b V^2 N_c}{\pi W_c^2} \tag{11.21}$$

Considering the ratio of N_{icn}/N_{idn}, the minimum number of cycles for the formation of craters is determined as N_{icn}/N_{idn} times the value of N_c or N_{idn}/N_{icn} times value of q_n. Hence the net energy per unit area for the formation of a crater is determined using Eqns. (11.9) and (11.21), and given as

$$q_{net} = \frac{N_c Q_c^2}{\pi C_b W_c^2} \times \frac{N_{idn}}{N_{icn}} \tag{11.22}$$

11.4.8 Duration/Number of Cycles Before Initiation of Craters on Track Surfaces

Instant localized thermal stresses due to thermal transients on the roller tracks of races caused by instantaneous leakage of accumulated charge energy by a leaky bearing capacitor depend on instant rise in temperature at the localized spots leading to the formation of craters in due course [14–16]. Once a small crater is formed, formation of other craters follows at various asperity contacts. Release of material from the crater/craters accelerates the process by bridging the film thickness between races and rolling-elements at different locations causing an increase in the number/diameter of craters. Simultaneously more number of craters could be initiated/could appear, depending on the operating parameters and the condition of a bearing. This is in general a random phenomenon.

On comparing the net electric energy per unit area, q_{net}, given in Eqn. (11.22) and the total energy per unit area, q_{cr}, required for appearance of a crater on the track surfaces of the bearing as given in Eqn. (11.5), the minimum number of cycles/duration, a bearing would have taken before the appearance of a crater is determined as

$$N_{\mathrm{c}} = \frac{\pi C_{\mathrm{b}} \partial_{\mathrm{c}}^2 W_{\mathrm{c}}^3}{4EQ_{\mathrm{c}}^2} \times \frac{N_{\mathrm{icn}}}{N_{\mathrm{idn}}} \qquad (11.23)$$

For M number of craters of diameter W_{c} to appear simultaneously on the track surfaces, the number of cycles will be M times that of N_{c}. Also, from this equation it is evident that the number of cycles required to form a crater depends on square of its diameter [since $\partial_{\mathrm{c}}^2 \propto W_c^{-1}$ as per Eqn. (11.4)]. Obviously there is a preference for small craters, although rather arbitrarily a bigger size (1 mm or so) is taken as the critical size for determining the life of the bearing.

11.5 Experimental Investigations

The investigations on the passage of a current through bearings show that the threshold voltage varies with operating parameters and lubricant characteristics. A low-resistivity lubricant (10^5 ohm m) having a viscosity of 99.92 cSt at 40 °C and 9.68 cSt at 100 °C, and a relative permittivity (ξ_{r}) of 2.5 begins to conduct electricity at a voltage of 1.12 V and allows a current of 50 A (AC) to pass through the NU 326 roller bearing tested on roller bearing test rig operating at 1100 rev. min^{-1} under radial loads of 10^4 N and 5×10^3 N [6]. The bearing shows the ratio of capacitive reactance to active resistance to be 14×10^2. It behaves like a conductor, allowing a current to pass without exhibiting a threshold voltage phenomenon and showing a capacitive response [2]. The bearing fails by initiation of slip bands leading to a corrugation pattern on the bearing surfaces. In contrast, a roller bearing using high-resistivity lubricant (10^9 ohm m) having a relative permittivity (ξ_{r}) of 2.5, indicates a ratio of capacitive reactance to active resistance of 0.14 and behaves like a capacitor until the threshold voltage is reached and indicates a threshold voltage phenomenon.

The charge accumulation on the track surfaces and threshold voltage have been established experimentally [2, 4]. The craters initiation on the track surfaces of the NU 330 bearing have been established at different operating parameters due to the leakage of charge energy. Figure 11.2 shows the enlarged view of the crater on the bearing track surface.

11.6 Data Deduction and Discussion on Energy & Time Required for the Formation of Craters

From the test data and dimensional parameters of the NU 330 bearing as given in Tables 11.1 and 11.2, the time/cycles required for the appearance of craters on the track surfaces have been determined using Eqn. (11.23) for a typical case where a diameter of the crater is found as 1 mm. The ratio of N_{icn}/N_{idn} in Eqn. (11.23) is

Table 11.1. Experimental and theoretical data for the initiation of craters on the roller track of races of the NU 330 bearing using high-resistivity lubricant due to stored energy of charge leakage between the races and rolling-elements

Dimensional and Operational Parameters of the NU 330 Bearing			
d	42 mm	E	$210 \times 10^9\ \mathrm{N\,m^{-2}}$
D	235 mm	ξ	$2.5 \times 8.85419 \times 10^{-12}\ \phi\,\mathrm{m}^{-1}$
K	3	δ	$0.020\ \mathrm{mm}^{-1}$
L	45 mm	β	$0.029\ \mathrm{mm}^{-1}$
N	14	ϕ	10^9 ohm m
n	$1100\ \mathrm{rev.min}^{-1}$	∂_c	$700 \times 10^6\ \mathrm{N\,m^{-2}}$
P	10^5 N		
W_c	1 mm		

Table 11.2. Experimental and theoretical data

n (rpm)	450	750	1000	1200
$Q_c \times 10^{-12}$ (C)	7.62	19.71	21.20	54.5
$C_b(p\phi)$ (ref. [5])	190.4	131.41	106.4	94.0
V (mV) (ref. [5])	40	150	200	580
N_c	1.51×10^{10}	3.8×10^9	1.09×10^9	1.5×10^8
t	>3 years	>3 years	>2 years	84 days

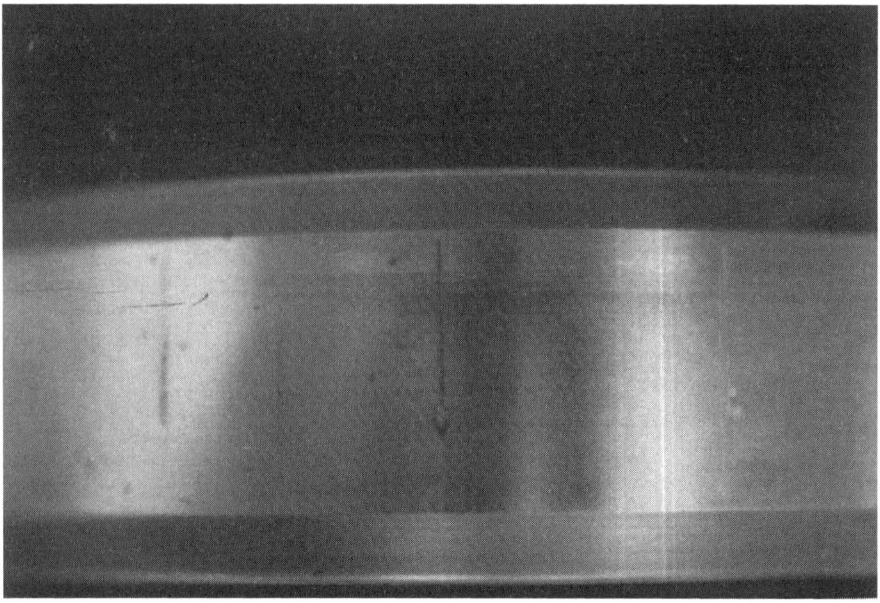

Fig. 11.1. Craters on the damaged inner-race track surface of a roller bearing by instantaneous leakage of charge energy

taken approximately 2.5 considering an average bearing-to-shaft voltage (V/e) as 0.66 [12]. Typical damage of the bearing through craters formation by spark erosion is shown in Figs. 11.1 and 11.2. Figure 11.3 shows the metallurgical structure of a surface damaged by spark erosion. Stereo microscope examination of the defective region shows the removal of material and evidence of it having been melted. Polishing and etching of the defective region shows a light etching tendency compared to the unaffected region which is etched darkly. The spark erosion region shows an untempered martensitic structure with a very high micro-hardness >1100 HV as compared to the tempered martensite structure with micro-hardness of ~600 HV as region away from the area of spark erosion (Fig. 11.3). The duration, thus determined theoretically, has been ascertained with the investigations of the case history for the formation of craters due to charge leakage through the bearings.

11.6.1 Energy for the Appearance of Craters on the Track Surfaces

The energy per unit area q_{cr} required for the appearance of a crater on the track surface of races of a rolling-element bearing is directly proportional to the square of the critical stress limit (∂_c), diameter of a crater, and inversely proportional to Young's

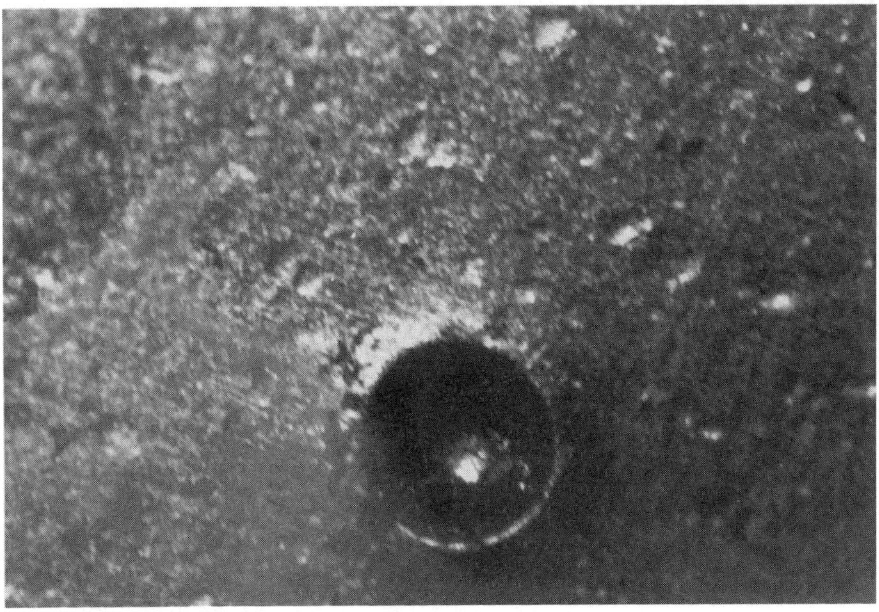

Fig. 11.2. Enlarged view of a crater on the track surface of the inner race of a rolling-element bearing

modulus of elasticity of the bearing material [Eqn. (11.5)]. Also, the electrical energy, q_{net}, required for the crater formation is proportional to the square of charge accumulation on the bearing surfaces, and inversely proportional to the bearing capacitance and square of the crater diameter. Furthermore, it is a function of number of cycles required for the crater formation and the ratio of the cycles for discharge of the accumulated energy to that of the stored energy of charge on the track surfaces [Eqn. (11.22)].

It may be pointed out that the electrical energy of the charge accumulation may dissipate at a number of points in a single shaft revolution depending on several causes, including instability, vibration, surface finish, etc. [16], and initiate craters of different sizes and intensities on the track surfaces. A similar pattern gets repeated in each shaft revolution, which results in the formation of identifiable craters in due course [17].

By the process, as analyzed, the formation of localized craters by spark erosion is continued, and the mass transfer from the developed craters deteriorate the condition of the bearing and lead to form more number of different sized craters much faster by the fusion of the tiny material particles released from the craters. At some locations an increased area is fused by spark erosion in due course, whereas initially, in general, the systematic formation of craters takes place (refer Fig. 11.1).

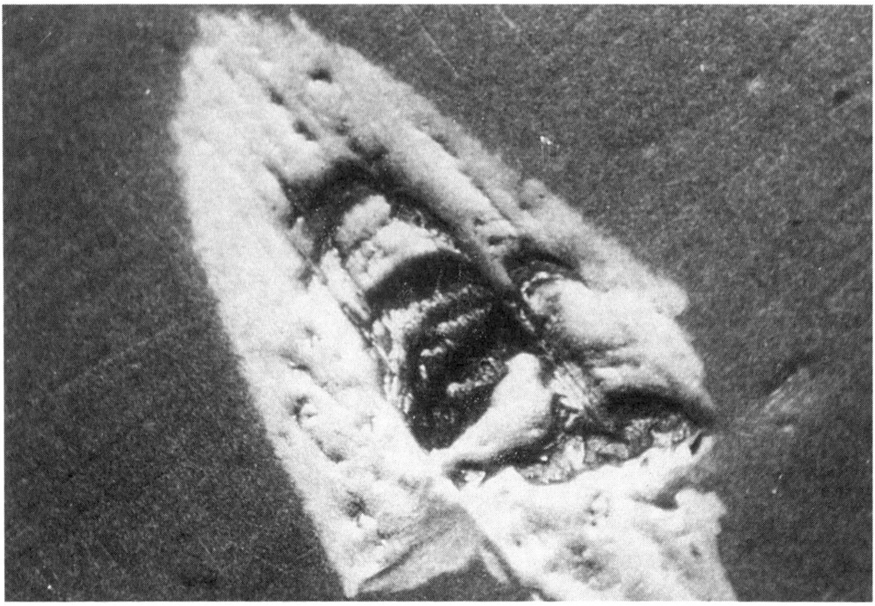

Fig. 11.3. Enlarged view of the spark erosion showing the cavity and molten like appearance

11.6.2 Cycles/Time Required for the Appearance of Craters on the Track Surfaces

The cycles/time N_c required for appearance of craters on the track surfaces is proportional to bearing capacitance, the square of critical stress limit for the appearance of craters, the cube of a crater's diameter, the ratio of cycles for accumulation and dissipation of charge energy, and inversely proportional to the square of the charge accumulation on bearing surfaces, and is a function of properties of the bearing material [Eqn. (11.23)]. Also, it is a function of number of craters, and there are chances for formation of small craters although arbitrary.

The value of cycles for the appearance of a crater of 1 mm diameter at the spatial location of the energy due to temporal build up vary between 1.51×10^{10} to 1.5×10^{8} cycles at different levels of bearing voltages and charge accumulation under different parameters of operation (Table 11.2). However, mechanical loading and temperature-rise due to frictional losses also contribute to the reduction in bearing life, but their effect is much less compared to the effect of dissipation of stored energy on the bearing surfaces. From Table 11.2, it is evident that from an increase in bearing voltage of 200–580 mV, the bearing life reduces from 1.09×10^{9} to 1.5×10^{8} cycles, i.e. from 750 to 84 days only. This suggests that the bearing deterioration is a steady process more resembling wear than a catastrophic fracture event. Below 200 mV shaft voltage the bearing life is more than 10^{10} cycles. This indicates that the safe limit of shaft voltage for a rolling-element bearing in general is less that 200 mV. The same has been the recommendation of the reputed manufacturers of the Power Plant Equipments. Furthermore, this has been ascertained by the investigation of operational data and the case history of various damaged bearings of different machines.

It may be noted from the analysis that the life expectancy of a bearing before the craters are initiated will be higher if the threshold voltage of operation is lower. This is because there is less rise in contact temperature. In addition, a decrease in charge accumulation and an increase in bearing capacitance result in a higher number of cycles before the initiation of craters. Particularly, the decrease in charge accumulation increases the bearing life more significantly, since the life is inversely proportional to the square of charge accumulation [Eqn. (11.23)]. Also, if the time/number of cycles, the bearing capacitance and the diameter of craters are known, then the charge accumulation/shaft voltage can be established theoretically by using Eqn. (11.23).

11.7 Conclusions

From the above analysis, the following conclusions are drawn [1, 17]:

1. The energy per unit area for the appearance of a crater on the track surfaces of a rolling-element bearing depends on the material properties, the critical/limiting level of contact stresses and the diameter of the crater.
2. The electrical energy per unit area for the appearance of a crater is a function of charge accumulation, the bearing capacitance, the crater's diameter, the number of cycles and the ratio of cycles for dissipation of the charge energy to that of the accumulation.
3. The number of cycles required for the appearance of a crater depends on the bearing capacitance, the charge accumulation, the crater's diameter, the critical/limiting level of contact stresses, the material properties and the number of cycles for accumulation of charge energy to that of the dissipation.
4. The decrease in the charge accumulation and the increase in the bearing capacitance results in a higher number of cycles before initiation of craters on the track surfaces.
5. The safe limit of shaft voltage for rolling-element bearings is less than 200 mV. Any increase in which reduces the life of a bearing very fast.

REFERENCES

1. Prashad, H. (2001). Appearance of craters on track surfaces of rolling-element bearings by spark erosion. *Tribol. Int.*, **34**, 39–47.
2. Prashad, H. (1987). Effects of operating parameters on the threshold voltage and impedance response on non-insulated rolling-element bearings under the action of electric current. *Wear*, **117**, 223–240.
3. Prashad, H. (1989). Analysis of the effects of an electric current on contact temperature, contact stresses and slip band initiation on the roller tracks of roller bearings. *Wear*, **131**, 1–14.
4. Prashad, H. (1988). Investigations on corrugated pattern on the surfaces of roller bearings operated under the influence of electrical fields. *Lubric. Eng.*, **44**(8), 710–718.
5. Prashad, H. (1988). Theoretical evaluation of impedance, capacitance, and charge accumulation on roller bearings operated under electrical fields. *Wear*, **125**, 223–239.
6. Prashad, H. (1988). Determination of time span for the appearance of flutes on the track surface of rolling-element bearings under the influence of electric current. *Tribol. Trans.*, **41**(1), 103–109.
7. Prashad, H. (1988). Diagnosis of deterioration of lithium base greases used in rolling-element bearings by X-ray diffractrometry. *Tribol. Trans.*, **32**(2), 205–214.
8. Prashad, H. (1999). Determination of magnetic flux density on the surface of rolling-element bearings as an indication of the current that has passed through them – an investigation. *Tribol. Int.*, **32**, 455–467.
9. Czichos, Host (1957). *Tribology Series 1*, Elsevier Sequoia, UK.
10. Prashad, H. (1990). Theoretical analysis of the effects of instantaneous charge leakage on roller track of roller bearings lubricated with high-resistivity lubricants. ASME Transaction, *J. Tribol.*, **112**, 37–43.
11. Prashad, H. (1987). The effect of cage and roller slip on the measured defect frequency response on rolling-element bearings. *ASLE Trans.*, **3**(1), 360–367.
12. Prashad, H. (1992). Theoretical analysis of capacitive effect of roller bearings on repeated starts and stops of a machine operating under the influence of shaft voltages. ASME, *J. Tribol.*, **114**, 818–822.
13. Sydney, S. G. (1960). *Electricity and Magnetism*, Longmans, Green and Co., London.
14. Patir, N. and Cheng, H. S. (1978). Effect of surface roughness orientation on the central film thickness in EHD contacts. Proceedings of 5th Leeds-Lyon Symposium.
15. Christensen, H. and Tonder, K. (1969/1970). Tribology of rough surfaces stochastic models of hydrodynamic lubrication. SINTEF REPORT, Trondheim.
16. Leenders, P. and Houpert, L. (1988). Study of lubricant film in rolling bearings; effect of roughness. Leeds-Lyon Symposium, Paper XXI(i), 629–638.
17. Prashad, H. and Venugopal, K. (2002). Formation of craters on the track surfaces of rolling-element bearings due to spark erosion. *BHEL J.*, **23**(1), 34–47.

Nomenclature

a	ratio of potential difference across bearing to shaft voltage ($V/e < 1$)
A	area of craters
C_b	equivalent capacitance of bearing
C_{ir}	capacitance between inner race and a roller
C_{or}	capacitance between outer race and a roller
d	diameter of rolling-element
D	pitch diameter
e	shaft voltage
E	Young's modulus of elasticity
h_o	minimum film thickness
I	current passing through bearing
K	number of rolling-elements in the loaded zone
L	length of rolling-element
M	number of craters
n	rev.min^{-1}
N	number of rolling-elements in bearing
N_c	number of shaft revolutions before initiation of craters
N_{icn}, N_{idn}	number of revolutions to accumulate and discharge of the accumulated charges on the track surfaces, respectively
p	positions of plane of action of loading ($p = 1, 2, 3, \ldots$)
P	resultant load on bearing
q	electric energy per unit area dissipated in a cycle for craters formation on the track surfaces
q_n	electrical energy per unit area dissipated in N_c cycles for appearance of craters
q_{net}	net electrical energy required for appearance of craters considering the ratio of shaft revolutions to accumulate and discharge of the accumulated energy in N_c cycles
q_{cr}	energy per unit area required for formation of craters on the track surfaces
Q	net stored electrical energy of charge between races and rolling elements
Q_c	charge accumulation on track surfaces of bearing
Q_{ir}	charge accumulation between inner race and rolling-element
Q_{or}	charge accumulation between outer race and rolling-elements
r	radius of rolling element
R_i	track radius of inner race
R_o	track radius of outer race

R_b	equivalent resistance of bearing
t	time required for appearance of craters on track surfaces
U	net energy required for appearance of a crater on track surfaces
U_1, U_2	net energy required for appearance of a crater when $\partial^* = 1$ and when $\partial^* = 0$, respectively
V	potential drop across bearing
W_c	diameter of crater
W_{ir}, W_{or}	width of contact of rolling-element on inner race and outer race, respectively
β	$(R_i + r)/(2R_i r)$ inner race constant
δ	$(R_o - r)/(2R_o r)$ outer race constant
∂_c	critical stress level for appearance of craters on the track surfaces
∂^*	∂/∂_c ratio of existing stress to the critical stress
∂	existing stress
ϕ	oil resistivity (ohm m)
ξ	permittivity (dielectric constant of lubricant) in $\phi\,m^{-1}$
ξ_r	relative permittivity of lubricant

Chapter 12

APPEARANCE AND EFFECT OF LOCALIZED ELECTRIC CURRENT IN ROLLING-ELEMENT BEARINGS

12.1 General

The diagnosis and cause analysis of rolling-element bearing failure have been well studied and established in literature. Failure of bearings due to unforeseen causes were reported as: puncturing of bearings insulation; grease deterioration; grease pipe contacting the motor base frame; unshielded instrumentation cable; the bearing operating under the influence of magnetic flux, etc. These causes lead to the passage of electric current through the bearings of motors and alternators and deteriorate them in due course. But, bearing failure due to the localized electric current between the track surfaces of races and the rolling-elements has not been hitherto diagnosed and analyzed.

This chapter explains the cause of generation of localized current in presence of shaft voltage. Also, it brings out the developed theoretical model to determine the value of localized current density depending on the dimensional parameters, the shaft voltage, the contact resistance, the frequency of rotation of shaft and rolling-elements of a bearing. Furthermore, failure caused by a flow of localized current has been experimentally investigated.

12.2 Introduction

The causes listed in Chapters 1 and 2 develop a magnetic flux, which closes in the circumference over the yoke and induces voltage on the shaft as the machine rotates.

This results in a localized current at each bearing rather than a potential difference between the shaft ends. A current path, however, along shaft, bearings and frame results in a potential difference between the shaft ends [1–3].

It has been reported that at a certain threshold voltage, depending on the resistivity of the lubricant and operating conditions, current flows through the bearing [4]. Also, it was established by Busse et al. [5] that the dielectric strength of the lubricant is the ability to withstand voltage without breakdown. Prashad [6] studied that the flow of circular current in the inner race leaks through the rolling-elements to the outer race by following a path of least resistance and establishes a field strength leading to the development of magnetic flux on the track surface of races and rolling-elements. Also, this has been experimentally investigated.

Studies were reported on the causes and control of electric currents in bearings by Morgan and Wyllie [7]. Andreason [8] determined flow of current through lubricated contacts. Simpson and Crump [9] studied the effect of electric current on bearing life. The effect of operating parameters on an impedance response of a rolling-element bearing has been analyzed by Prashad [4, 10]. The deterioration of lubricants used in non-insulated bearings investigated by Komatsuzaki [11], Prashad [12], and Remy and Magnin [13]. Investigations were also carried out to evaluate the pitting mechanism on the lubricated surfaces under AC electric current by Lin et al. [14], and by Chiou et al. [15].

The surveys of the failure of rolling-element bearings indicate various causes including that of failure due to corrugations by Winder and Wolfe [16] and Prashad [17, 18]. The mechanism of corrugations formed on the track surfaces and related investigations were reported by Prashad and co-workers [3, 19]. In addition to the well-established causes, the bearing may also fail due to the causes generally unforeseen during design and operation. These causes have been discussed by Prashad [20].

This chapter presents with the help of case study the failure of bearings of motors by the effect of localized electric current between the track surface of races and rolling-elements under the conditions when the other causes of bearing failure have been ruled out by all possible investigations. Also, the principle for generation of localized current has been established and analyzed along with the development of theoretical model to determine its value.

12.3 Bearing Arrangement and Nature of Bearing Failure

Failure of bearings was detected in a few motors. These motors rated 2100 kW, operating at 1494 rpm, have a three-bearing arrangement. The motors are used to drive primary air fans.

At the non-drive end (NDE) of the motor, the bearing type NU 228 is used and is insulated. Figure 12.1 shows the arrangement of the bearing in the motor. At the drive-end (DE) of the motor, roller bearing type NU 232 and ball bearing type 6332 are used for taking both radial and axial loads simultaneously. All the bearings are grease-lubricated, and the motors are designed for continuous operation with periodic re-lubrication. The bearings are also instrumented for continuous measurement of the temperature during operation.

After commissioning, a few motor bearings both on NDE and DE failed prematurely. The nature of the failure was due to the formation of corrugations and flutings on the roller tracks of races besides corrosion on the raceways and rolling-elements, irrespective of the make of the bearings used. Typical corrugation pattern on the track surface of a failed bearing is shown in Chapter 5. However, the depth of corrugations and degree of development of corrugations and flux density were different on various failed bearings. The original color of the grease in between the rolling-elements was also found to change and became black.

12.4 Investigations, Observations and Data Collection

Various investigations pertaining to shaft voltage and vibrations were carried out on bearings of different motors, both on NDE and DE, and various other related aspects were inspected. The shaft voltages were measured between NDE and DE shaft, DE shaft to ground, and NDE shaft to ground. The vibration levels were measured in microns and mm s^{-1} at both NDE and DE bearings. The dimensional accuracy of the new bearings were checked and found to be in line with the specifications. Inspection of the dimensional accuracy and the metallurgical examinations of the components of the failed bearings also revealed consistency with the specifications. Various failed bearings were examined. Samples of the fresh grease and the used grease from the rolling-elements of the failed bearings were collected and analyzed.

None of the following causes of the bearing failure could be established, which might have led to the passage of current though bearings (Fig. 12.1).

I. Grease pipe was not in contact with the base frame in any of the motors.
II. Current path through bearings via the path through the collection of leaked grease and the base frame was not detected.
III. All the instrumentation cables were completely shielded.
IV. Insulation of the bearings was fully intact in all the failed motor bearings. This was established by the value of high resistance measured across the bearing insulation.
V. Passage of current through connecting the bolts, the nuts, etc. could not be established in any of the failed motor bearings.

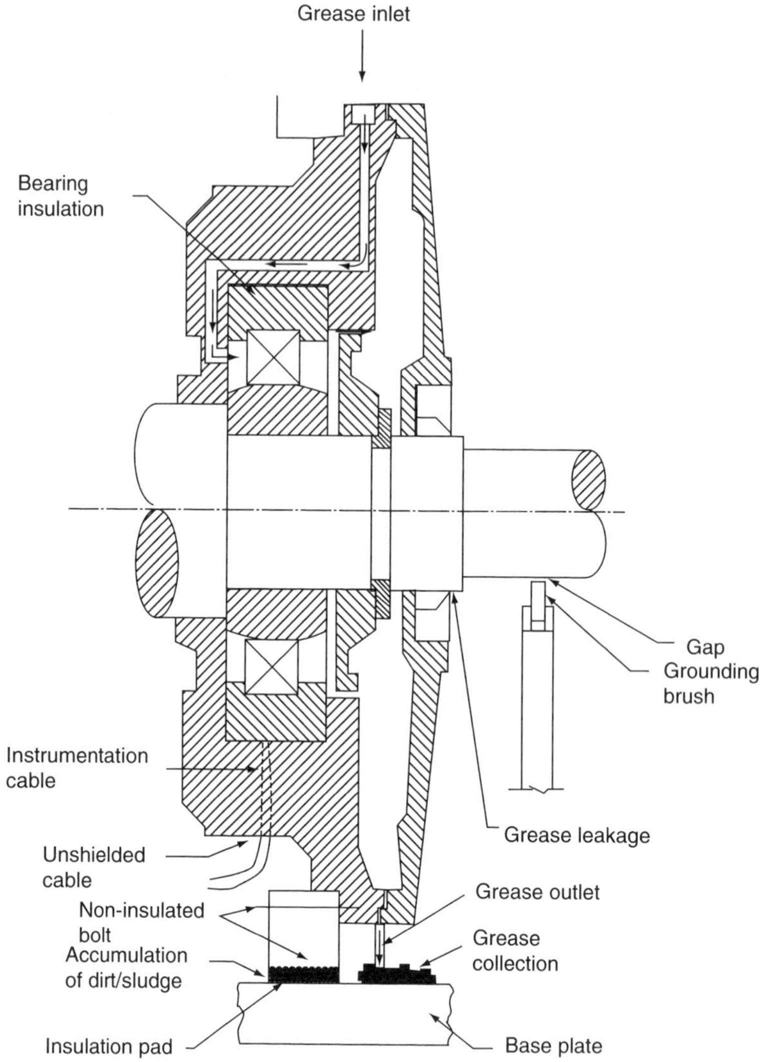

Fig. 12.1. Schematic showing bearing arrangement with lubricant flow path, unshielded instrumentation cable and other unforeseen causes leading to the passage of electric current

The shaft voltage at the site on these failed motor bearings was measured as 1600 mV before and after the replacement of failed bearings. The shaft voltage at the works during the test run was also recorded as high as 500 mV for these motors as compared to the other motors. The difference of shaft voltage measured at the sites and the works may be attributed to the used analogue/digital devices besides the coupled mode of operation and other unforeseen causes. For other normal running motors, shaft voltage as measured was less then 200 mV. The failure of NU 228 bearings was reported at the site on these motors after a run of approximately 300 h. A grounding brush was not used in all such motors as a practice. Furthermore, there was no adequate space available on the shaft to install the grounding brush.

12.5 Theoretical Model and Approach to Determine the Flow of Localized Current in a Bearing

In a rolling-element bearing using low-resistivity lubricant (10^5ohm m) in the presence of shaft voltage, current can flow through the bearings. This occurs in all such cases when current finds a low resistance path to flow as brought out in Section 12.4. But when the current does not find a way to pass through any of the paths as listed in Section 12.4, and the shaft voltage is more than the threshold voltage/safe voltage (>200 mV), a substantial level of localized current between the rolling-elements and the track surface of the inner race may appear which may deteriorate the bearing and disintegrate the lubricant in course of time causing flux density distribution, corrosion and blackening of the track surface as reported by Prashad [20]. This reduces the fatigue life of bearings and lead to premature failure.

The existing shaft voltage on the rotating shaft induces the voltage on the track surface of rolling-elements. Since the rolling-element rotates at much higher frequency than the shaft rotating speed [19, 21], this leads to generate higher voltage on the surface of rolling-elements as per principles of electric and electromagnetic fields [22]. Thus, potential difference between the inner race and rolling-elements is developed. The revolution of rolling-elements along the track surface further affects the induced voltage on the track surface of the rolling-elements besides their rotation around their own axis. This makes the analytical model quite complex. However, the potential difference between rotating shaft/inner race and rolling-elements lead to the passage of localized current at the contact surface depending on the contact resistance between the contact zone of track surface of the inner race and the rolling-elements. The induced voltage phenomenon also occurs between the track surface of the outer race and the rolling-elements. Furthermore, the flow of localized current between the inner race and the rolling-elements depending on the different points of contacts during rotation turns into circular current in the inner race and rolling-elements quite frequently in the course of operation. The proximity contact of rolling-elements with the outer race may lead to a flow of current through the outer race for a short duration. This flow of partial/full circular current establishes the field strength leading to the development of residual magnetic flux on the track surface of races and rolling-elements in due course [6].

It may be imagined that current enters the track surface of the inner race due to the developed potential drop between rolling-elements and track surface of the inner race in the loaded zone at the surfaces of contact points between them, in a distributed form. The current flows around and outward until it concentrates at one or several rolling-contacts. As the rolling-elements orbit, the current carrying elements travel. Current then flows through each of the conducting rolling-elements, between two diametrically opposite contact areas. As the rolling-elements rotate with respect to the contact areas, these currents gradually might sweep out 360° around a major circle. If there was only one conducting rolling-element at a time, then two arc currents would travel in the race towards its contact areas one clockwise and the other counterclockwise. Depending on the resistance these two currents may or may not be equal in magnitude. If more than one rolling-element contacts, then, of course, the situation is more complex. In the theoretical model, current flow in one direction in the inner race is assumed for analysis. The flow and break of flow of a localized electric current continues depending on condition and regime of operation.

12.5.1 Determination of Speed of Rolling-Elements

Under pure rolling, the absolute velocity at a point located on the circumference of a rolling-element is equal to the circumferential speed of the inner race. Simultaneously, on an opposite point (at 180°) of the same rolling-element contacts the stationary outer race. The absolute speed of each point on the rolling-element is a combination of the speed of the rolling-element around its axis and the speed of the set of rolling-elements around the bearing axis. The absolute velocities of the two points 180° apart on the diameter of the rolling-element are parallel and vary linearly between the velocities at the points of contact on the inner and outer race.

With the above analysis, the rotating speed frequency of the rolling-elements has been determined as [19]

$$f_{\mathrm{b}} = 0.5 f_{\mathrm{s}} \left(\frac{D}{d} \right) \left[1 - \left(\frac{d}{D} \right)^2 \cos^2 \alpha \right] \tag{12.1}$$

The ratio of the speed of rotation of rolling-elements and shaft speed is given as

$$V_{\mathrm{bs}} = \frac{f_{\mathrm{b}}}{f_{\mathrm{s}}} = 0.5 \left(\frac{D}{d} \right) \left[1 - \left(\frac{d}{D} \right)^2 \cos^2 \alpha \right] \tag{12.2}$$

For a radial cylindrical roller bearing, $\alpha = 0$

So,

$$V_{bs} = 0.5\left(\frac{D^2 - d^2}{dD}\right) \tag{12.3}$$

12.5.2 Induced Voltage on Rolling-Elements

As the shaft-voltage (E) develops and changes with time, the voltage on the track surface of the inner race, E_{ir}, accordingly changes. This is because the inner race is press fitted on the shaft, and it is considered integral with the shaft. The voltage on the track surface of rotating inner race, E_{ir}, induces the voltage on the rolling-elements rotating around their own axis and moving along the track surface of the inner and outer races. This makes the phenomenon of induced voltage very complex.

The rotating rolling-elements loop may be considered as a type of alternating current generator or in other words, rolling-elements develop voltage depending on the speed of rotation. The voltage in the absolute term is more on those surfaces of rolling-elements, which are at right angles to the existing field because of shaft voltage on track surface of inner race and zero on the parallel surfaces to the field. The developed voltage on rolling-elements broadly depends on the area of such loop formed by rotating rolling-elements and not on their shapes. The voltages, thus developed on the surfaces of rolling-elements, (E_{ir}), depend on the rotating speed of the rolling-elements besides other electromagnetic factors. Considering the ratio of the speed of rolling-elements with respect to the shaft speed, V_{bs}, the voltage on the rolling-elements (E_r) is taken as the product of V_{bs} and shaft voltage (E) using principles of electric current theory [22]. This leads to determine the potential difference between rolling-elements and track surface of inner race as

$$E_{rir} = E_r - E_{ir} \tag{12.4}$$

On using Eqn. (12.3), the potential difference E_{rir} is determined as (where $E_{ir} = E$, and $E_r = EV_{bs}$)

$$E_{rir} = EV_{bs} - E \tag{12.5}$$

or

$$E_{rir} = E\left(\frac{D^2 - d^2 - 2Dd}{2Dd}\right) \tag{12.6}$$

Furthermore, development of same induced voltage as on rolling-elements occurs on the track surface of outer race under the influence of voltage existing on surfaces of rolling-elements. So, the potential difference between the rolling-elements and outer race will be of insignificant value as the outer race of bearing is stationary [22].

12.5.3 Bearing Resistance

The resistance between the track surface of the inner race and rolling-elements depends on bearing kinematics, lubricant film thickness, width of contact between track surface and rolling-elements and the number of rolling-elements in loaded zone. For the bearing using low-resistivity lubricant (10^5 ohm m), the resistance (R_c) of the bearing as determined is found to vary between 0.1 and 0.5 ohm [4].

12.5.4 Localized Electric Current in Bearings

Localized current between the track surface of the inner race, and rolling-elements is determined using Eqn. (12.6) and the contact resistance (R_c), and is given as

$$I_b = \frac{E_{rir}}{R_c} = \frac{E}{R_c}\left(\frac{D^2 - d^2 - 2Dd}{2Dd}\right) \tag{12.7}$$

12.6 Field Strength on the Track Surface of Races and Rolling-Elements

Due to the presence of localized electric current in the partial/complete arc of the inner race and rolling-elements, the field strength develops on the track surfaces of rolling-elements and inner race in due course. The field strength on these elements can be determined separately as reported by Prashad [6].

12.6.1 Field Strength on the Track Surface of Inner Race

The field strength (H_{irr}) on the track surface of the inner race due to flow of current in rolling-elements has been analyzed in Chapter 9, and is determined as [6]

$$H_{irr} = \frac{2\pi I_b\left(R^2 - R_{ir}^2\right)}{R^3} \tag{12.8}$$

12.6.2 Field Strength on Rolling-Elements

On rotation, rolling-elements change polarity, and the outer race being stationary, the field strength on the surface of rolling-elements depends on the flow of circular current in inner race and is given as [6] (refer Chapter 9)

$$H_{rir} = \frac{2\pi I_b R_{ir}(R^2 - R_{ir}^2)}{R^4} \tag{12.9}$$

The chances of development of significant field strength on the track surface of outer race due to the flow of intermittent local current in the track arcs of the inner race and rolling-elements are quite remote, and hence it is not considered in the analysis. However, sometimes development of traces of flux density on the outer-race track surface can not be ruled out.

12.7 Magnetic Flux Density

The field strength on the track surface of the inner race and rolling-elements of a bearing gives rise to the development of magnetic flux density on these surfaces (refer Chapter 9). These can be determined analytically as reported by Prashad [6].

12.7.1 Magnetic Flux Density on the Track Surface of Inner Race

The magnetic flux density on the track surface of the inner race of a bearing, B_{ir}, due to field strength, H_{irr}, in an oil medium of relative permeability, U_r with respect to free space is given by [6]

$$B_{ir} = UU_r H_{ir} = 4\pi \times 10^{-7} U_r H_{irr} \quad \text{(in tesla)} \tag{12.10}$$

which is determined (in gauss) using Eqn. (12.8) as

$$B_{ir} = \frac{78.96 \times 10^{-3} U_r I_b (R^2 - R_{ir}^2)}{R^3} \tag{12.11}$$

12.7.2 Magnetic Flux Density on Track Surface of Rolling-Elements

On using Eqns. (12.9) and (12.10), the residual flux density on a few rolling-elements (in gauss) is determined as

$$B_r = \frac{78.96 \times 10^{-3} U_r I_b R_{ir} (R^2 - R_{ir}^2)}{R^4} \tag{12.12}$$

It may be noted that the residual magnetization of steel, remaining after an alternating magnetizing current is switched off, bears no simple theoretical relationship to the magnetization during passage of a direct current of the same effective value, as established.

12.8 Determination of Time Span for the Appearance of Flutes on the Track Surfaces

Instant thermal stresses due to thermal transients on the roller track of races caused by the roller contact under the influence of electric current depends on the instant rise in temperature. As the temperature-rise stabilizes, the contact thermal stresses increase and affect the fatigue life. The duration, the bearing, would have taken after the slip bands formation before the appearance of flutes/corrugations on the track surface of inner race is determined as [23]

$$t = \frac{(\pi \partial_{\mathrm{corr}})^2 (BD_{\mathrm{ir}} + BD_{\mathrm{or}} + LdN)(BD_{\mathrm{ir}}\Delta_{\mathrm{ir}}^{-1} + BD_{\mathrm{or}}\Delta_{\mathrm{or}}^{-1} + LdN\Delta_{\mathrm{r}}^{-1})}{2E_{\mathrm{rir}} I_{\mathrm{b}} Y} \tag{12.13}$$

12.9 Data Analysis

All the failed bearings both on NDE and DE ends of the three motors were examined. In these bearings the damage as reported in Section 12.3 was studied. Investigations were carried out on the NU 228 bearings and the flux density distribution on their track surfaces was detected using magnetic flux monitor similar to procedure reported by Prashad [6]. The pitch of corrugations on the track surfaces of the NU 228 bearing was measured. The theoretical time span for the formation of corrugations after the slip band formation was determined using Eqn. (12.13). The dimensional and operating parameters of bearing type NU 228 and various determined values of measured/analytical parameters are given in Tables 12.1 and 12.2, respectively.

Table 12.1. Dimensional and operating parameters of NU 228 bearing

B	42 mm
R_{ir}	79 mm
R_{oro}	125 mm
R_{or}	116 mm
R	97.5 mm
d	37 mm
D	195 mm
D_{ir}	158 mm
D_{or}	232 mm
L/B	1
N	14
n	1500 rev. min^{-1}

Table 12.2. Experimental and theoretical data

E	1.6 V
R_c	0.20 ohm
E_{rir}	2.46 V
E_{rir}/E	1.54
I_b	12.30 A
Y	210×10^9 N m^{-2}
B_{ir}	3.6 gauss (theoretical)
B_{ir}	5 gauss (maximum on measurement)
B_r	3 gauss (experimental on a few rolling-elements)
B_r	2.7 gauss (theoretical)
Δ_{ir}	0.25 mm
Δ_{or}	Not detected
Δ_r	Not detected
∂_{corr}	700×10^6 N m^{-2}
t	107.45 h
t_1	300 h

12.10 Results and Discussion on the Localized Current, the Flux Density, Time Span and Mechanism of Bearing Failure

12.10.1 Potential Difference between Rolling-Elements and Track Surface of Inner Race, and Flow of Localized Current Between Them

The potential difference between the rolling-elements and track surface of inner race of an insulated bearing working under the influence of shaft voltage depends on the shaft voltage, the pitch diameter and diameter of rolling-elements as per the derived relations in Eqn. (12.6). The bearing type NU 228, working under the influence of shaft voltage of 1.6 V, develops potential difference of 2.46 V between the rolling-elements and the track surface of the inner race. The ratio of this potential difference to the shaft voltage, E_{rir}/E_r, has been worked out as 1.54 (Table 12.2). This ratio depends on the speed of rotation of rolling-elements to the shaft speed and is a function of the dimensional parameters of a roller bearing, Eqn. (12.3). Furthermore, localized electric current between the track surface of the inner race and rolling-elements depends on the potential difference (E_{rir}) and contact resistance (R_c). It is a function of the shaft voltage, E, the contact resistance, R_c and dimensional parameters of a bearing [Eqn. (12.7)]. The intensity of local current, theoretically has been determined as 12.30 A for the existing shaft voltage of 1.6 V and resistance, R_c, of 0.2 ohm (Table 12.2). The localized electric current damages the track surfaces of the bearing in due course and initiates to form corrugations similar to the figures shown in Chapter 5. On the contrary, for the shaft voltages of 500 and 200 mV, the value of local current has been determined as 3.84 and 1.54 A, respectively. These low values of current do not affect the bearing during operation. In addition, the outer race being stationary, significant potential difference is not generated between the outer race and rolling-elements, and thus current does not pass between them. However, momentarily circular current, through an arc of the outer race can flow as and when proximity contact of the outer race and rolling-elements takes place in a sector due to instabilities/vibrations during operation.

12.10.2 Residual flux Density Distribution on the Track Surface of Inner Race and Rolling-Elements

Flow of current between the track surface of inner race and rolling-elements due to the existing potential drop leads to generate the residual flux density distribution on the track surface of the inner race and rolling-elements in due course. The flux density on the track surface of the inner race has been determined theoretically as 3.6 gauss, using Eqn. (12.8). The maximum value of flux density was determined as 5 gauss by measurement (Table 12.1). On a few rolling-elements the flux density was determined and found to be maximum of 4 gauss. Theoretically, the flux density was determined as 2.7 gauss on rolling-elements using Eqn. (12.9). The presence of flux density by measurement and theoretical investigations indicates that the localized current has passed through the bearing arc partially [6]. This has created corrugations/flutes on the inner-race track surface of the bearings and deteriorated the lubricant, which has corroded the surfaces of bearings and led to failure by the reduction of fatigue life [10, 12, 24].

12.10.3 Assessment of Time Span Before the Appearance of Flutes on the Track Surfaces

As a result of the shearing of atomic planes within the crystals, some crystals on the roller track of the inner or outer race develop slip bands. The slip bands are formed in the sub-surface prior to the appearance of flutes/corrugations on the track surfaces. The formation of slip bands is initiated by shear stress caused by operating parameters, and is accelerated by the passage of electric current, the corrosion and oxidation of track surfaces. Furthermore, the formation of slip bands depends on lubricant characteristics and quality of a bearing. The following may take place before the initiation of corrugations:

1. Generation of persistent slip bands (PSB);
2. Crack form along PSB and initiated from the tip of slip bands, and
3. Formation and propagation of flutes/corrugations on the track surface.

After the cracks/slip bands under the track surfaces are initiated, the process that governs the propagation leading to the formation of corrugations/flutes is considered by the continuum theory of Griffith [10, 23].

The time span, a bearing takes before the formation of flutes on the track surfaces after the slip bands formation depends on the dimensions of the track surface, the pitch of corrugations, the number of rolling-elements, the potential difference between the track surface and rolling-elements, the intensity of local current and the properties of the bearing material. The pitch of corrugations on the rolling-elements and outer race were not found and hence the values of Δ_r^{-1}, Δ_{or}^{-1} are negligible, Eqn. (12.13). This time span, t, for the initiation of corrugations is determined as 107.45 h, using Eqn. (12.13), without considering the time for the formation of slip bands (Table 12.2). The net time for the development of slip bands including that of flute formation on track surface of the inner race after commissioning is found to be approximately 300 h according to site data as given in Table 12.2. This might match practically with the actual time of initiation of flutes after the formation of slip bands, as determined, analytically.

12.10.4 Bearing Failure Under Localized Current

Electric current damage of a bearing is of two types. In the first type, when the low-resistivity lubricant (<10^5 ohm m) is used in the roller bearings, a silent discharge occurs through the bearing elements under the influence of electric current. This decomposes the used lubricant and corrodes the surfaces of the bearings, which lowers the fatigue life of the bearings as discussed by Prashad [3, 12]. Furthermore, passage of current through a bearing increases its operating temperature. Subsequently, thermal stresses are increased and surface heating takes place, which leads to low-temperature tempering of the track surfaces. This accelerates the formation of slip bands and corrugations on the track surfaces in due course [10, 18]. Magnetic flux density is also developed on the track surfaces, and original structure of the lubricant undergoes changes [12, 6].

In the second type of bearing failure, when the high-resistivity lubricant (>10^9 ohm m) is used in the roller bearings, an accumulation of charges occurs on the track surfaces until it reaches a threshold critical value when breakdown takes place. This leads to damage of the track surface caused by arcing. This is accompanied by mass transfer and elevated local temperature on the asperity of the contact surfaces [4, 25].

12.11 Conclusions

Based on the above analysis, the following conclusions are drawn [26]:

1. Under the influence of higher shaft voltage, a rolling-element bearing using low-resistivity lubricant, may deteriorate due to the development of potential difference between track surface of inner race and rolling-elements. This leads to the passage of localized current depending on the contact resistance between them.
2. The potential difference between the track surface of inner race and rolling-elements develops because of the higher frequency of rotation of rolling-elements as compared to the shaft speed. The potential difference, thus developed, depends on the shaft voltage and bearing kinematics.
3. The ratio of potential difference between the track surfaces of inner race and rolling-elements to that of the shaft voltage is a function of pitch and rolling-elements diameters of a bearing. For the NU 228 bearing, it is determined as 1.54.
4. The development of magnetic flux density on the track surface of the inner race and rolling-elements indicates flow of locally generated current between them.
5. The time of the appearance of flutes on the track surface can be estimated by, the bearing kinematics, the existing potential difference between the track surface of inner race and rolling-elements, the value of localized current, the properties of bearing material together with measured values of pitch of the corrugations on the track surfaces.
6. The failure of a bearing by the localized current can be avoided by limiting the shaft voltage to a maximum of 200 mV. In case a higher value of shaft voltage persists, it is preferable to ground a brush. However, sometimes by altering the path of a flow of circular current through dismantling and reassembling of the bearing as a means of modulation of track surfaces under interaction, the effect of localized current can be minimized.
7. Study of localized current in rolling-element bearings is a complex phenomenon and needs solution of complex analytical models and investigations.

REFERENCES

1. Prashad, H. (1994). Investigations of damaged rolling-element bearings and deterioration of lubricants under the influence of electric current. *Wear*, **176**, 151–161.
2. Bradford, M. (1984). Prediction of bearing wear due to shaft voltage in electrical machines. Technical Report No. 84–007, ERA Technology Limited, England, 49–53.

3. Prashad, H. (1988). Investigations on corrugated pattern on the surface of roller bearings operated under the influence of electrical fields. *STLE Lubric. Eng.*, **44**(8), 710–718.
4. Prashad, H. (1987). Effects of operating parameters on the threshold voltages and impedance response of non-insulated rolling-element bearings under the action of electric current. *Wear*, **117**, 223–240.
5. Busse, D., Erdman, J., Kerkman, R. J., et al. (1997). System electrical parameters and their effects on bearing currents. *IEEE Trans. Ind. Appl.*, **33**(2), 577–584.
6. Prashad, H. (1999). Determination of magnetic flux density on the surfaces of rolling-element bearings as an indication of the current that has passed through them—an investigation. *Tribol. Int.*, **32**, 455–467.
7. Morgan, A. W. and Wyllie, D. (1969–1970). A survey of rolling bearing failures. *Proc. Inst. Mech. Eng.*, **184**(Pt 3F), 48–56.
8. Andreason, S. (1968). Passage of electric current through rolling bearings. *Ball Bearing J.*, **153**, 6–12.
9. Simpson, F. E. and Crump, W. J. J. (1963). Effects of electric currents on the life of rolling contact bearings. Proceedings of the Lubrication and Wear Convention Bournmouth, Institution of Mechanical Engineers, London, Paper 27, 296–304.
10. Prashad, H. (1989). Analysis of the effects of an electric current on contact temperature contact stresses and slip band initiation on the roller tracks of roller bearings. *Wear*, **131**, 1–14.
11. Komatsuzaki, S. (1987). Bearing damage by electrical wear and its effects on deterioration of lubricating grease. *STLE Lubric. Eng.*, **43**(1), 25–30.
12. Prashad, H. (1989). Diagnosis of deterioration of lithium greases used in rolling-element bearings by X-ray diffractometry. *STLE Tribol. Trans.*, **32**(2), 205–214.
13. Remy, M. and Magnin, A. (1996). Rheological and physical studies of lubricating greases before and after use in bearings. ASME, *J. Tribol.*, **118**, 681–686.
14. Lin, C. M., Chiou, Y. C. and Lee, R. T. (2001). Pitting mechanism on lubricated surface of Babbitt alloy/bearing steel pair under A.C. electric field. *Wear*, **249**, 133–142.
15. Chiou, Y. C. Lee, R. T. and Lin, C. M. (1999). Formation criterion and mechanism of electrical pitting on the lubricated surface under A.C. electrical field. *Wear*, **236**, 62–72.
16. Winder, L. R. and Wolfe, O. J. (1968). Valuable results from bearing damage analysis. *Metal Progr.*, **4**, 52–59.
17. Prashad, H. (1996). Diagnosis of failure of rolling-element bearings of alternators – a study. *Wear*, **198**, 46–51.
18. Prashad, H. (1998). The effect of current leakage on electro-adhension forces in rolling-friction and magnetic flux density distribution on the surface of rolling-element bearing. ASME, *J. Tribol.*, **110**, 448–455.
19. Prashad, H., Ghosh, M. and Biswas, S. (1985). Diagnostic monitoring of rolling-element bearing by high-frequency resource technique. *ASLE Trans.*, **28**(4), 439–448.
20. Prashad, H. (1999). Diagnosis and cause analysis of rolling-element bearing failure in electrical power equipment due to current leakage. *STLE Lubric. Eng.*, **55**(5), 30–35.

21. Prashad, H. (1987). The effect of cage and roller slip on the measured defect frequency reference of rolling-element bearings. *ASLE Trans.*, **30**(3), 360–366.
22. Starling, S. G. (1960). *Electricity & Magnetism*, 7th ed., Longmans, Green and Co., New York.
23. Prashad, H. (1998). Determination of time span for appearance of flutes on track surface of rolling-element bearings under the influence of electrical current. *Tribol. Trans.*, **41**(1), 103–109.
24. Prashad, H. and Murthy, T. S. R. (1988). Behavior of greases in statically bounded conditions and when used in non-insulated anti-friction bearings under the influence of electrical fields. *STLE Lubric. Eng.*, **44**(3), 239–246.
25. Prashad, H. (1990). Theoretical analysis of the effects of instantaneous charge leakage on roller track of roller bearings lubricated with high-resistivity lubricants. ASME, *J. Tribol.*, **112**, 37–43.
26. Prashad, H. (2002). Diagnosis of rolling-element bearings failure by localized current between track surfaces of races and rolling-elements. ASME, *J. Tribol.*, **124**, 468–473.

Nomenclature

B	width of track surface
B_{ir}	magnetic flux density on track surface of inner race
B_{or}	magnetic flux density on track surface of outer race
B_{r}	magnetic flux density on rolling-elements
d	diameter of rolling-element
D	pitch diameter
D_{ir}	outside diameter of inner race
D_{or}	inner diameter of outer race
E	shaft voltage
E_{r}	voltage on rolling-elements
E_{ir}	voltage on track surface of rotating inner race
E_{rir}	potential difference between rolling-elements and track surface of inner race
f_{s}	shaft rotational frequency
f_{b}	rolling-element frequency
H_{irr}	field strength on track surface of inner race due to flow of current in rolling-elements
H_{rir}	field strength on surface of rolling-elements due to flow of current in inner race
I_{b}	bearing current
L	length of rolling-element
n	rpm

N	number of rolling-elements in a bearing
R	pitch radius of bearing
R_c	bearing resistance
R_{ir}	track radius of inner race
R_{or}	track radius of outer race
R_{oro}	outside radius of outer race
t	time required for appearance of flutes after slip bands formation on track surfaces
t_1	net time of bearing operation after commissioning for inspection
V_{bs}	ratio of speed of rolling-elements and shaft speed (f_b/f_s)
Y	Young's modulus of elasticity
U_r	relative permeability oil with respect to free space (=1)
U_o	permeability of free space (4×10^{-7} henry m^{-1})
α	contact angle
∂_{corr}	stress for flute appearance from opening of the tip of flute from an existing slip band on the track surface
Δ_{ir}, Δ_{or}, Δ_r	pitch of corrugation on inner race, outer race and rolling-element, respectively.

Chapter 13

ALTERNATIVE APPROACHES TO DETERMINATION OF STIFFNESS OF ROLLING-ELEMENT BEARINGS

13.1 General

An alternative approach has been developed to determination of the effective stiffness of rolling-element bearings through evaluation of electrical parameters under different operating conditions. These parameters have been theoretically evaluated using width of deformation on the track surface of races, minimum film thickness, lubricant characteristics and bearing geometry. Besides this, by applying another developed alternative approach also, based on the inverse of the rate of change of width of bearing deformation on the track surface of races with applied load, the effective stiffness of a roller bearing has been assessed approximately.

The effective stiffness has been determined at different loads and speeds of operation using electrical analogy, and by approximate method using width of deformation under different loads. The values of effective stiffness by these developed alternative procedures, thus determined, have been compared with that of the values determined by existing conventional procedure, and found to have matching trend.

13.2 Introduction

In rotating rolling-element bearings, the actual surface area of contact depends on surface roughness and primarily comprises of 'asperity contacts'. The bearing performance is improved if the film thickness is 2–3 times more than the average roughness depth. The minimum film thickness between races and rolling-elements offers maximum capacitance and minimum capacitive reactance, depending on the permittivity of the lubricant. On the contrary, the active resistance offered by a bearing is minimum at the minimum film thickness; however, it is primarily governed by the resistivity of the lubricant. The interaction between races and rolling-elements in presence of oil film is like a resistor-capacitor (RC) circuit, with the oil film offering an impedance to current flow [1].

In general, the effect of electric current on contact temperature, slip bands initiation, and life of rolling contact bearings has been investigated [2]. In addition, analysis has been made on corrugation patterns, and the theoretical evaluation of capacitance, resistance, impedance and instant charge accumulation on the surfaces of rolling-element bearings operated under the influence of electric current has been reported [3, 4].

In most cases, it is possible to represent a mechanical system by an electrical circuit. This analogy between the electrical and mechanical systems exists because of the similarity of the differential equations, which represent them [5]. Using this analogy, equivalent dynamic coefficients of two-lobe and three-lobe journal bearings have been determined [6, 7].

This chapter discusses and gives the theoretical procedure to determine the effective stiffness of rolling-element bearings based on the evaluated electrical parameters under different operating conditions. Damping of a bearing has been assessed by electrical analogy. The effective stiffness has also been determined approximately on considering the reverse of rate of change of width of deformation on the track surface of races with applied load on a rolling-element bearing. The effective stiffness of a rolling-element bearing, thus determined, by these developed approaches has been compared with that of the existing technique [8, 9].

13.3 Theoretical Analysis to Determination of Stiffness of Roller Bearing

13.3.1 Determination of Width of Deformation

Rollers of different diameters of the same material in contact under load form theoretically a rectangular surface of deformation of a length equal to the length of the rollers. Width of deformation on these rollers has been determined in Chapter 5, and is a function of material characteristics, their dimensions and load on the rollers and is given as [3]

$$W_{\mathrm{ir}} = 2.15\left[\frac{Pd(D-d)}{pN_{\mathrm{z}}ELD}\right]^{0.5} \tag{13.1}$$

and

$$W_{\text{or}} = 2.15\left[\frac{Pd(D+d)}{pN_zELD}\right]^{0.5} \tag{13.2}$$

13.3.2 Determination of Bearing Capacitance and Resistance

The capacitance and resistance have been determined between track surface of inner race and a roller, as well as track surface of outer race and a roller in Chapter 6, and given as [4]

$$C_{\text{ir}} = \frac{2\xi L}{(\beta_{\text{ir}}h_{\text{o}})^{1/2}} \tan^{-1}\left[0.5\, W_{\text{ir}}\left(\frac{\beta_{\text{ir}}}{h_{\text{o}}}\right)^{1/2}\right] \tag{13.3}$$

and

$$C_{\text{or}} = \frac{2\xi L}{(\beta_{\text{or}}h_{\text{o}})^{1/2}} \tan^{-1}\left[0.5 W_{\text{or}}\left(\frac{\beta_{\text{or}}}{h_{\text{o}}}\right)^{1/2}\right] \tag{13.4}$$

as well as

$$R_{\text{i}} = \frac{\rho(\beta_{\text{ir}}h_{\text{o}})^{1/2}}{2L\tan^{-1}\left[W_{\text{ir}}/2(\beta_{\text{ir}}/h_{\text{o}})^{1/2}\right]} \tag{13.5}$$

and

$$R_{\text{o}} = \frac{\rho(\beta_{\text{or}}h_{\text{o}})^{1/2}}{2L\tan^{-1}\left[W_{\text{or}}/2(\beta_{\text{or}}/h_{\text{o}})^{1/2}\right]} \tag{13.6}$$

where

$$\beta_{\text{ir}} = 0.5\,[r^{-1} + R_{\text{ir}}^{-1}] \quad \text{and} \quad \beta_{\text{or}} = 0.5\,[r^{-1} - R_{\text{or}}^{-1}] \tag{13.7}$$

The minimum film thickness (h_{o}) is determined by the Grubin's formula. The equivalent capacitance and resistance of a roller bearing are determined from the bearing's equivalent circuit. It is evident that the equivalent circuit of inner as well as outer races with rollers are RC circuits in parallel, whereas inner and outer-race circuits are in series to each other, considering the flow of current from shaft to the bearing housing as shown in Fig. 13.1. Resistance and capacitance between inner race and a roller as well as outer race and a roller are considered in parallel to each other for evaluation of impedance, capacitive reactance and equivalent capacitance of races and a roller (refer Fig. 13.1).

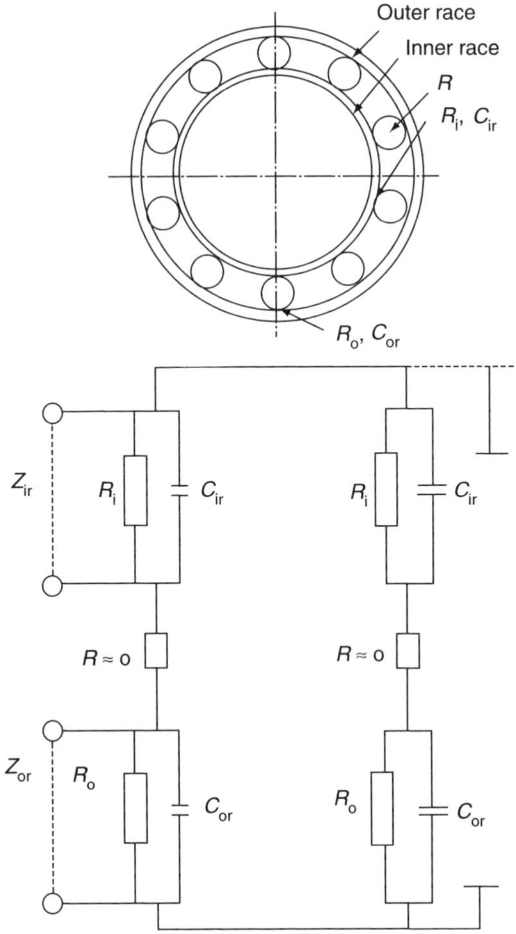

Fig. 13.1. Electrical equivalent bearing circuit

The following equations used for evaluation of equivalent capacitance and resistance:

$$Z_{\mathrm{ir}} = \frac{jX_{\mathrm{ir}}R_{\mathrm{i}}}{R_i + jX_{\mathrm{ir}}} \tag{13.8}$$

and

$$Z_{\mathrm{or}} = \frac{jX_{\mathrm{or}}R_{\mathrm{o}}}{R_{\mathrm{o}} + jX_{\mathrm{or}}} \tag{13.9}$$

where

$$X_{\text{ir}} = (wC_{\text{ir}})^{-1} \quad \text{and} \quad X_{\text{or}} = (wC_{\text{or}})^{-1} \tag{13.10}$$

Since

$$Z_{\text{r}} = Z_{\text{ir}} + Z_{\text{or}} \tag{13.11}$$

which gives

$$R_{\text{r}} = \frac{X_{\text{ir}}^2 X_{\text{or}}^2 (R_{\text{i}} + R_{\text{o}}) + R_{\text{i}} R_{\text{o}} (X_{\text{or}}^2 R_{\text{i}} + X_{\text{ir}}^2 R_{\text{o}})}{(R_{\text{i}} R_{\text{o}} - X_{\text{ir}} X_{\text{or}})^2 + (X_{\text{or}} R_{\text{i}} + X_{\text{ir}} R_{\text{o}})^2} \tag{13.12}$$

and

$$X_{\text{r}} = \frac{R_{\text{i}}^2 R_{\text{o}}^2 (X_{\text{ir}} + X_{\text{or}}) + X_{\text{ir}} X_{\text{or}} (R_{\text{i}}^2 X_{\text{or}} + R_{\text{o}}^2 X_{\text{ir}})}{(R_{\text{i}} R_{\text{o}} - X_{\text{ir}} X_{\text{or}})^2 + (X_{\text{or}} R_{\text{i}} + X_{\text{ir}} R_{\text{o}})^2} \tag{13.13}$$

This leads to determine capacitance between races and a roller as

$$C_{\text{r}} = (wX_{\text{r}})^{-1} \tag{13.14}$$

Assuming that the rollers in the loaded zone (N_{z}) of a bearing under inner and outer races are equally loaded, the equivalent capacitance of a bearing is given as [4]

$$C_{\text{b}} = N_{\text{z}} C_{\text{r}} \tag{13.15}$$

Similarly, equivalent resistance of a bearing is given as [4]

$$R_{\text{b}} = \frac{R_{\text{b}}}{N_{\text{z}}} \tag{13.16}$$

13.3.3 Stiffness of Roller Bearing by Electrical Analogy

An electrical circuit with components including inductance, resistance and capacitance under the influence of voltage correspond to equation representing a mechanical system with damping and stiffness characteristics with a single degree of freedom [5, 6]. These equations imply a voltage–force analogy, in which, capacitance (C_{b}) in farads and resistance (R_{b}) in ohms of a bearing correspond to equivalent stiffness^{-1}(K_{qe})$^{-1}$ in m N^{-1} and damping in N s m^{-1}, respectively [6]. Also, bearing stiffness due to bending moment and radial force between rollers and outer race [since, K_{r} and K_{m} are parallel to each other (refer Fig. 13.2)], and between rollers and inner race is given as

$$N_{\text{z}} K_{\text{rm}} = N_{\text{z}}(K_{\text{r}} + K_{\text{m}}) = K_{\text{qe}}, \quad \text{and} \quad N_{\text{z}} K_{\text{r}} = K_{\text{qe}} \tag{13.17}$$

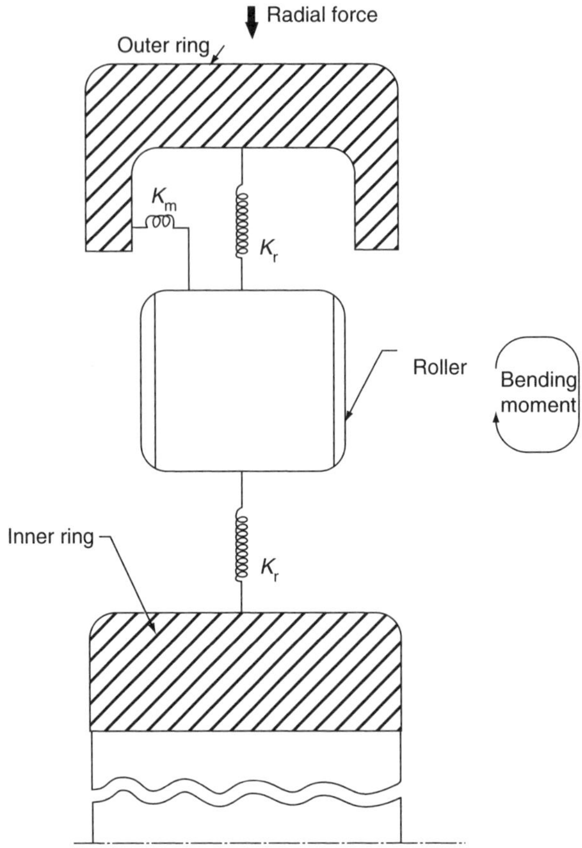

Fig. 13.2. Dynamic model of a roller bearing

In case of a rolling bearing, resistance between races and rollers (R_b) is less because of very small oil film thickness existing between rolling-elements and track surface of races in the loaded zone. Furthermore, periodic asperity contacts between the races and rolling-elements under different conditions of operation in rotation reduce the bearing resistance. Consequently, by electrical analogy, rolling bearings show poor damping characteristics. On the contrary, stiffness characteristics play a very significant role in such bearings. This is because of variation in capacitance, which in turn vary inverse of the film thickness between the rolling-elements and track surface of races (Chapter 6).

It may be noted that the electrical analogy is geometrically based on the solutions for charge distribution along potential and isostatic surfaces, with an assumed analogy between force and voltage. The solution satisfies continuity. Ratios of solutions eliminate the need for heuristic data such as resistivity, dielectric constant, etc.; thus any value will suffice, as these values are not required for comparisons provided the lubricant conditions remain constant. Solutions to electrical equations provide potential lines and iso-surfaces. Hence, the electrical analogy can be used as an alternative to conventional analysis, as a potential tool for relative comparison in addition to the absolute [6].

13.3.4 Effective Stiffness and Dynamic Model of a Roller Bearing

A roller bearing under rotation is generally loaded by radial force and bending moment. Consequently, bearing rings have displacement in radial direction and to the direction of bending moment, and have stiffness (K_r and K_m) due to the forces acting in these directions. Figure 13.2 shows the dynamic model of a roller bearing. Under the effect of K_r and K_m, roller bearing dynamic model become very complex for analysis. Furthermore, mutual interaction of K_r and K_m makes it mathematically more complex. However, K_m, being very small in a roller bearing, can be neglected for determination of effective stiffness (K_{qef}). Hence from Eqn. (13.17) for outer race and rollers, it is evident that

$$K_r N_z = K_{qe} \quad (\text{since } K_m = 0) \tag{13.18}$$

Since, rollers and races are in series to each other for the flow of current from one race to another through rollers, so the effective bearing stiffness (K_{qef}), i.e. the product of radial stiffness and number of rollers in the loaded zone ($K_r N_z$), considering the net effect of radial force on the inner and outer races through rollers can be determined as

$$2(K_r N_z)^{-1} = 2(K_{qe})^{-1} = (K_{qef})^{-1}, \quad \text{or} \quad K_{qef} = K_{qe}/2 \tag{13.19}$$

13.3.5 Assessment of Approximate Stiffness of Roller Bearings

The stiffness, a common design parameter for oil film bearing, is inverse of the rate of change of bearing deflection with applied load. The maximum elastic contact deformation and width of contact deformation is dependent on the rolling-element load as discussed in Chapter 5 [3, 9]. The principal bearing deflections, which affect the stiffness, are radial, axial and moment loads applied simultaneously require complex analysis. Considering the width of deflections/deformations on the track surface of inner and outer races by p positions of plane of loading, the stiffness of inner and outer races can be determined, approximately, by differentiating Eqns. (13.1) and (13.2), which yield

$$K_{\text{ir}} = \frac{dP}{dW_{\text{ir}}} = 0.43 \times \frac{W_{\text{ir}} p N_z ELD}{d(D - d)} \tag{13.20}$$

and

$$K_{\text{or}} = \frac{dP}{dW_{\text{or}}} = 0.43 \times \frac{W_{\text{or}} p N_z ELD}{d(D + d)} \tag{13.21}$$

Since K_{ir} and K_{or} are in series and rolling-elements are in parallel to each other with respect to the applied load. Thus, the approximate effective bearing stiffness can be determined as

$$K_{\text{qwf}} = \frac{Z K_{\text{ir}} K_{\text{or}}}{K_{\text{ir}} + K_{\text{or}}} \tag{13.22}$$

13.3.6 Stiffness by Conventional Approach

In lieu of the more rigorous approach to bearing deflection, Palmgren [9] has developed series of formulas to calculate bearing deflection for specific conditions of loading. For a radial roller bearing for slow and moderate speed with line contact at each raceway, the radial deflection is given as [8]

$$\delta_{\text{r}} = \frac{8.71 \times 10^{-7} P_{\max}^{0.9}}{L^{0.8} \cos\alpha} \tag{13.23}$$

where

$$P_{\max} = \frac{5P}{Z\cos\alpha} \tag{13.24}$$

This gives roller bearing stiffness as [9]

$$K_{\text{qp}} = \frac{dP}{d\delta_{\text{r}}} = 3 \times 10^5 Z^{0.9} L^{0.8} P^{0.1} \cos^{1.9}\alpha (1\text{bin.}^{-1}) \tag{13.25}$$

13.3.7 Theoretical Data at Different Operating Parameters

The equivalent capacitance of the NU 330 roller bearing has been computed using Eqns. (13.3)–(13.15) for different parameters of operation given in Table 13.1. The effective stiffness at different loads and speeds of the bearing has also been computed using electrical analogy and given in Table 13.1. The number of rolling-elements in the loaded zone has been taken considering the load distribution in the bearing [8]. The number of rolling-elements in the loaded zone vary depending on

Table 13.1. Capacitance, resistance and stiffness of roller bearing type NU 330 at different operating parameters

Speed (rpm)		450			750			1200	
P (N)	4500	7500	10^4	4500	7500	10^4	4500	7500	10^4
W_{ir} (mm)	0.163	0.210	0.243	0.163	0.210	0.243	0.163	0.210	0.243
W_{or} (mm)	0.195	0.252	0.291	0.195	0.252	0.291	0.195	0.252	0.291
$h_o \times 10^{-4}$ (mm)	22.4	21.2	20.70	32.5	30.8	30.1	45.7	43.4	42.3
$C_{ir}(p\phi)$	72.4	98.7	116.9	49.9	67.9	80.4	35.5	48.2	57.9
$C_{ir}(p\phi)$	72.4	98.7	116.9	49.9	67.9	80.4	35.5	48.2	57.9
$C_{or}(p\phi)$	86.6	118.4	140.1	59.7	81.4	96.5	42.5	57.8	69.6
$X_{ir} \times 10^7$ (ohm)	4.4	3.2	2.7	6.4	4.6	4.0	8.97	6.6	5.6
$X_{or} \times 10^7$ (ohm)	3.68	2.6	2.3	5.33	3.83	3.30	7.5	5.62	4.6
$R_i \times 10^7$ (ohm)	30.68	22.4	18.96	44.4	32.6	27.5	62.3	45.93	38.6
$R_o \times 10^7$ (ohm)	25.5	18.6	15.8	37.0	27.17	22.9	52.1	38.27	32.2
$R_r \times 10^7$ (ohm)	1.1	0.79	0.71	1.6	1.17	1.02	2.3	1.7	1.01
$R_b \times 10^7$ (ohm)	0.13	0.13	0.14	0.15	0.15	0.15	0.16	0.17	0.15
$X_r \times 10^7$ (ohm)	7.92	5.69	4.93	11.48	8.11	6.97	16.14	12.93	9.98
$C_r(p\phi)$	40.2	56.0	64.6	27.7	39.3	45.7	19.73	24.6	31.87
$C_b(p\phi)$	361.8	336.0	323.0	304.7	314.4	319.9	276.2	295.2	286.8
$K_{qe} \times 10^9 (\mathrm{N\,m^{-1}})$	2.76	2.98	3.1	3.28	3.18	3.13	3.62	3.39	3.49
$K_{qef} \times 10^9 (\mathrm{N\,m^{-1}})$	1.38	1.49	1.55	1.64	1.59	1.57	1.81	1.70	1.75
$K_{qp} \times 10^9 (\mathrm{N\,m^{-1}})$		1.8 (at 4500 N)			1.92 (at 7500 N)			1.97 (at 10^4 N	
$K_{qwf} \times 10^9 (\mathrm{N\,m^{-1}})$		0.4 (at 4500 N)			0.5 (at 7500 N)			0.6 (at 10^4 N)	

Kinematics: $d = 42\,\mathrm{mm}$, $R_{ir} = 96.5\,\mathrm{mm}$, $R_{or} = 138.5\,\mathrm{mm}$, $D = 235\,\mathrm{mm}$, $L = 45\,\mathrm{mm}$, $p = 1$, $Z = 14$, $\beta_{ir} = 0.020\,\mathrm{mm}^{-1}$, $\beta_{or} = 0.029\,\mathrm{mm}^{-1}$, $\xi = 2.5 \times 8.85419 \times 10^{-12} \phi\,\mathrm{m}^{-1}$, $\rho = 10^9$ ohm m, $n = 50$ Hz, $R = 0$ ohm (K_{qp} and K_{qwf} are without speed consideration).

load, bearing clearance and operating speed. The stiffness values, computed by the developed approximate procedure given in Eqns. (13.20)–(13.22), are shown in the same table along with the stiffness values, determined by the conventional approach using Eqns. (13.23)–(13.25).

13.3.8 Data Analysis and Discussion

13.3.8.1 Effective Stiffness by Electrical Analogy

The equivalent stiffness depends on inverse of the equivalent capacitance of a roller bearing, which in turn is a function of capacitive reactance (X_{ir} and X_{or}), resistance (R_i and R_o) and capacitance (C_{ir} and C_{or}) between a roller and track surface of inner and outer races, and number of rolling-elements in the loaded zone considering load distribution, clearance and operating speed [Eqns. (13.3)–(13.15)]. C_{ir} and C_{or} depend on lubricant permittivity, minimum film thickness, roller's radius and length, roller track radius of respective inner and outer races, and width of deformation on the track surface of respective races [Eqns. (13.3), (13.4) and (13.7)]. Similarly, R_i and R_o depend on the same parameters along with lubricant resistivity except that of the permittivity of the lubricant as shown in Eqns. (13.5)–(13.7). The equivalent stiffness (K_{qe}) is a complex function of stiffness due to radial load and bending moment (K_r and K_m), and depends on the dynamic model of a bearing (Fig. 13.2). The effective stiffness (K_{qef}) is established as 1/2 that of the equivalent stiffness (K_{qe}) as shown in Eqn. (13.19).

The effective stiffness (K_{qef}) of the NU 330 roller bearing vary between 1.38×10^9 and $1.81 \times 10^9\,\mathrm{N\,m^{-1}}$ under different combination of loads varying between 4500 and 10^4 N, and speeds varying between 450 and 1200 rpm, as shown in Table 13.1. With speed variation from 450 to 1200 rpm at 7500 and 10^4 N loads, the effective stiffness (K_{qef}) changes from 1.49×10^9 to $1.76 \times 10^9\,\mathrm{N\,m^{-1}}$, and from 1.55×10^9 to $1.75 \times 10^9\,\mathrm{N\,m^{-1}}$, respectively (Table 13.1). At constant speed with increase in load, effective stiffness is slightly changed. This occurs because of change of film thickness with load [8]. At 750 and 1200 rpm, with increase in load from 4500 to 10 000 N, K_{qef} changes from 1.64×10^9 to $1.57 \times 10^9\,\mathrm{N\,m^{-1}}$, and 1.81×10^9 to $1.75 \times 10^9\,\mathrm{N\,m^{-1}}$, respectively, as shown in Table 13.1. In general, variation of stiffness is non-linear with operating parameters. This occurs due to the variation of film thickness and lubricant characteristics with change in operating parameters.

From Table 13.1, it is evident that the values of equivalent bearing resistance (in ohms), i.e. bearing damping in $\mathrm{N\,s\,m^{-1}}$ [7] under different operating parameters is approximately 10^3 times less as compared to the bearing stiffness without considering the periodic asperity contacts between races and rolling-elements.

13.3.8.2 Stiffness by Developed Approximate Procedure Using Width of Deformation

The effective stiffness of a roller bearing by the developed approximate procedure depends on the bearing dimensional parameters, the material properties, the bearing load and the number of rolling-elements in the loaded zone and the total number of rolling-elements in a bearing as shown in Eqns. (13.20)–(13.22). The operating speed of bearing has not been considered for determination of stiffness by this method. The values of the effective stiffness (K_{qwf}) vary from 0.40×10^9 to $0.6 \times 10^9\,\mathrm{N\,m^{-1}}$ as the load changes from 4500 to 10^4 N as against 1.38×10^9 to $1.55 \times 10^9\,\mathrm{N\,m^{-1}}$ under similar load conditions, but at 450 rpm of rotation, using electrical analogy approach. The difference in stiffness values, thus determined, is attributed to the effect of operating parameters on the film thickness and load distribution in a bearing.

13.3.8.3 Stiffness by Conventional Procedure

The stiffness of a roller bearing under radial load as worked out by Palmgren [9] [Eqn. (13.25)] is a function of the number and length of rollers, and load acting on them. Under different loads varying between 4500 and 10^4 N, stiffness of NU 330 roller bearing vary between 1.8×10^9 and $1.97 \times 10^9\,\mathrm{N\,m^{-1}}$. Under similar load conditions at 450 rpm of rotation using developed electrical analogy approach, stiffness vary between 1.38×10^9 and $1.55 \times 10^9\,\mathrm{N\,m^{-1}}$ (Table 13.1). However, at 1200 rpm at 4500 N load, the stiffness is determined as $1.81 \times 10^9\,\mathrm{N\,m^{-1}}$. The difference in stiffness values, determined by these methods can be attributed to the effect of operating parameters and load distribution. The speed of operation of a bearing affects the dynamic load and film thickness, which in turn influence the bearing stiffness. However, in general, the values of stiffness determined by the conventional method and electrical approach match within close limits. Also, the procedure given by Palmgren and the developed approximate method using width of deformation indicate identical behavior pattern of the change of effective stiffness under identical conditions (Table 13.1).

13.3.9 Conclusions

The major conclusions based on the above analysis are [10]:

1. The inverse of capacitance in (farads) under various operating parameters can be taken as equivalent stiffness ($\mathrm{N\,m^{-1}}$) of a roller bearing.
2. The effective stiffness is a function of equivalent stiffness, and depends on dynamic model including that of the stiffness due to radial force and bending moment.

3. The capacitance/stiffness^{-1} of a roller bearing is a function of lubricant permittivity and resistivity, minimum film thickness, length and radius of roller, roller track radius, width of deformation on track surface and number of rolling-elements in the loaded zone.
4. The effective stiffness determined by the approximate method is a function of bearing dimensional parameters, bearing load, number of rolling-elements in a bearing and number of rolling-elements in the loaded zone.
5. The effective stiffness of a rolling-element bearing changes with operating parameters and load characteristics and its distribution.
6. In general, under identical operating conditions, the values of stiffness determined by conventional method and developed electrical approach match within proximity limits.
7. The developed approximate method using width of deformation and conventional method indicate identical pattern of the change of effective stiffness under identical operating conditions.

Nomenclature

C_r	equivalent capacitance between races and a roller
C_b	equivalent capacitance of roller bearing
C_{ir}, C_{or}	capacitance between inner race and a roller, and outer race and a roller, respectively
d	diameter of rolling-element
D	pitch diameter of roller bearing
E	Young's modulus of elasticity
h_o	minimum film thickness
K_{ir}	stiffness between inner-race track surface and a roller
K_{or}	stiffness between outer-race track surface and a roller
K_r	stiffness between races and a roller due to radial force
K_m	stiffness between races and a roller due to bending moment
K_{rm}	stiffness between races and a roller due to radial force and bending moment
K_{qe}	equivalent bearing stiffness by electrical analogy
K_{qef}	effective stiffness of roller bearing by electrical analogy
K_{qwf}	approximate effective stiffness determined by width of deformation
K_{qp}	stiffness by conventional existing procedure
L	length of roller
n	frequency
N_z	number of rollers in the loaded zone of bearing

p	plane of action of loading ($p = 1, 2, 3, \ldots$)
P	applied load
P_{max}	maximum radial load
r	radius of roller
R	resistance of rolling-element
R_{ir}, R_{or}	track radius of inner race and outer race in a radial plane, respectively
R_i	resistance between inner race and a roller
R_o	resistance between outer race and a roller
R_r	equivalent resistance between races and a roller
R_b	equivalent resistance of roller bearing
w	$2\pi n$
W_{ir}, W_{or}	width of deformation on inner and outer race, respectively
X_{ir}	capacitive reactance between inner race and a roller
X_{or}	capacitive reactance between outer race and a roller
X_r	equivalent capacitive reactance between races and a roller
Z	number of rolling-elements in a bearing
Z_{ir}	impedance between inner race and a roller
Z_{or}	impedance between outer race and a roller
Z_r	impedance between races and a roller
α	contact angle
β_{ir}	bearing inner race constant
β_{or}	bearing outer race constant
δ_r	deflection/contact deformation
ρ	resistivity of lubricant (ohm m)
ξ	permittivity/dielectric constant of lubricant ($\phi\,m^{-1}$)
ξ_r	relative permittivity of lubricant

13.4 Theoretical Analysis to Determination of Stiffness of Ball Bearing

13.4.1 Determination of Width of Deformation

Balls of different diameters of the same material in contact under load form a surface of deformation and have a specific radius of contact area. Width of deformation on these balls has been determined in Chapter 5, and is a function of material characteristics, their dimensions and load on the balls [11, 12].

The width to be covered by deformation/corrugation by p positions of plane of action of loading by the resultant load P by balls on the track surface of inner and outer races is determined in Chapter 5, and given as [12]

$$W_{\text{ib}} = 1.76\left[\frac{Pd(D-d)}{pN_zE(D-2d)}\right]^{1/3} \tag{13.26}$$

and

$$W_{\text{ob}} = 1.76\left[\frac{Pd(D+d)}{pN_zED}\right]^{1/3} \tag{13.27}$$

The width of deformation on the ball track of inner race and outer race is separated by a minimum film thickness (h_o). This forms a parallel plate capacitor between the track surface of races and a ball in presence of the lubricant of known dielectric constant (ξ). Also, resistance exists between track surface of races and a ball, depending on the resistivity of the used lubricant.

13.4.2 Determination of Bearing Capacitance and Resistance

The capacitance of parallel-plate capacitor formed by inner and outer race, and a ball of a ball bearing is maximum at the point of contact at the minimum film thickness, and reduces on both sides of the contact. On the contrary, resistance is minimum at the point of contact at the minimum thickness. The capacitance and resistance have been determined between track surface of inner race and a ball, as well as between the track surface of outer race and a ball [13] in Chapter 6, and are given as

$$C_{\text{ib}} = 2\pi\xi\beta_1\phi_1^{-1}\text{Log}_e\left(\frac{1+\phi_1 W_{\text{ib}}^2}{4h_o}\right) \tag{13.28}$$

and

$$C_{\text{ob}} = 2\pi\xi\beta_2\delta_1^{-1}\text{Log}_e\left(\frac{1+\delta_1 W_{\text{ob}}^2}{4h_o}\right) \tag{13.29}$$

as well as

$$R_{\text{ib}} = \frac{\rho\phi_1}{2\pi\beta_1 \log_e(1+\phi_1 W_{\text{ib}}^2/4h_o)} \tag{13.30}$$

and

$$R_{\text{ob}} = \frac{\rho\delta_1}{2\pi\beta_2 \log_e(1+\delta_1 W_{\text{ob}}^2/4h_o)} \tag{13.31}$$

where

$$\beta_1 = \left(\frac{\phi_1}{\phi_2}\right)^{1/2}, \quad \beta_2 = \left(\frac{\delta_1}{\delta_2}\right)^{1/2}$$

and

$$\begin{aligned} \phi_1 &= 0.5[r^{-1} + r_{\mathrm{i}}^{-1}], \quad \phi_2 = 0.5[r^{-1} - r_{\mathrm{ib}}^{-1}] \\ \delta_1 &= 0.5[r^{-1} - r_{\mathrm{o}}^{-1}], \quad \delta_2 = 0.5[r^{-1} - r_{\mathrm{ob}}^{-1}] \end{aligned} \tag{13.32}$$

and

$$r_{\mathrm{ib}} = \frac{d(D-d)}{D}, \; r_{\mathrm{ob}} = \frac{d(D+d)}{D+2d}$$

The minimum film thickness (h_{o}) is determined by the Grubin's formula [8, 9]. The equivalent capacitance and resistance of a ball bearing are determined from the bearing's equivalent circuit. It is evident that the equivalent circuit of inner as well as outer races with balls is RC circuits in parallel, whereas inner and outer-race circuits are in series to each other, considering the flow of current from shaft to the bearing housing, as shown in Fig. 13.3. Resistance and capacitance between inner race and a ball as well as outer race and a ball are considered in parallel for evaluation of impedance, capacitive reactance and equivalent capacitance of races and a ball (Fig. 13.3).

$$Z_{\mathrm{ib}} = \frac{jX_{\mathrm{ib}}R_{\mathrm{ib}}}{R_{\mathrm{ib}} + jX_{\mathrm{ib}}} \tag{13.33}$$

and

$$Z_{\mathrm{ob}} = \frac{jX_{\mathrm{ob}}R_{\mathrm{ob}}}{R_{\mathrm{ob}} + jX_{\mathrm{ob}}} \tag{13.34}$$

where

$$X_{\mathrm{ib}} = (wC_{\mathrm{ib}})^{-1} \quad \text{and} \quad X_{\mathrm{ob}} = (wC_{\mathrm{ob}})^{-1} \tag{13.35}$$

Since

$$Z_{\mathrm{b}} = Z_{\mathrm{ib}} + Z_{\mathrm{ob}} \tag{13.36}$$

which gives

$$R_{\mathrm{b}} = \frac{X_{\mathrm{ib}}^2 X_{\mathrm{ob}}^2 (R_{\mathrm{ib}} + R_{\mathrm{ob}}) + R_{\mathrm{ib}} R_{\mathrm{ob}} (X_{\mathrm{ob}}^2 R_{\mathrm{ib}} + X_{\mathrm{ib}}^2 R_{\mathrm{ob}})}{(R_{\mathrm{ib}} R_{\mathrm{ob}} - X_{\mathrm{ib}} X_{\mathrm{ob}})^2 + (X_{\mathrm{ob}} R_{\mathrm{ib}} + X_{\mathrm{ib}} R_{\mathrm{ob}})^2} \tag{13.37}$$

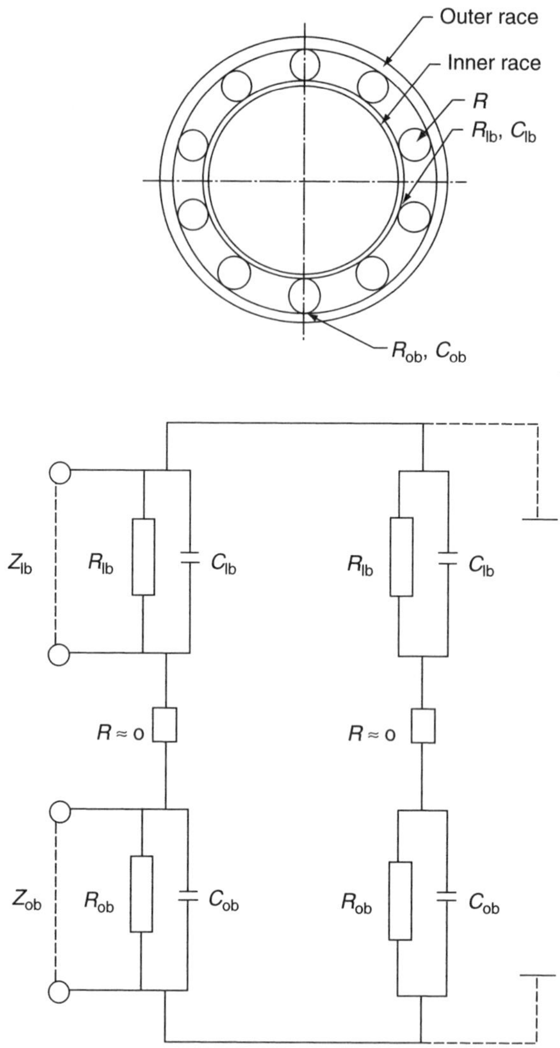

Fig. 13.3. Electrical equivalent ball bearing circuit

and

$$X_b = \frac{R_{ib}^2 R_{ob}^2 (X_{ib} + X_{ob}) + X_{ib} X_{ob} (R_{ib}^2 X_{ob} + R_{ob}^2 X_{ib})}{(R_{ib} R_{ob} - X_{ib} X_{ob})^2 + (X_{ob} R_{ib} + X_{ib} R_{ob})^2} \tag{13.38}$$

This leads to determination of capacitance between races and a ball as

$$C_b = (w X_b)^{-1} \tag{13.39}$$

Assuming that the balls in the loaded zone (N_z) of a bearing under inner and outer races are equally loaded, the equivalent capacitance of a bearing is given as [13]

$$C_r = N_z C_b \tag{13.40}$$

Similarly, equivalent resistance of a bearing is given as [12]

$$R_r = \frac{R_b}{N_z} \tag{13.41}$$

13.4.3 Determination of Stiffness of Ball Bearing by Electrical Analogy

An electrical circuit with components including inductance, resistance and capacitance under the influence of voltage correspond to equation representing a mechanical system with damping and stiffness characteristics with a single degree of freedom [5, 6]. These equations imply a voltage–force analogy, in which capacitance (C_r) in farads and resistance (R_r) in ohms of a bearing correspond to equivalent stiffness^{-1}(K_{qe})$^{-1}$ in m N^{-1} and damping (N s m^{-1}), respectively [6]. Equivalent stiffness (K_{qe}) can be expressed as the sum of radial stiffness, axial stiffness, bending moment between each race and a ball with the product of the number of balls in the loaded zone, i.e. stiffness of ball bearing due to radial and axial forces, and bending moment is given as

$$N_z K_{rma} = N_z(K_r + K_m + K_a) = K_{qe} \tag{13.42}$$

(Since K_r, K_m and K_a are parallel to each other, as shown in Fig. 13.4.)

In the case of a rolling bearing, resistance between races and balls is negligible because of very small oil film thickness existing between rolling-elements and track surface of races in the loaded zone [9]. Furthermore, periodic asperity contacts between the races and balls under different conditions of operation in rotation reduce the bearing resistance. Consequently, by electrical analogy, rolling-element bearings show poor damping characteristics. On the contrary, stiffness characteristics play a very significant role in such bearings. This is because of variation in capacitance, which in turn varies inversely with the film thickness between the rolling-elements and track surface of races [13].

13.4.4 Effective Stiffness and Dynamic Model of a Ball Bearing

A ball bearing under rotation is generally loaded by radial force, axial force and bending moment. Consequently, bearing rings have displacement in radial, axial and direction of the bending moment alongwith stiffness (K_r, K_a, K_m) due to the forces acting in these directions. Figure 13.4 shows dynamic model of a ball bearing. Under

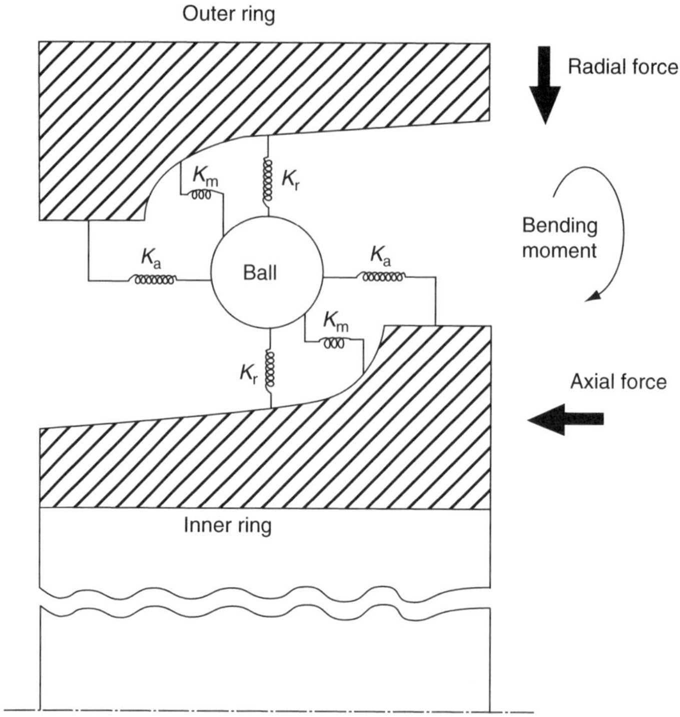

Fig. 13.4. Dynamic model of a ball bearing

the effect of K_r, K_a and K_m, ball bearing dynamic model becomes very complex for analysis. Furthermore, mutual interaction of K_r, K_a and K_m in these directions between inner race and balls, outer race and balls make it mathematically more complex as shown in Fig. 13.4 projecting the dynamic model of a ball bearing. Also, it is evident that K_r, K_a and K_m are in parallel to each other, and can be considered approximately equal in close proximity under the effect of small film thickness and stable operating conditions (Fig. 13.3). Hence, from Eqn. (13.42), it is evident that

$$N_z K_{rma} = 3N_z K_r = K_{qe} \quad (\text{since } K_r = K_m = K_a) \tag{13.43}$$

From Fig. 13.3, it is evident that K_r, K_m and K_a between inner race and a ball, and similarly between outer race and a ball are in series to each other, considering the flow of current between the races. However, individually, K_r, K_m and K_a are in parallel to each other. The effective stiffness (K_{qef}) of a ball bearing considering the net interaction between balls and inner and outer races due to the radial force, i.e.

product of radial stiffness and number of rolling-elements in the loaded zone ($N_z K_r$) can be determined as [Eqn. (13.43)]

$$3(K_{qe})^{-1} + 3(K_{qe})^{-1} = 2(N_z K_r)^{-1} = (K_{qef})^{-1}, \text{ or } K_{qef} = K_{qe}/6 \tag{13.44}$$

13.4.5 Assessment of Approximate Effective Stiffness of Ball Bearings

The stiffness, a common design parameter for oil film bearing, is inverse of the change of bearing deflection with applied load. The maximum elastic contact deformation and width of contact deformation is dependent on the rolling-element load [12, 9]. The principal bearing deflections which affect the stiffness are radial and axial loads applied simultaneously, besides bending moment, require complex analysis. Considering the width of deflections/deformations on the track surface of inner and outer races by p positions of plane of loading, the stiffness of inner and outer races and a ball can be determined, approximately, by differentiating Eqns. (13.26) and (13.27), which yield

$$K_{ib} = \frac{dp}{dW_{ib}} = \frac{0.55 W_{ib}^2\, pN_z E(D - 2d)}{d(D - d)} \tag{13.45}$$

and

$$K_{ob} = \frac{dP}{dW_{ob}} = \frac{0.55 W_{ob}^2\, pN_z ED}{d(D + d)} \tag{13.46}$$

Since K_{ib} and K_{ob} are in series and rolling-elements are in parallel to each other with respect to the applied load on shaft, the approximate net/effective bearing stiffness can be determined as

$$K_{qw} = \frac{ZK_{ib}K_{ob}}{K_{ib} + K_{ob}} \tag{13.47}$$

13.4.6 Determination of Bearing Stiffness by Conventional Approach

In lieu of the more rigorous approach to bearing deflection, Palmgren [9] has developed series of formulas to calculate bearing deflection for specific conditions of loading. For slow and moderate speed, deep-groove and angular-contact ball bearing subjected to radial load, which causes only radial deflection, that is [8]

$$\delta_r = \frac{1.58 \times 10^{-5} P_{max}^{2/3}}{d^{1/3} \cos\alpha} \tag{13.48}$$

where

$$P_{\text{max}} = \frac{5P}{Z\cos\alpha} \tag{13.49}$$

This gives ball bearing effective stiffness as [9]

$$K_{\text{qp}} = \frac{dP}{d\delta_{\text{r}}} = 0.32 \times 10^5 z^{2/3} d^{1/3} p^{1/3} \cos^{5/3}\alpha \ (1\,\text{bin.}^{-1}) \tag{13.50}$$

13.4.7 Data Deduction and Discussion

The equivalent capacitance of the 6326 ball bearing has been computed using Eqns. (13.28)–(13.40) for different parameters of operation given in Table 13.2. The equivalent (K_{qe}) and effective stiffness (K_{qef}) at different loads and speeds of the bearing has also been computed [Eqn. (13.44)] using electrical analogy, and are given in Table 13.2. The number of balls in the loaded zone has been taken considering the load distribution in the bearing [8]. The number of balls in the loaded zone varies, depending on load, bearing clearance and operating speed. The effective stiffness values, computed by the developed approximate procedure given in Eqns. (13.45)–(13.47), are shown in Table 13.2 alongwith the stiffness values, determined by the conventional approach using Eqns. (13.48)–(13.50).

13.4.7.1 Determination of Effective Stiffness by Electrical Analogy

The equivalent stiffness of a ball bearing depends on inverse of its equivalent capacitance, which in turn is a function of capacitive reactance (X_{ib} and X_{ob}), resistance (R_{ib} and R_{ob}) and capacitance (C_{ib} and C_{ob}) between a ball and track surface of inner and outer races, and the number of rolling-elements in the loaded zone considering load distribution, clearance and operating speed [Eqns. (13.28)–(13.40)]. C_{ib} and C_{ob} depend on lubricant permittivity, minimum film thickness, ball's radius, ball track radius of respective inner and outer races, and width of deformation on the track surface of respective races [Eqns. (13.28), (13.29) and (13.32)]. Similarly, R_{ib} and R_{ob} depend on the same parameters as C_{ib} and C_{ob} along with lubricant resistivity except that of the permittivity of the lubricant as shown in Eqns. (13.30)–(13.32). The equivalent stiffness (K_{qe}) is a complex function of K_{r}, K_{a} and K_{m}, and depends on the dynamic model of a ball bearing (Fig. 13.4). The difference among these values is not very significant under the influence of small film thickness and stabilized conditions of operation. Considering this, the effective

Table 13.2. Capacitance, resistance, capacitive reactance and stiffness of ball bearing type 6326 at different operating parameters

Speed(rpm)	1000		1500	
P (N)	11.7×10^3	70×10^3	95×10^3	120×10^3
W_{ib} (mm)	1.20	2.40	2.89	2.97
W_{ob} (mm)	1.15	2.29	2.77	2.84
$h_o \times 10^{-4}$ (mm)	23.02	28.17	27.83	27.31
C_{ib} ($p\phi$)	15.87	25.74	29.17	29.81
C_{ob} ($p\phi$)	14.02	24.57	28.49	29.21
$X_{ib} \times 10^7$ (ohm)	20.1	12.4	10.9	10.7
$X_{ob} \times 10^7$ (ohm)	22.7	13.0	11.2	10.9
$R_{ib} \times 10^7$ (ohm)	139.3	85.9	75.8	74.2
$R_{ob} \times 10^7$ (ohm)	157.9	90.1	77.7	75.8
$R_b \times 10^7$ (ohm)	6.1	3.6	3.1	3.03
$R_r \times 10^7$ (ohm)	0.87	0.9	1.03	1.01
$X_r \times 10^7$ (ohm)	41.8	25.0	21.7	21.12
C_b ($p\phi$)	7.62	12.7	14.7	15.1
C_r ($p\phi$)	68.58	101.6	88.5	90.6
$K_{qe} \times 10^9$ (N m^{-1})	14.5	9.84	11.3	11.03
$K_{qef} \times 10^9$ (N m^{-1})	2.4	1.64	1.88	1.84
$K_{qp} \times 10^9$ (N m^{-1})	0.86	1.56	1.73	1.87
$K_{qw} \times 10^9$ (N m^{-1})	0.12	0.42	0.60	0.66

Kinematics: $d = 46$ mm, $r_i = 80$ mm, $r_o = 126$ mm, $D = 206$ mm, $p = 1$, $Z = 9$, $\delta_1 = 0.018\,\text{mm}^{-1}$, $\delta_2 = 0.0089\,\text{mm}^{-1}$, $\xi = 2.5 \times 8.85419 \times 10^{-12}\,\phi\,\text{m}^{-1}$, $\phi_1 = 0.028\,\text{mm}^{-1}$, $\phi_2 = 0.007745\,\text{mm}^{-1}$, $\beta_1 = 1.42$, $\beta_2 = 1.9013$, $\rho = 10^9$ ohm m, $n = 50$ Hz, $R = 0$ ohm (K_{qp} and K_{qw} are without speed consideration).

stiffness (K_{qef}) is established as one-sixth of the equivalent stiffness (K_{qe}) as shown in Eqn. (13.44).

The effective stiffness of the 6326 ball bearing varies between 1.64×10^9 and $1.88 \times 10^9\,\text{N m}^{-1}$ under different combinations of loads varying between 70×10^3 and 120×10^3 N, and speeds varying between 1000 and 1500 rpm, as shown in Table 13.2. At 1000 rpm at 11.7×10^3 and 70×10^3 N loads, the effective stiffness (K_{qef}) changes from 2.4×10^9 to $1.64 \times 10^9\,\text{N m}^{-1}$, and at 1500 rpm at 95×10^3 and 120×10^3 N loads, the effective stiffness changes from 1.88×10^9 to $1.84 \times 10^9\,\text{N m}^{-1}$ (Table 13.1). At constant speed with increase in load, effective stiffness is slightly changed. This occurs because of decrease in film thickness with increase in load [8]. In general, variation of stiffness is non-linear with operating parameters. This occurs due to the variation of film thickness and lubricant characteristics with change in operating parameters.

From Table 13.2, it is evident that the values of equivalent bearing resistance (in ohms), i.e. bearing damping in N s m^{-1} [7] under different operating parameters are approximately 10^3 times less as compared to the bearing stiffness without considering the periodic asperity contacts between races and rolling-elements.

13.4.7.2 Determination of Stiffness by Developed Approximate Procedure Using Width of Deformation

The effective stiffness of a ball bearing by the developed approximate procedure depends on the bearing dimensional parameters, material properties, bearing load and the number of rolling-elements in the loaded zone as shown in Eqns. (13.45)–(13.47). The operating speed of bearing has not been considered for determination of stiffness by this method. The values of the effective stiffness (K_{qw}) vary from 0.42×10^9, 0.60×10^9 to 0.66×10^9 N m^{-1} as the load changes from 70×10^3, 95×10^3 to 120×10^3 N as against 1.64×10^9, 1.88×10^9 to 1.84×10^9 N m^{-1} under similar load conditions, but at 1000 to 1500 rpm of rotation, using electrical analogy approach (Table 13.2). The difference in stiffness values, thus determined, is attributed to influence of operating parameters on the film thickness and load distribution in the bearing.

13.4.7.3 Determination of Stiffness by Conventional Procedure

The stiffness of a ball bearing under radial load as worked out by Palmgren [9] [Eqn. (13.50)] is a function of the number and diameter of balls, and load acting on them. Under different varying loads of 70×10^3, 95×10^3 and 120×10^3 N, stiffness of 6326 ball bearing varies as 1.56×10^9, 1.73×10^9 and 1.87×10^9 N m^{-1}, respectively. Under similar load conditions at 1000 to 1500 rpm of rotation using developed electrical analogy approach, effective stiffness varies as 1.64×10^9, 1.88×10^9 and 1.84×10^9 N m^{-1} (Table 13.2). The difference in stiffness values, determined by these methods can be attributed to the effect of speed. The speed of operation of a bearing affects the dynamic load and film thickness, which in turn influences the bearing stiffness. However, in general, the values of stiffness determined by the conventional method (K_{qp}) and electrical approach (K_{qef}) match closely as shown in Table 13.2. Also, the procedure given by Palmgren and the developed approximate method using width of deformation indicate identical behavior pattern of the change of effective stiffness under identical conditions (Table 13.2).

13.4.8 Conclusions

Besides conclusions given in Section 13.3.9 for a roller bearing, additional major conclusions for ball bearing are [14]:

1. The effective stiffness is a function of equivalent stiffness, and depends on the dynamic model of a ball bearing, particularly on radial and axial stiffness including that of the bending moment.
2. The equivalent capacitance/stiffness^{-1} of a ball bearing is a function of the lubricant permittivity, the minimum film thickness, the diameter of ball, the ball track radius, the width of deformation on track surfaces and the number of balls in the loaded zone.
3. By electrical analogy, it is established that the effective stiffness of a ball bearing changes with operating speed and load characteristics, and it is approximately one-sixth of the equivalent stiffness determined by electrical analogy.

13.5 Conclusions

The technique developed, and further development and extension of the methodology of electrical analogy has potential to determine effective stiffness more reliably under complex loading pattern and faster in different situations, as compared to the conventional technique for a rolling-element bearing.

REFERENCES

1. Prashad, H. (1987). Effects of operating parameters on the threshold voltages and impedance response on non-insulated rolling-element bearings under the action of electrical currents. *Wear*, **117**, 223–240.
2. Prashad, H. (1998). Determination of time span for appearance of flutes on the track surface of rolling-element bearings under the influence of electric current. Presented at WTC in London during 8–12 September, 1997. Also, published by STLE in *Tribol. Trans.*, **41**(1), 103–109.
3. Prashad, H. (1988). Investigations on corrugated pattern on the surface of roller bearings operated under the influence of electrical fields. *Lubric. Eng.*, **44**(8), 710–718.
4. Prashad, H. (1988). Theoretical evaluation of impedance, capacitance, and charge accumulation on roller bearings operated under electrical fields. *Wear*, **125**, 223–239.
5. Pujara, K. K. (1972). *Vibration for Engineers*, Dhanpat Rai and Sons, Delhi.
6. Prashad, H. (1996). Evaluation of dynamic coefficients of two-lobe journal bearings using an electrical analogy approach. ASME, J. *Tribol.*, **118**, 657–662.

7. Prashad, H. (1997). Assessment of dynamic coefficients of three-lobe journal bearings through evaluation of electrical parameters – a new approach. *BHEL J.*, **18**(1), 40–48.
8. Harris, A. T. (1984). *Rolling Bearing Analysis*, John Wiley and Sons, New York.
9. Palmgren, A. (1959). *Ball and Roller Bearing Engineering*, 3rd ed., Burbank, Philadelphia, 49–51.
10. Prashad, H. (2004). Determination of stiffness of roller bearings—an alternative approach. *IE(I) J.-MC*, **84**, 186–192.
11. Prashad, H. (1990). Analysis of the effects of electrical current on contact temperature, residual stresses leading to slip bands initiation and formation of corrugation pattern on ball tracks of ball bearings. *BHEL J.*, **11**(1), 39–47.
12. Prashad, H. (1991). Theoretical and experimental investigations on the pitch and width of corrugations pattern on the surface of ball bearings. *Wear*, **143**, 1–14.
13. Prashad, H. (1993). Theoretical determination of impedance, resistance, capacitive reactance and capacitance of ball bearings. *BHEL J.*, **14**(2), 40–48.
14. Prashad, H. (2003). Alternative approaches to determination of stiffness of ball bearings. *BHEL J.*, **24**(3), 59–70.

Nomenclature

C_b	equivalent capacitance between races and a ball
C_r	equivalent capacitance of ball bearing
C_{ib}, C_{ob}	capacitance between inner race and a ball, and outer race and a ball, respectively
d	diameter of rolling-element
D	pitch diameter of ball bearing
E	Young's modulus of elasticity
h_o	minimum film thickness
K_r	stiffness between race and a ball due to radial force
K_a	stiffness between race and a ball due to axial force
K_m	stiffness between race and a ball due to bending moment
K_{rma}	stiffness between races and a ball due to radial and axial forces, and bending moment
K_{ib}, K_{ob}	stiffness of inner and outer-race track surfaces and a rolling-element, respectively
K_{qe}	equivalent bearing stiffness by electrical analogy
K_{qef}	effective stiffness of ball bearing
K_{qw}	approximate effective stiffness by width of deformation
K_{qp}	effective stiffness by existing procedure
N_z	number of balls in the loaded zone of bearing
n	frequency (H_Z)

p	plane of action of loading ($p = 1, 2, 3, \ldots$)
P	applied load
P_{max}	maximum radial load
r	radius of ball
r_i, r_o	track radius of inner race and outer race, respectively
r_{ib}, r_{ob}	common deformation radius bisecting r and r_i, r and r_o, respectively
R_{ib}	resistance between inner race and a ball
R_{ob}	resistance between outer race and a ball
R_b	equivalent resistance between races and a ball
R_r	equivalent resistance of ball bearing
R	resistance of ball
w	$2\pi n$
W_{ib}, W_{ob}	width of deformation on inner and outer race, respectively
X_{ib}	capacitive reactance between inner race and a ball
X_{ob}	capacitive reactance between outer race and a ball
X_b	equivalent capacitive reactance between races and a ball
Z	number of rolling-elements in a bearing
Z_{ib}	impedance between inner race and a ball
Z_{ob}	impedance between outer race and a ball
Z_b	impedance between races and a ball
β_1, β_2	functions of inner and outer race constants
δ_1, δ_2	bearing outer race constants
δ_r	deflection/contact deformation
ϕ_1, ϕ_2	bearing inner race constants
ρ	resistivity of lubricant (ohm m) contact angle
ξ	permittivity/dielectric constant of lubricant ($\phi\,m^{-1}$)
ξ_r	relative permittivity of lubricant

Chapter 14

FAILURE DIAGNOSIS AND INVESTIGATIONS OF ROLLING-ELEMENT BEARINGS DUE TO UNFORESEEN CAUSES

14.1 General

This chapter highlights the investigations pertaining to the diagnosis of rolling-element bearings of motors and alternators failed due to the causes generally unforeseen during design and operation. However, in general the diagnosis of the failure of the bearings has been well established in literature. The unforeseen causes by failure diagnosis have been established. These generally unforeseen causes found to lead to the passage of electric current through bearings of the motors and alternators and deteriorate them in due course.

The vibration and shaft-voltage data, characteristics of the grease used in various rolling-element bearings, and other related aspects were analyzed and unforeseen causes of failure of bearings diagnosed.

14.2 Introduction

14.2.1 Causes of Shaft Voltages and Flow of Current Through Bearings

Phenomenon of shaft voltages exists in electrical machines, which causes flow of current through the bearings depending on the resistance of the bearing circuits. This has been discussed in Chapters 2 and 12 in details [1–15].

At a certain threshold voltage, depending on the resistivity of the lubricant and operating conditions, current flows through the bearing [4]. Thus, the flow of circular current in the inner race leaks through the rolling-elements to the outer race by following a path of least resistance and establishes a field strength leading to the development of magnetic flux on the track surface of races and rolling-elements [5] as discussed in Chapter 9.

Studies have been carried out by various authors on the causes and control of electric currents in bearings [6], flow of current through lubricated contacts [7], the effect of electric current on bearing life [8], the effect of operating parameters on an impedance response [4], and the deterioration of lubricants used in non-insulated bearings [10–12]. These have been discussed in Chapters 3–10.

The surveys of the failure of rolling-element bearings indicate various causes including that of failure due to corrugations [16, 13, 14]. The mechanism of formation of pattern of corrugations on the track surfaces and related investigations on roller bearing surfaces have been explained in Chapter 5.

This chapter deals the investigations pertaining to generally unforeseen causes, which lead to the premature failure of rolling-element bearings in motors and alternators.

14.3 Bearing Arrangement and Nature of Bearing Failure

The motors with rated capacity 2100 kW operating at 1494 rpm have three bearing arrangements. The motors are used to drive primary air fans of 500-MW generators.

On the non-drive end (NDE) of the motor, an insulated bearing NU 228 is used. On the drive-end (DE) of the motor, roller bearing type NU 232 and ball bearing type NU 6326 are used for taking both radial and axial loads simultaneously. All the bearings are grease lubricated and the motors are designed for continuous operation with the periodic re-lubrication. Bearings are also instrumented for continuous measurement of the temperature during operation. Figure 12.1 shows the schematic diagram of NDE bearing arrangement in the motor.

The alternators were designed with the output voltage of 24 V with 4.5-kW rating. Each alternators uses bearing type NU 311 at the drive-end and 6309 at the non-drive end. The alternators are used for charging batteries and operation of lighting system.

After commissioning, a few motors and alternators, bearings were failed prematurely. The nature of the failure was the formation of corrugations and flutings on the roller tracks of the races besides corrosion on the raceways and rolling-elements, irrespective of the make of the bearings used. However, the depth of corrugations and degree of development of corrugations were different. The color of the grease was also found to have got blackened.

14.4 Investigations, Observations and Data Collection

The investigations pertaining to shaft voltage, vibration and shock pulse levels were carried out on various motors, both at non-drive end and drive-end, and various other related aspects were inspected.

14.4.1 Measurement of Shaft Voltages

The shaft voltages were measured between non-drive end and drive-end shaft, drive-end shaft to ground, and non-drive end shaft to ground. Table 14.1 shows shaft-voltage data after removing the grounding brush. Table 14.2 indicates frequency analysis of different shaft voltages after spectrum analysis of the shaft voltages recorded on tape recorder using shaft probes. Shaft voltages were also measured on retaining the grounding brush and keeping the brush intact (Fig. 12.1).

14.4.2 Measurement of Vibration Levels and Spectrum Analysis

Vibration levels were measured in mm s^{-1} on both non-drive end and drive-end bearings using B and K vibration-measuring instruments. Table 14.3 shows the vibration levels in horizontal, vertical and axial directions.

14.4.3 Inspection of Dimensional Accuracy and Metallurgical Examination

The dimensional accuracy of new bearings were checked and found to be in line with the specifications. Also, the dimensional accuracy and metallurgical examinations of the components of the failed bearings revealed consistency with the specifications.

Table 14.1. Overall values of shaft voltages of different motors without grounding brush (in V)

Motor	Non-drive end to drive-end shaft	Non-drive end to ground	Drive-end shaft to ground
A	1.004	1.004	1.0
B	0.64	0.030	0.65
C	1.14	1.15	0.47
D	0.85	0.85	0.58
E	0.65	0.80	0.82
F	0.74	0.91	0.40

Table 14.2. Frequency analysis of shaft voltages (V) of different motors

	Frequency (Hz)									
	Non-drive end to drive-end					Non-drive end to ground				
Motor No.	50	195	1150	1445	Over all	50	195	1150	1445	Over all
A	0.04	–	0.10	0.33	–	0.13	0.03	0.28	0.55	0.74
B	0.67	–	0.07	0.17	0.81	0.008	0.002	0.009	0.014	0.02
C	0.96	0.15	0.19	0.46	1.44	0.062	0.18	0.31	0.65	0.91
D	–	–	–	–	–	0.77	0.1	0.21	0.65	1.22
E	0.18	0.029	0.11	0.37	0.525	0.06	0.026	0.123	0.47	0.53
F	0.87	0.0212	0.053	0.19	0.978	0.55	–	0.076	0.22	0.76

Table 14.3. Vibration levels of different motors in mm s^{-1} (rms)

	Non-drive end			Drive-end			
Motor	H	V	A	H	V	A	Remarks
A	0.7	1.2	0.7	1.5	1.5	0.8	H, V, A are horizontal, vertical and axial, levels of vibration.
B	0.7	0.65	0.7	2.0	1.6	1.1	
C	0.8	0.8	0.65	1.2	1.1	0.8	
D	0.6	0.6	0.55	1.0	0.8	0.7	
E	0.65	0.35	0.35	0.8	0.9	0.5	As per ISO 10816 Part 1–5 bearings are in good condition up to vibration levels of 1.8 mm s^{-1}, and above this, in satisfactory condition.
F	0.6	1.0	1.0	2.0	1.5	1.0	

14.4.4 Analysis of Lubricants from Failed Bearings

Various failed bearings were examined. Sample of fresh grease and used grease from the rolling-elements of the failed bearings were collected and analyzed.

14.4.5 Study of Bearing Location in Alternators

In certain original design of alternators, it was found that the rolling-element bearing type NU 311, at the drive-end was mounted in the housing as shown in

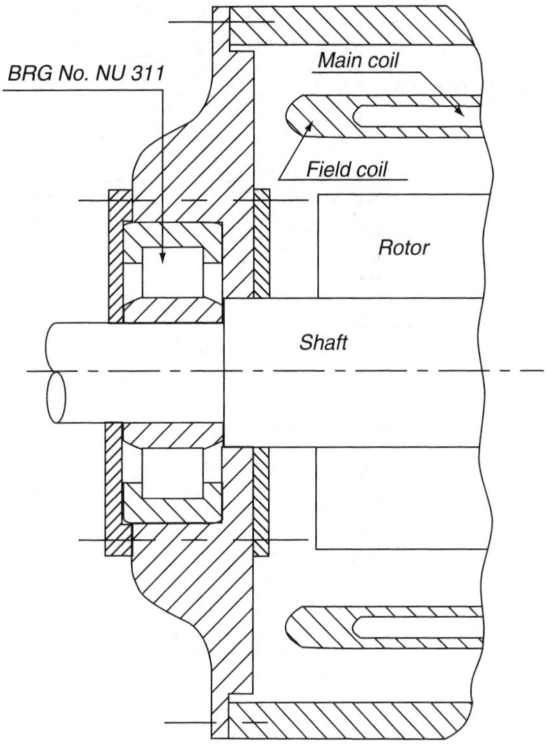

Fig. 14.1. Design configuration showing location of bearing, field and main coils of original alternator

Fig. 14.1. In the modified design of the alternators for the compact design, bearings were provided in the projection of the end cover (Fig. 14.2). But the projection was towards the inside of the alternator, because of which the bearing was located almost under the influence of the field coil, as shown in Fig. 14.2. The bearing elements were found to have been magnetized. The magnetic flux density on the track surface of some bearings was found to be as high as 40 gauss by gauss meter. In general, flux density found to vary between 10 and 40 gauss. Also, the stray voltages of 2–6 and 6–20 V were measured under 'no-load' and 'load' conditions, respectively.

14.4.6 Checking for Grease Pipe Contacting the Base Frame

In a few motors, it was found the outlet grease pipe was either in direct contact with the base plate or passing through the accumulation of grease outlet from the bearing (Fig. 12.1).

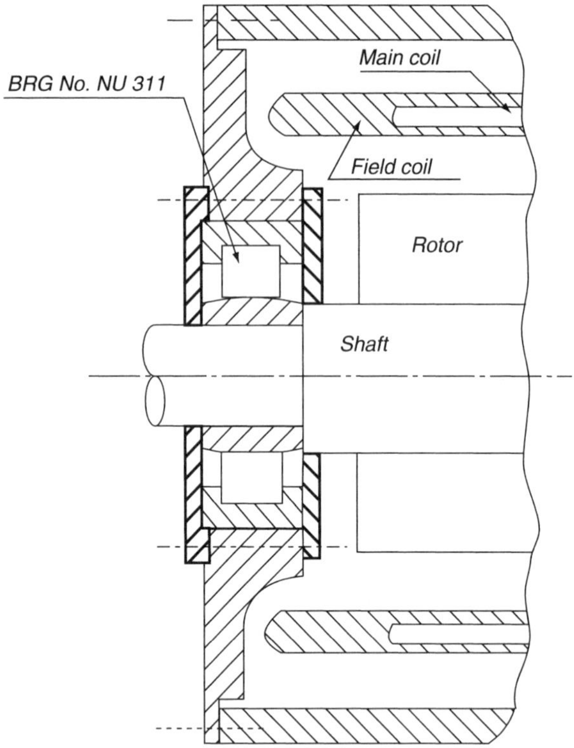

Fig. 14.2. Design configuration showing location of bearing, field and main coils of modified alternator

14.4.7 Checking of Grease Leakage Through Seals

Excess of grease leaking through seals got collected just beside the bearing housing, and thus the contact between the bearing housing and the base plate was made through the contaminated grease even though the bearing pedestal was insulated from the base plate, as shown in Fig. 12.1. Collection of dirt/sludge etc. between the corners of the base at bearing pedestal and the base plate was also detected, particularly in the thermal power houses.

14.4.8 Checking for Unshielded Instrumentation Cables

Instrumentation cable used for measurement of bearing temperature sometimes gets unshielded due to various unforeseen reasons, and the cable comes partially in contact with the base plate/bearing housing. A few of such cases have been found and investigated. In fact in the present case study, unshielded instrumentation cables were established to be the cause of bearing failure, after the investigations.

14.4.9 Checking for Improper Contact of Grounding Brush with Shaft

Due to prolonged operation, dust collection, improper gap between brush and shaft, and improper maintenance, the grounding brush loses the grip with the rotating shaft. This was found in various locations.

14.4.10 Checking for Damage of Bearing Insulation

The prolong operation of bearings under vibration, leads to the ageing of insulation, depending on quality of insulating material. This results in the breakage/crack of the bearing insulation. Consequently, it generates a resistance-free path of bearing current, which leads to catastrophic failure of the bearings (Fig. 12.1). This has been detected in a few motor bearings.

14.4.11 Checking for Passage of Current Through Connecting Bolts, Nuts, etc.

If the connecting bolts, nuts, joints of bearing pedestal and base plate are not properly shielded, a current is found to pass through the bearing even if the bearing insulation pads are properly maintained (Fig. 12.1). A few such cases have been found and investigated.

14.5 Results and Discussion

14.5.1 Shaft Voltages and Their Frequencies

The measurement of shaft voltages with and without the grounding brush indicated that the gap between the shaft and grounding brush was not set precisely. Also, the brush was not maintained and cleaned properly. That is why the brush was not able to ground the shaft voltage properly. Table 14.1 indicates that the different levels of shaft voltages exist in all the motors because of the various causes as brought out and explained in Chapter 2. The minimum voltage of motor 'B' between NDE and ground was measured as 0.030 V as against 0.64 V between NDE and DE shaft, and 0.65 V between DE and ground (Table 14.1). This (minimum voltage of 0.030 V) indicates that the insulation at NDE bearing of motor 'B' was bridged on the path. This was confirmed on dismantling the bearings of motor 'B', and breakage of the insulation was detected. The maximum voltage of motor 'C' between NDE and ground was measured as 1.15 V. The drive-end shaft to ground voltage of 0.4 V of motor 'F' indicates the passage of feeble current through the bearings as compared to motors 'A', 'C', 'D' and 'E'. The investigations indicated the partial unshielded instrumentation cable contacting the bearing housing was the source of the passage of current through the bearing of the motor 'F'.

The frequency analysis of the shaft voltage signals, shown in Table 14.2, indicates that the major component of shaft voltage consists of the magnitude of the slot passing frequency at 195, 1150 and 1445 Hz, and line passing frequency component of 50 Hz. The voltage of slot passing frequency component at 1445 Hz between NDE and DE varies between 0.17 and 0.37 V, and that between NDE and ground varies between 0.014 and 0.65 V. The voltage at line frequency component of 50 Hz, between NDE and DE, and between NDE and ground varies between 0.04 and 0.96 V, and 0.008 and 0.77 V, respectively.

The presence of different magnitudes at the above frequencies of the shaft voltage is attributed to the magnetic asymmetry at unequal air gap created by the static and dynamic eccentricity. This may occur partially due to the deviation in quality control norms, and may be controlled to some extent by high precision erection of the machines [5].

14.5.2 Vibration Analysis

The vibration levels and noise emission indicate the bearing condition. For an unhealthy rolling-element bearing, levels of high frequency components and overall vibration levels rise considerably. Initially, incipient damage like, microspalls on the track surface of rolling-element bearings, generates high frequency vibrations. However, when the defects in the roller track increase and grow in size, then the magnitude of low frequency components increases [16].

From the magnitude of vibration levels, it is evident that the overall vibration levels in axial and radial directions were within acceptable limits as per ISO 10816, Part 1–5 (Table 14.3). The presence of the principal slot harmonic frequency components and their side bands between 1300 and 1600 Hz existed in all the motor bearings with varying higher levels in axial direction. This indicates the existence of varying degree of magnetic asymmetry of air gaps due to static and dynamic eccentricity in all the motors. This originates different levels of shaft voltages [1–3] as shown in Tables 14.1 and 14.2, and led to bearing failure as a result of the passage of electric current through bearings due to the various unforeseen causes as brought out in Section 14.3 of this chapter (Chapters 3 and 5).

The overall vibration levels in all the motors were in normal limits except in motors 'B' and 'F', where overall vibration levels were rose to 2 mm s^{-1}, as shown in Table 14.3. Moreover, the major constituents of the spectrum of motor 'B' and 'F' were found to exist in low frequency range, which indicates that in these motor bearings, the incipient damage had already set. Shock pulse levels of motors 'B' and 'F' also indicated initial incipient stage of damage of bearings, whereas condition of all other motor bearings was normal but necessitated immediate re-lubrication.

14.5.3 Passage of Current Through Bearings

Whenever a grease outlet pipe and the base plate are in contact, either directly or through the accumulation of grease outlet from the pipe, this makes closed circular path of current flow through the rotating shaft and the bearings. This happens even if the bearing insulation and/or insulation pads are intact. Similarly, accumulation of grease leakage from bearing seals on the base plate and pedestal makes the passage for a flow of feeble current. Furthermore, accumulation of excessive dirt/sludge between the edge of the base of bearing pedestal and base plate creates the path for a current flow even in the presence of insulation pads. This happens when the bearing is not insulated in the housings, as shown in Fig. 12.1.

Under the condition of improper contact of grounding brush with shaft, the brush is not able to make the path of least resistance for grounding the shaft current. The current then tend to pass through the bearing or tend to create localized loop in the bearing, depending on the bearing impedance. This leads to deterioration of the bearing condition and causes the bearing failure in due course. This happens when the bearing housing and bearing pedestal are not properly insulated.

Sometimes, apart from the puncturing of the bearing insulation, unshielded instrumentation cable contacting the base plate also create the path of the least resistance for the flow of electric current through the bearings. This was established as the cause of failure of motor bearings 'F'.

14.5.4 Magnetic Flux Density

The presence of magnetic flux density along with the corrugation pattern and corrosion on the bearing surfaces indicate the damage due to electric current as explained in Chapter 9 [14, 15]. The presence of corrugation pattern without significant flux density distribution indicates plastic deformation of the bearing surfaces by mechanical loading, accompanied by the flexibility of the supporting structure, and was influenced by the frequency of rotation of rolling-elements in the inner race.

From Fig. 14.2, it is evident that the non-insulated bearing NU 311 was located under the field winding in the alternator design. This makes the NU 311 bearing of alternator permanently magnetized, and magnetic flux density develops on the bearing elements [5]. If the bearing had been located away from the influence of the field coils, as shown in Fig. 14.1, magnetic flux density would not have been developed on the bearing elements [14].

When the outer race of the bearing is magnetized and the inner race and rolling-elements rotating inside the outer race, voltage is generated by electromagnetic principles and the flow of current starts through the inner race and rolling-elements depending on the impedance of the oil film thickness and threshold voltage phenomenon [4]. This was confirmed by the stray voltages of 2–6 and 6–20 V of the alternator measured under 'no-load' and 'load' conditions, respectively. This shows dissymmetry of the magnetized field, rotor eccentricity apart from the other causes given in references [3, 5]. However, in addition to other manufacturing errors, the major contribution to the higher voltage between the shaft and the body is attributed to the bearing location under the influence at the field coil, which magnetized the bearing and damaged it due course as explained in Chapter 9 [5, 14, 15].

14.5.5 Analysis of Failed Bearings

Corrugation and ridges were found on the track surface of the races of all the failed bearings, besides the corrosion on the track surfaces. The track surface of the bearings got corroded because of the decomposition of the grease, and formation of corrugations on the track surfaces by low-temperature tempering and Hertzion pressure on the raceways [1, 3], which finally results in the reduction of bearing fatigue life and failure in due course as dealt in Chapter 5 [8, 9].

14.5.6 Effects of Bearing Current on Lubricant

The zinc additive, i.e. zinc dithiophosphate or zinc dialkyldithiophosphate (ZDTP), used as multifunctional additive in the grease, under rolling friction protects the rubbing metal surfaces and contributes to friction and wear reduction, which depends partly on the amount of additive on these surfaces. Decomposition of ZDTP in the lithium base greases under the influence of electric fields leads to the formation of lithium zinc silicate ($Li_{3.6}Zn_{0.2}SiO_2$) in the presence of a high relative percentage of free lithium and silica impurity in the grease under high temperature in the asperity contacts along with the formation of gamma lithium iron oxide (γ-$LiFeO_2$). During the process, lithium hydroxide is also formed which, corrodes the bearing surfaces. And the original structure of lithium stearate changes to lithium palmitate. On the contrary, these changes are not detected under pure rolling friction.

The used grease taken from the motor bearing 'B' showed above changes, similar to those described in references [10, 11] and Chapter 3.

14.5.7 Process of Bearing Failure Under the Influence of Leakage Current

When the current leaks through the roller bearing in which low-resistivity (10^5 ohm m) grease has been used, a 'silent' discharge passes through the bearing elements. This creates a magnetic flux density distribution on the bearing surfaces. Also, when a bearing is located and operates under the influence of magnetic field, voltage is generated and current flows through the bearing depending on the bearing impedance and the threshold voltage phenomenon. Under these conditions, this leads to the electrochemical decomposition of the grease in the initial stages, which, in turn, corrodes the bearing surfaces [10] as explained in Chapter 3. And, then, gradual formation of flutings and corrugations on the surfaces [3, 7, 13] occurs as found in the bearings of motors 'B' and 'F'. Subsequently, wear increases and the bearing fails.

14.6 Conclusions

From the above investigations and analysis, the following are concluded [17, 18]:

(i) Current passes through the bearing because of puncturing of bearing housing insulation, grease outlet pipe touching motor base frame, and improper contact of grounding brush with the shaft.

(ii) Current can also pass through a bearing on such occasions when un-shielded instrumentation cable touches the bearing, even if the bearing insulation is healthy.

(iii) If the bearing is not insulated in the housing, accumulation of dirt/sludge between the pedestal and base plate creates the path for the flow of leakage current through the bearing even if pedestal insulation pads are healthy.

(iv) Under the influence of field coils, a roller bearing develops magnetic flux. During operation of a bearing under the influence of magnetic field, voltage is generated. Also, stray voltage on the bearing is developed.

(v) Current passes through a bearing, depending on the bearing impedance and threshold voltage phenomenon.

(vi) Levels of measured voltage between bearing and ground, and between bearing and shaft indicate the condition of the bearing insulation. Vibration levels indicate the bearing condition.

(vii) A varying magnetic flux density of 10–40 gauss is developed on the track surface of races of bearings, depending on a flow of current.

(viii) In the case of bearings using grease of low-resistivity, failure occurs under the 'silent' electric discharge due to chemical decomposition, and the formation of flutings and corrugations on the bearing surfaces.

(ix) To avoid the bearing failure under the influence of electric current, bearings should not be located under the influence of field coils in the alternators.

(x) Bearings should be properly insulated and shielded for any possible means of leakage of electric current.

By the detection of the flux density on the bearing surfaces along with the corrugated pattern on the tracks surfaces, by analysis of deterioration of greases used in the bearings, and by measurement of stray and shaft voltages, the failure of bearings by the passage of electric current can be established. And the extent of deterioration is ascertained by the vibration analysis. Also, the bearings can be shielded against any unforeseen causes for precluding any possible means of leakage of current, for ensuring their trouble-free operational life.

REFERENCES

1. Prashad, H. (1994). Investigations of damaged rolling-element bearings and deterioration of lubricants under the influence of electric current. *Wear*, **176**, 151–161.
2. Bradford, M. (1984). Prediction of bearing wear due to shaft voltage in electrical machines, ERA Technology Limited.
3. Prashad, H. (1988). Investigations on corrugated pattern on the surface of roller bearings operated under the influence of electrical fields. *Lubric. Eng.*, **44**(8), 710–718.
4. Prashad, H. (1987). Effects of operating parameters on the threshold voltages and impedance response of non-insulated rolling-element bearings under the action of electric current. *Wear*, **117**, 223–240.
5. Prashad, H. (1996). Magnetic flux density distribution on the track surface of rolling-element bearings – an experimental and theoretical investigation. *Tribology Transaction*, **39**(2), 386–391.
6. Morgan, A. W. and Whillie, D. (1969–1970). A survey of rolling bearing failures. *Proc. I. Mech. E.F.*, **184**, 48–56.
7. Anderson, S. (1968). Passage of electric current through rolling bearings. *Ball Bearing J.*, **153**, 6–12.
8. Simpson, F. E. and Crump, J. J. (1963). Effects of electric currents on the life of rolling contact bearings. Proceedings of Lubrication and Wear Convention, Bournmouth, 1963, Institution of Mechanical Engineers, London, Paper 27.
9. Prashad, H. (1989). Analysis of the effects of an electric current on contact temperature, contact stresses and slip band initiation on the roller tracks of roller bearings. *Wear*, **131**, 1–14.
10. Prashad, H. (1989). Diagnosis of deterioration of lithium greases used in rolling-element bearings by X-ray diffractometry. *Tribol. Trans.*, **32**(2), 205–214.

11. Komatsuzaki, S. (1987). Bearing damage by electrical wear and its effects on deterioration of lubricating grease. *Lubric. Eng.*, **43**, 25–30.
12. Remy, M. and Magnin, A. (1996). Rheological and physical studies of lubricating greases before and after use in bearings. ASME, *J. Tribol.*, **118**, 681–686.
13. Winder, L. R. and Wolfe, O. J. (1968). Valuable results from bearing damage analysis. *Met. Proc.*, 52–59.
14. Prashad, H. (1996). Diagnosis of failure of rolling-element bearings of alternators – a study. *Wear*, **198**(1–2), 46–51.
15. Prashad, H. (1988). The effect of current leakage on electro-adhension forces in rolling-friction and magnetic flux density distribution on the surface of rolling-element bearing. ASME, *J. Tribol.*, **110**, 448–455.
16. Prashad, H. et al. (1985). Diagnostic monitoring of rolling-element bearing by high-frequency resource technique. *ASLE Trans.*, **28**(4), 39–448.
17. Prashad, H. (1999). Investigations and diagnosis of failure of rolling-element bearings due to unforeseen causes—a case study. *BHEL J.*, **20**(1), 59–67.
18. Prashad, H. (1999). Diagnosis and cause analysis of rolling-element bearings failure in electric power equipments due to current passage. STLE, *J. Lubric. Eng.*, **5**, 30–35.

Chapter 15

ELECTRICAL PARAMETERS OF A CYLINDRICAL HYDRODYNAMIC JOURNAL BEARING AND REDUCTION IN ITS LIFE UNDER THE INFLUENCE OF DIFFERENT LEVELS OF SHAFT-VOLTAGES

15.1 Introduction

For the reliable performance of a hydrodynamic journal bearing an adequate amount of oil is required for the bearing surfaces under relative motion. This forms the oil film thickness between the journal and the bearing surface. The quantity of oil needed depends on the heat generated that needs to be transferred within a bearing. There exists a considerable difference between the oil temperature in the region of load-carrying oil film – in the zone of minimum film thickness – and the outlet oil temperature from a bearing [1]. Some of the lubricant is drawn into a bearing clearance by rotation of a journal and the remaining lubricant is forced into the bearing by delivery supply pressure.

It is well established that the temperature rise of the exit oil from the load-carrying oil film of a bearing is comparatively less when increasing rpm at constant load compared to increasing load at constant rpm. This is because of a decrease in eccentricity ratio and corresponding increase in film thickness (by increasing rpm at constant load). This causes more heat transfer by increased flow of oil from load-carrying oil film, and may increase the exit oil temperature from this region. On the contrary, at constant speed and increasing load, the eccentricity ratio increases and film thickness decreases. Under these conditions, the quantity of exit oil flowing from the loaded zone is reduced. Thus, the exit oil temperature from the load-carrying oil film may increase. But, by mixing of fresh oil from the supply groove, the exit oil temperature may reduce and also may add to increase temperature inhomogeneity in the unloaded zone of a bearing [2]. Also in both the above conditions, the profile of the film thickness in the loads carrying zone will be quite different.

It is evident that in hydrodynamic journal bearings, the zone of minimum film thickness i.e. load-carrying oil film varies along the circumference of a bearing-through the bearing length. This forms a capacitor of varying capacitance between the journal and the bearing depending on permittivity of the lubricant, circumferential length of load-carrying oil film, bearing length, eccentricity ratio and the clearance ratio of a bearing. Besides that, the load-carrying oil film offers resistance depending on operating parameters and resistivity of the lubricant. Thus, the load-carrying oil film forms a resistor-capacitor (RC) circuit and offers impedance to a current flow in a hydrodynamic journal bearing. So, analyzing the RC circuit the behavior of a journal bearing can be predicated.

This chapter brings out the theoretical studies undertaken to determine the capacitance and active resistance of the interacting surfaces of a journal and bearing in the load-carrying oil film under different operating parameters. Also, the bearing capacitive reactance has been determined to analyze the effect of permittivity and the role of capacitance on the behavior of a hydrodynamic journal bearing. Besides this, optimum variation of capacitance with rpm at different *L*/*d* ratio has been computed, which can be utilized to determine the optimum value of clearance/diameter ratio under different eccentricity ratios to determine the safe load-carrying capacity. Also, this chapter brings out the theoretical studies establishing the reduction in bearing life under the influence of different levels of shaft voltages before initiation of craters on the bearing surfaces as compared to the bearings operating without the influence of shaft voltages.

15.2 Theoretical Determination of Electrical Parameters of Cylindrical Hydrodynamic Journal Bearings

15.2.1 Capacitance

Hydrodynamic theories consider the idealized case of smooth surfaces and continue to predict finite film thickness under conditions of high load and low speed. Furthermore, for each revolution of the shaft, part of the circumference of the journal passes through a zone of load-carrying oil film. This forms a parallel plate capacitor of varying thickness between the interacting surfaces of the journal and the bearing. However, the capacitance of parallel plate capacitor is maximum at minimum film thickness and reduces on both sides as the film thickness is increased.

The film thickness (h) at an arbitrary point C (Fig. 15.1) on the bearing surface at a distance X from the minimum film thickness (h_o) along the bearing circumference at an attitude angle θ is given as:

$$h = C_r[1 - \varepsilon \cos\theta] \tag{15.1}$$

And the minimum film thickness is obtained when $\theta = 0$ and given as:

$$h_o = C_r[1 - \varepsilon] \tag{15.2}$$

The change in capacitance (ΔC) between two rectangular plates of width $\mathrm{d}X$ situated between $X + \mathrm{d}X$ and X along X-axis (along the bearing circumference) of bearing length L (along Y-axis) located at the attitude angles varying between $\theta + \mathrm{d}\theta$ and θ (between points C and D) and separated by a film thickness varying between $h + \Delta h$ and h is given by:

$$\Delta C = 2\xi \frac{[(X + \mathrm{d}X)L - XL]}{(2h + \Delta h)} \tag{15.3}$$

The change in film thickness (Δh) is very small in comparison to $\mathrm{d}X$, and hence can be neglected. Thus the change in capacitance is given as:

$$\Delta C = \frac{\xi d XL}{h} \tag{15.4}$$

From Fig. 15.1, it is evident that

$$\mathrm{d}X = \frac{d\theta d}{2} \tag{15.5}$$

On substituting Eqns. (15.5) and (15.1) in Eqn. (15.4), the change in capacitance is given as:

$$\Delta C = \frac{L\xi d d\theta}{[2C_r(1 - \varepsilon \operatorname{Cos}\theta)]} \tag{15.6}$$

The maximum change in capacitance depends on the attitude angle and it is along the film thickness. So, on integrating between $-\theta$ and $+\theta$ on both sides of minimum film thickness (i.e. $\theta = 0$), the equivalent bearing capacitance can be determined as:

$$C_b = \left(\frac{L\xi}{\psi}\right) \int_{-\theta}^{\theta} \frac{d\theta}{(1 - \varepsilon \cos\theta)} \tag{15.7}$$

where

$$\psi = \frac{2C_r}{d} \tag{15.8}$$

and hence

$$C_b = \left[\frac{4L\xi}{\psi(1-\varepsilon^2)^{\frac{1}{2}}}\right]\left\{\tan^{-1}\ \tan 0.5\theta\left[\frac{(1+\varepsilon)}{(1-\varepsilon)}\right]^{\frac{1}{2}}\right\} \tag{15.9}$$

Theoretically, boundary limit for the length of load-carrying oil film is not available. However, it has been established that the angle $\pi/2$ (or $l = \pi d/4$), made by corresponding arc of load-carrying oil film at the bearing center can be used with reasonable accuracy for eccentricity ratio ranging between 0.6 and 0.8 [1, 3]. Thus, the maximum change in capacitance occurs in limits between $-\pi/4$ and $\pi/4$ from the

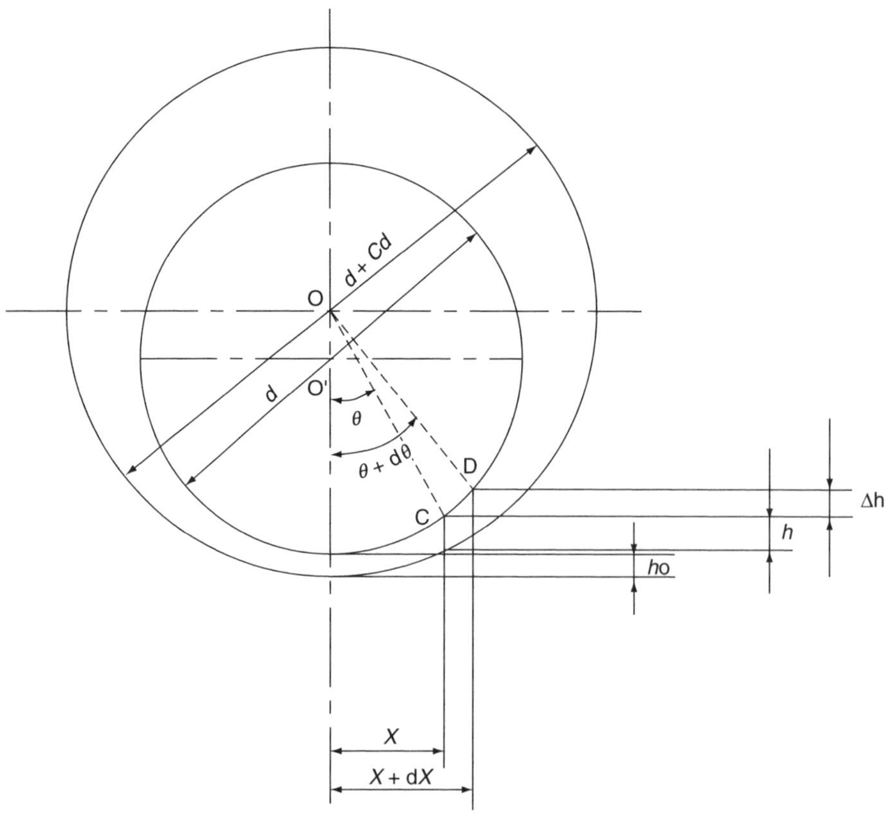

Fig. 15.1. Journal configuration in a hydrodynamic cylindrical journal bearing under rotation

location of minimum film thickness where θ is taken as zero (Fig. 15.1). Hence, equivalent bearing capacitance is determined as:

$$C_b = \frac{\left\{4\delta\xi d \tan^{-1} 0.41 \left[\frac{(1+\varepsilon)}{(1-\varepsilon)}\right]^{1/2}\right\}}{\psi(1-\varepsilon^2)^{1/2}} \tag{15.10}$$

15.2.2 Active Resistance

The change in conductance/resistance^{-1}(ΔF_b) between two rectangular plates of width dX situated between (X + dX) and X along X-axis (along the bearing circumference) of bearing length L (along Y-axis) located at the attitude angles varying between (θ + dθ) and θ and separated by a film thickness varying between $h + \Delta h$ and h (Fig. 15.1), is given as:

$$\Delta F_b = \frac{dXL}{\rho h} \quad [\Delta h \text{ being negligible}] \tag{15.11}$$

The maximum change in conductance/resistance^{-1} depends on the attitude angle and it is along the film thickness. So, on integrating between $+\theta$ and $-\theta$ on both sides of the minimum film thickness (i.e. $\theta = 0$) similar to Eqn. (15.7) and using Eqns. (15.1), (15.5) and (15.8) the equivalent bearing conductance is determined as:

$$F_b = \frac{1}{R_b} = \int_{-\theta}^{\theta} \frac{(L/\psi\rho)\mathrm{d}\theta}{(1-\varepsilon\cos\theta)} \tag{15.12}$$

and hence

$$R_b = \frac{\left[\rho\psi(1-\varepsilon^2)^{\frac{1}{2}}\right]}{4L\tan^{-1}\ \tan 0.5\theta\left[\frac{(1+\varepsilon)}{(1-\varepsilon)}\right]^{\frac{1}{2}}} \tag{15.13}$$

Similar to the capacitance the maximum change in conductance/resistance^{-1} occurs in the limits from $\pi/4$ to $-\pi/4$ from the minimum film thickness where θ is taken as zero. Thus the equivalent bearing resistance of a hydrodynamic journal bearing with the reasonable accuracy is determined as:

$$R_b = \frac{\left[\rho\psi(1-\varepsilon^2)^{\frac{1}{2}}\right]}{4\delta d\tan^{-1} 0.41\left[\frac{(1+\varepsilon)}{(1-\varepsilon)}\right]^{\frac{1}{2}}} \tag{15.14}$$

In operation, the actual circumferential contact in the zone of load-carrying oil film between the journal and a bearing depends on surface roughness and irregular oil film, and comprises primarily the high 'points' contact. Also, the area of contact is governed by vibrations of the oil film under different conditions of operation. In case of a stationary journal, the actual area of contact is relatively large and consists mainly of quasi-metallic surfaces of the journal and a bearing, the resistance of which is measured as 0.4–0.5 ohms. Instantaneously, resistance under operating conditions (R_b) can be taken as equivalent to the high 'points' contact resistance under static conditions (R_s). This is because for a fraction of a second (10^{-3} s – duration to pass zone of load-carrying oil film – as shown in Table 15.1), the journal and bearing are treated as stationary at the high points during rotation of the journal. This can also be explained due to the temporary lack of slip at the asperities due to elastic deformation between the journal and bearing. Also, the surface effect due to shaft voltage on the Babbitt liner is taken as up to the depth (H) of 0.001 mm [4].

15.2.3 Determination of Capacitive Reactance and Impedance of Hydrodynamic Journal Bearings

Capacitive reactance between the interacting surfaces of journal and the bearing in the load-carrying oil film is given as [5]:

$$X_{cb} = \frac{1}{wC_b} \tag{15.15}$$

where

$$w = 2\pi f \quad \text{and} \quad f = 50\,\text{Hz}$$

The equivalent circuit of a bearing consisting of a journal with a load-carrying oil film is considered as a series resistor-capacitor (RC) circuit. The impedance of bearing equivalent circuit is determined as [5]:

$$Z_b = R_b - jX_{cb} \tag{15.16}$$

And the ratio of capacitive reactance to active resistance is given as [Eqns. (15.10), (15.14) and (15.15)]:

$$\frac{X_{cb}}{R_b} = \frac{1}{w\rho\xi} \tag{15.17}$$

15.2.4 Determination of Stored Charge on Bearings

When a journal bearing is exposed to an electric current, the resultant/equivalent capacitance of the bearing opposes any change in the existing voltage and causes an electric charge to be stored. The stored charge (Q) on the bearing surfaces is determined as [5]

$$Q = VC_{\mathrm{b}} \tag{15.18}$$

15.2.5 Lubricant Characteristics and Bearing Behavior

When a bearing using a high-resistivity lubricant (10^{14} ohm cm) having viscosity 43–48 cSt at 40 °C and viscosity index 95 (min) with relative permittivity (ξ_{r}) of 2.2 operates under electric fields, the current does not begin to flow until the first threshold voltage is reached. At the first threshold voltage (the increase in voltage from zero to a specific maximum value with negligible current flow), the increase in current flow is instantaneous (Chapter 4). Beyond the first threshold voltage, even a further small increase in voltage results in a very high increase in the current similar to rolling-element bearings using high-resistivity lubricant.

15.3 Optimum Clearance Ratios and Safe Load-Carrying Capacity

For assessment of optimum value of clearance ratio at different eccentricity ratios, the following relations are used to determine S, ε, and p [1, 3].

$$S = \frac{2.08}{(1 - \varepsilon)} \tag{15.19}$$

where S, the loading coefficient is given by [3]:

$$S = \frac{2p\psi^2}{\eta W} \quad \text{and} \quad p = \frac{S\eta W}{2\psi^2} \tag{15.20}$$

Also, W and p are determined as:

$$W = \frac{2\pi n}{60}, \; p = \frac{P}{Ld} \tag{15.21}$$

The maximum safe load-carrying capacity per unit area (at 120 °C exit oil temperature from load-carrying oil film) at different operating speeds, clearance and eccentricity ratios has been worked out for the oil (viscosity 43–48 cSt at 40 °C and about 11–12 cSt at 90 °C), and shown in Fig. 15.6 – variation of safe load-carrying capacity with clearance ratio at different rpm of operation – of reference [3].

15.4 Theoretical Analysis of the Estimation of Bearing Life Under the Influence of the Shaft Voltages

For establishing the reduction in bearing life under the effect of shaft voltages, the following analysis has been carried out.

15.4.1 Reduction in Bearing Life

Instantaneous charge leakage, accumulated on the bearing surfaces, leads to rise in contact temperature between the journal and hydrodynamic journal bearing lubricated with a high-resistivity lubricant under the influence of different levels of shaft voltages. An analysis has been carried out for the rise of contact stresses by the instantaneous charge leakage, where stresses lead to the initiation of craters by the arcing effect, resulting in subsequent damage to the bearing liner. Charge leakage between high 'points' of journal and a bearing liner during momentary contact in the zone of load-carrying oil film has been used to establish heat generated and instantaneous temperature rise in each shaft rotation.

Using this temperature rise, the contact stresses are determined and a reduction in bearing life compared to the bearing operating without the influence of shaft voltage is established. Subsequently, the safe limit of a shaft voltage has been assessed for the reliable operation of a bearing operating under the influence of shaft voltage.

15.4.2 Determination of Width of Contact Between the Journal and Bearing

To determine the width of contact around the circumference between the journal and a hydrodynamic bearing, the transformation of relation of the width of contact between two rollers is used. Rollers of diameters d_1 and d_2 in contact under the action of load P forms a rectangular surface of deformation of a length equal to the length of rollers L. Width of contact on these rollers is given as [6]:

$$W_r = 2.15\left[\frac{Pd_1d_2}{EL(d_1+d_2)}\right]^{1/2} \tag{15.22}$$

To transform the above relation for a journal bearing, d_1 is taken as the journal diameter (d) and d_2 as bearing diameter ($C_d + d$). Thus, the width of contact on bearing circumference is determined as:

$$W_b = 2.15d\left[\frac{p\left(1+\frac{1}{\psi}\right)}{E}\right]^{\frac{1}{2}} \tag{15.23}$$

15.4.3 Duration of Circumferential Contact Between Journal and Bearing

Duration of circumferential contact between the journal and a bearing crossing the zone of load-carrying oil film is determined as W_b/V_n, and is given as:

$$t_{in} = \frac{0.68}{f_s}\left[\frac{p\left(1+\frac{1}{\psi}\right)}{E}\right]^{\frac{1}{2}} \quad (15.24)$$

15.4.4 Heat Generated and Instantaneous Temperature Rise of Bearing Liner by Instantaneous Charge Leakage in the Zone of Load-Carrying Oil Film Between the Journal and Bearing

When a bearing lubricated with high-resistivity oil is operated under the influence of shaft voltage, current does not flow through the bearing and charge gets accumulated on the journal surface until the shaft voltage is below the threshold voltage. However, slow passage of charge takes place through the bearing when the high 'points' on the journal and the bearing in the zone of load-carrying oil film come close or when the conducting particles bridge the oil film. These conducting paths are broken when high points become separated by greater thickness of oil film. Under such conditions, arcing results due to instantaneous charge leakage under different operating conditions by vibration effects due to various reasons. This gives rise to instantaneous minimum resistance between the journal and bearing, equivalent to that of bearing resistance under static conditions (R_s).

The instantaneous charge leakage between the journal and a bearing generates heat instantly. The electrical energy q due to charge leakage Q for a duration t_{in}, at the first threshold voltage V, through a contact resistance R_s between the journal and bearing having capacitance C_b, is given as:

$$q = I_b^2 R_s t_{in} \quad (15.25)$$

This energy influences the zone of load-carrying oil film (of width W_b, length L_b and depth H), and increases its temperature after certain number of cycles. Hence, by using Eqns. (15.18), (15.24) and (15.25) and neglecting loss of heat by oil flow and other means, the instantaneous temperature rise is determined as:

$$T_{in} = \frac{Q^2}{\pi C_b^2 f_s R_s L_b d H \rho_1 c} \quad (15.26)$$

15.4.5 Instantaneous Temperature Rise due to Charge Leakage in the Zone of Load-Carrying Oil Film Between the Journal and Bearing in each Shaft Rotation

The number of times the journal passes over the W_b in each shaft rotation is $\pi d/W_b$. Assuming that 50 per cent of the journal circumference has high 'points' which come in contact with the high 'points' of W_b – the width of contact on a bearing liner in the zone of load-carrying oil film, gives the instantaneous temperature rise at any single position on a bearing liner in each shaft rotation as T_{in} times the $\pi d/2W_b$. This is simplified using Eqns. (15.23) and (15.26), and given as:

$$T_n = \frac{Q^2}{4.30d}\left[\frac{p}{E}\left(1+\frac{1}{\psi}\right)\right]^{\frac{1}{2}} C_b^2 f_s R_s L_b H \rho_1 c \tag{15.27}$$

15.4.6 Thermal Stresses due to Thermal Transients on Bearing Liner

During the operation of a hydrodynamic journal bearing, from the initial-conditions, a transient thermal gradient due to charge leakage occurs first. If at an instant, bearing liner temperature is known, then the transient thermal gradients can be determined at various time intervals. In addition, the thermal stresses can be evaluated for any instant of time/point in time. The overall temperature rise of bearing liner due to charge leakage for a short interval (10^{-3} s) approaches zero. However, instantaneous rise in contact stresses on the high 'points' of the liner are determined as [6]:

$$\sigma_{in} = \frac{-\alpha E T_{in}}{(1-\mu)} \tag{15.28}$$

And initial contact stresses without the influence of shaft voltage and charge leakage are given as:

$$\sigma_o = \frac{-\alpha E T_o}{(1-\mu)} \tag{15.29}$$

The operating temperature of a bearing affects the contact stresses. However, the instantaneous temperature rise of liner due to charge leakage in each shaft rotation may set residual contact stresses, which are determined as:

$$\sigma_n = \frac{-\alpha E (T_n + T_o)}{(1-\mu)} \tag{15.30}$$

15.4.7 Determination of Decrease in Life Span of Bearing Liners due to the Initiation of Craters under Various Levels of Shaft Voltages

Due to charge leakage on the high 'points', the contact stresses increase in due course on the bearing liner. The ratio of these stresses as compared to the original stresses without the effect of shaft voltages is determined as [by Eqns. (15.29) and (15.30)]:

$$\frac{\sigma_n}{\sigma_o} = 1 + \frac{T_n}{T_o} \tag{15.31}$$

The ratio of σ_o/σ_n gives the indication of a decrease in the life of bearing liner under the effect of shaft voltages as compared to the original life. The ratio of percentage increase in stresses compared to the original stresses [$(\sigma_n - \sigma_o)/\sigma_o$ per cent] gives the percentage reduction in bearing liner life on the high 'points' before the initiation of craters.

15.5 Data of Theoretical Analysis

Values of C_b, R_b, X_{cb}, and Z_b for the hydrodynamic cylindrical bearings with various *L*/*d* and clearance ratios have been determined at different eccentricity ratios using Eqns. (15.10), (15.14), (15.15) and (15.16). Figure 15.2 shows variation of bearing capacitance (C_b) with clearance ratio (ψ) at different values of ε and δ. Figure 15.3 shows the variation of capacitance (C_b) with eccentricity ratio (ε) at different values of δ and ψ.

Similarly variation of resistance (R_b) with clearance and eccentricity ratios (ψ, ε) is shown in Figs. 15.4 and 15.5. Figure 15.6 shows variation of bearing capacitive reactance (X_{cb}) with clearance ratio (ψ) at different values of eccentricity (ε) and length/diameter (δ) ratios.

The optimum variation of capacitance with speed is shown in Fig. 15.7 for bearings with *L*/*d* (δ) of 0.7 and 1.5. However, for study of this variation the optimum values of clearance (ψ) and eccentricity (ε) ratios used for the determination of safe load-carrying capacity has been used [3].

By the field data analysis of various bearings operating under normal conditions, the stabilized bearing temperature has been taken as 50 °C. And σ_n has been determined for different values of shaft voltages varying from 0.25 V to 2 V to determine the ratio of σ_o/σ_n of a bearing operating under stabilized regime and having the parameters given in Table 15.1. The shaft voltages are taken less than the first threshold voltage to avoid the effect of chemical decomposition of the oil. For various bearing parameters for $L/d = 1$ and $d = 300$ mm see Table 15.1 [3].

Table 15.1. Instantaneous temperature rise, contact stresses, contact duration and reduction in bearing liner life due to charge leakage between the journal and hydrodynamic journal bearing under the influence of different levels of shaft voltages

Analytical values of temperature rise at different levels of charge accumulation and shaft voltages							
	(1)	(2)	(3)	(4)	(5)	(6)	(7)
V (V)	0.25	0.5	0.75	1.0	1.25	1.5	2.0
$Q \times 10^{-9}$(C)	6.25	12.5	18.75	25	31.25	37.5	50
T_n (°C)	0.67	2.68	6.0	10.72	16.75	24.12	42.88
σ_o/σ_n	0.99	0.95	0.89	0.82	0.75	0.67	0.54
% reduction in bearing life	1	5	11	18	25	33	46
Ratio of temperature rise as against 0.5 V	–	1	2.24	4.01	6.25	9	16
Ratio of reduction in bearing life as against 0.5 V	–	1	2.20	3.60	5	6.6	9.2

Bearing dimensions, operating parameters and analytical values $d = 300$ mm, $L = 300$, $L/d(\delta) = 1$, $n = 1500$ rpm, $f_s = 25$, $C_d/d(\delta) = 0.0015$, $\varepsilon = 0.82$, $p = 0.2\,\text{Kg mm}^{-2}$, $\rho = 10^{14}$ohm cm, $H = 0.001$ mm, $R_s = 0.4$ ohm, $\xi = 2.5 \times 8.85419 \times 10\,\phi\, m^{-1}$, $\xi_r = 2.5$, $T_o = 50\,°\text{C}$, $C_b = 25 \times 10^{-9}\phi$ [Eqn.(9.10)], $W_b = 117.84$ mm, $t_{in} = 5 \times 10^{-3}$ s, $L_b = 2\,\%$ of L in each contact (W_b) times the 50% of circumferential high 'points' contact in each shaft rotation i.e. $(2L/100)\,(\pi d/2W_b) = 23.99$ mm, $\rho_1 = 7.46 \times 10^3\,\text{kg m}^{-3}$, $c = 2.26 \times 10^2\,\text{ws kg}^{-1}\,°\text{C}^{-1}$, $\mu = 0.36$, $E = 40 \times 10^9\,\text{N m}^{-2}$.

15.5.1 Evaluation of Safe Load-Carrying Capacity by Determining Bearing Capacitance

For evaluating the safe load-carrying capacity, firstly the optimum value of capacitance of a bearing with given L/d ratio at the operating speed is determined by Fig. 15.7. The capacitance, thus determined, is used to evaluate the clearance ratio at the desired eccentricity ratio (Figs. 15.2 and 15.3). The values of clearance and eccentricity ratios, thus established, are used to work out the safe load-carrying capacity by Eqns. (15.19), (15.20) and (15.21).

15.6 Discussion on Theoretical Analysis

15.6.1 Bearing Capacitance

Bearing capacitance decreases with increase in clearance ratio. However, with increase in L/d (δ) at constant clearance ratio, the bearing capacitance increases

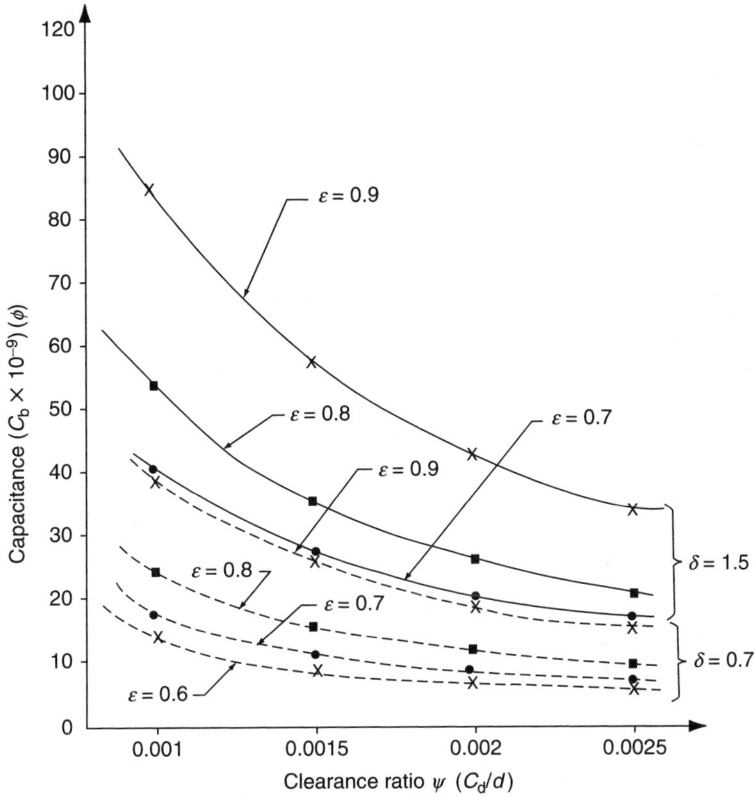

Fig. 15.2. Variation of bearing capacitance (C_b) with clearance ratio (ψ) at different eccentricity ratios (ε) for length/diameter ratio (δ) of 0.7 and 1.5 of bearing with diameter 300 mm

(Fig. 15.2). Similarly, capacitance increases with eccentricity ratio at constant *L/d* and clearance ratio (Fig. 15.3). The capacitance of 300 mm diameter bearing with different *L/d* ratio vary between 5.62 $n\phi$ to 84.51 $n\phi$ under different operating conditions (Fig. 15.2).

For the bearing with $L/d = 1.5$ and $C_d/d = 0.0025$, the change in capacitance is from 15.29 to 33.81 $n\phi$ as against 38.19 to 84.51 $n\phi$ for $C_d/d = 0.001$ and $L/d(\delta) = 1.5$ when eccentricity ratio is varied from 0.7 to 0.9 (Figs. 15.2 and 15.3). For the same eccentricity variation for the bearing with $C_d/d = 0.0025$ and $\delta = 0.7$, the change in capacitance is from 7.13 $n\phi$ to 15.78 $n\phi$ as against 17.82 $n\phi$ to 39.44 $n\phi$ for $C_d/d = 0.001$ and $\delta = 0.7$ (Fig. 15.2).

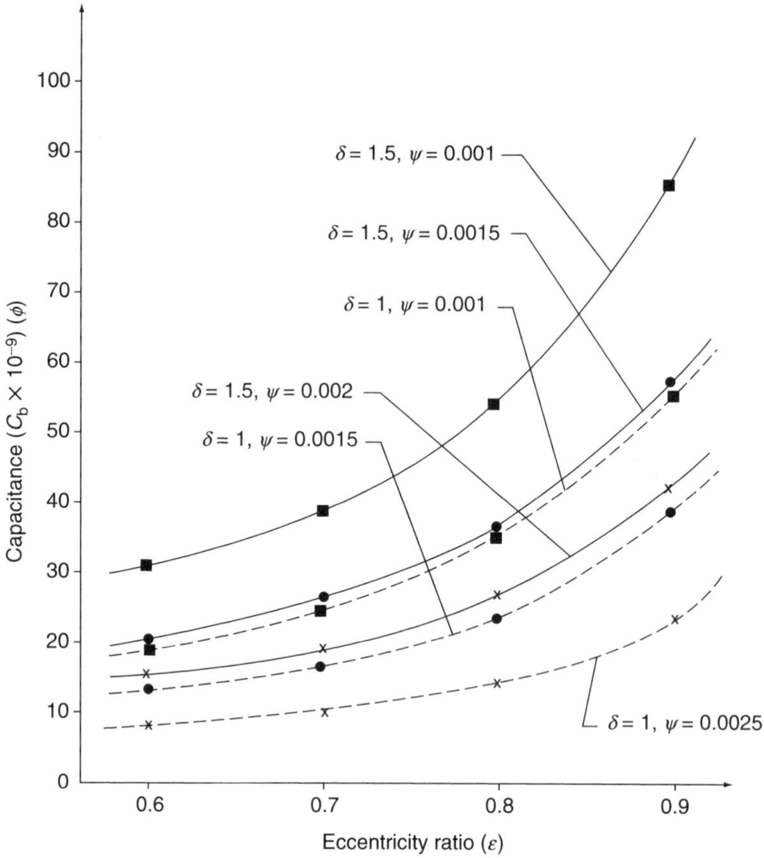

Fig. 15.3. Variation of bearing capacitance (C_b) with eccentricity ratio (ε) at length/diameter ratio (δ) of 1 and 1.5 with different clearance ratios (ψ) of 300 mm diameter bearing

15.6.2 Capacitive Reactance, Resistance and Impedance of Hydrodynamic Journal Bearings

The capacitive reactance of a hydrodynamic journal bearing is about 10^4 times less than that of its resistance under identical operating conditions. The capacitive reactance (X_{cb}) of a bearing using lubricant of resistivity 10^{14} ohm cm and relative permittivity of 2.2 is of the order of 10^5 ohms as against active resistance of 10^9 ohms (Figs. 15.4–15.6). Capacitive reactance of the bearing vary between 0.56×10^5 ohms and 5.66×10^5 ohms. It decreases with increase in eccentricity ratio for a bearing with constant C_d/d ratio. However, it increases with the bearing clearance.

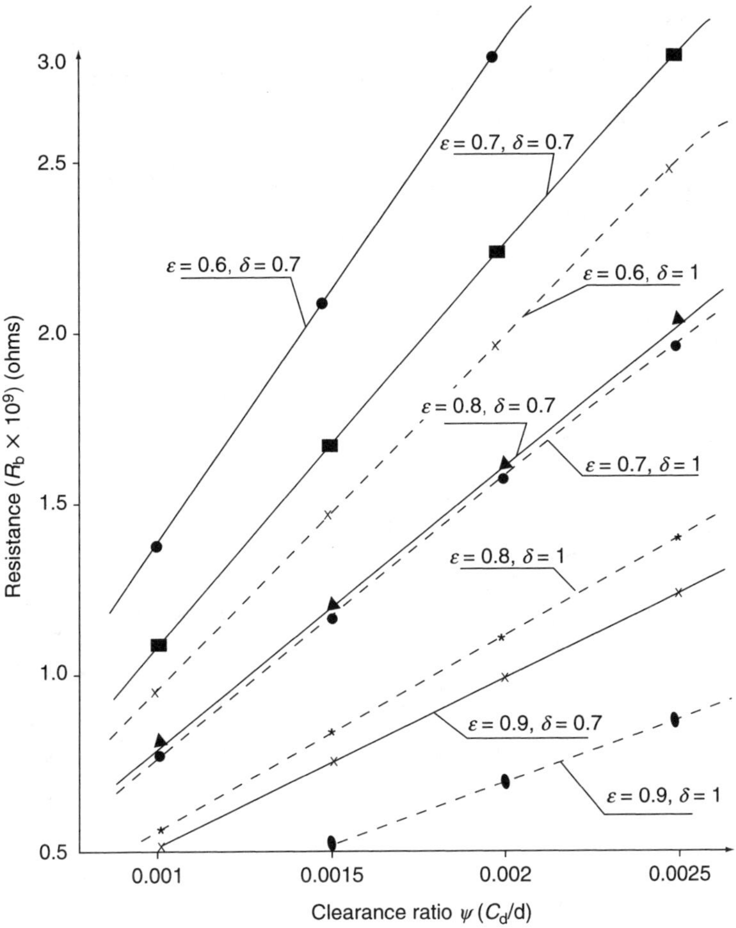

Fig. 15.4. Variation of bearing resistance (R_b) with clearance ratio (ψ) at different eccentricity ratios (ε) for $\delta = 0.7$ and 1 for bearing of diameter 300 mm

For $\delta = 0.7$ and $C_d/d = 0.001, X_{cb}$ decreases from 2.26×10^5 to 0.81×10^5 ohms as ε varies from 0.6 to 0.9. And X_{cb} increases from 0.56×10^5 to 1.41×10^5 ohms for the bearing with $\delta = 1$, $\varepsilon = 0.9$ as C_d/d varies from 0.001 to 0.0025 (Fig. 15.6).

On the contrary, capacitive reactance decreases with increase in *L*/*d* ratio under the fixed clearance and eccentricity ratio. For $C_d/d = 0.0025$ and $\varepsilon = 0.9$, the X_{cb} decreases from 2.02×10^5 to 1.41×10^5 ohms as $L/d(\delta)$ increases from 0.7 to 1 (Fig. 15.6).

The variation of resistance with ε, δ is in the same pattern as that of the capacitive reactance (Figs. 15.4 and 15.5). Resistance decreases with increase in

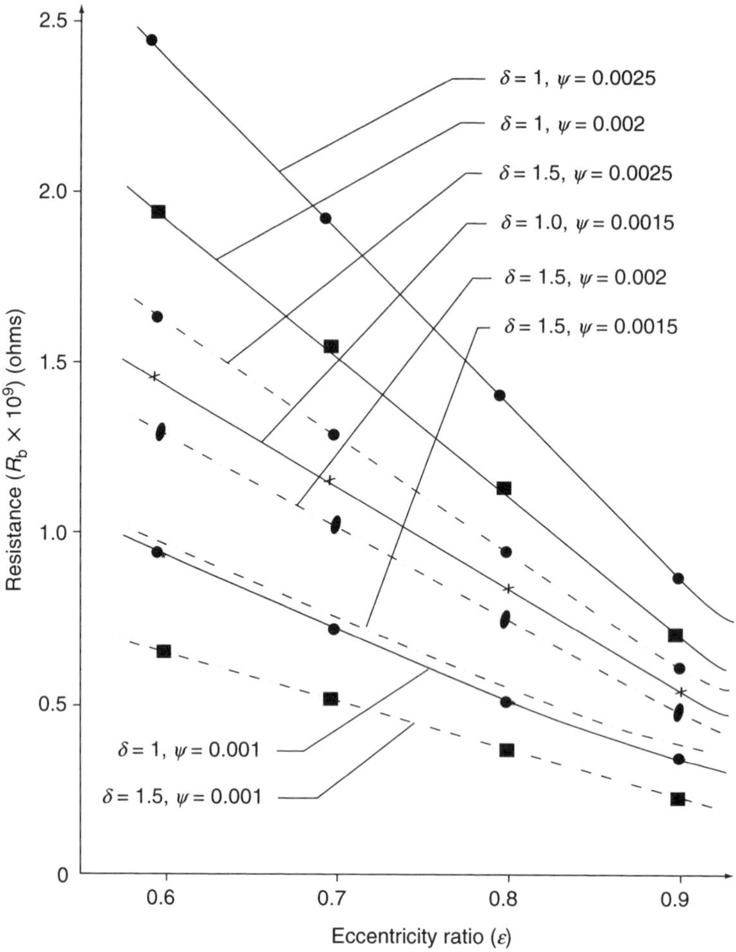

Fig. 15.5. Variation of bearing resistance (R_b) with eccentricity ratio (ε) at different clearance ratios (ψ) for $\delta = 1$ and 1.5 of 300 mm diameter bearing

eccentricity ratio for a bearing with constant C_d/d ratio. However, it increases with increase in C_d/d ratio. For $\delta = 1.5$ and $\psi = 0.002$, R_b decreases from 1.30×10^9 to 0.46×10^9 ohms as ε varies from 0.6 to 0.9 (Fig. 15.5). And R_b increases from 1.38×10^9 to 3.45×10^9 ohms for the bearing with $\delta = 0.7$ and $\varepsilon = 0.6$ as C_d/d increases from 0.001 to 0.0025 (Fig. 15.4). However, resistance decreases with increase in δ under the constant clearance and eccentricity ratios. For $\psi = 0.0025$ and $\varepsilon = 0.8$, the R_b decreases from 2.0×10^9 to 0.94×10^9 ohms as δ increases from 0.7 to 1.5 (Figs. 15.4 and 15.5).

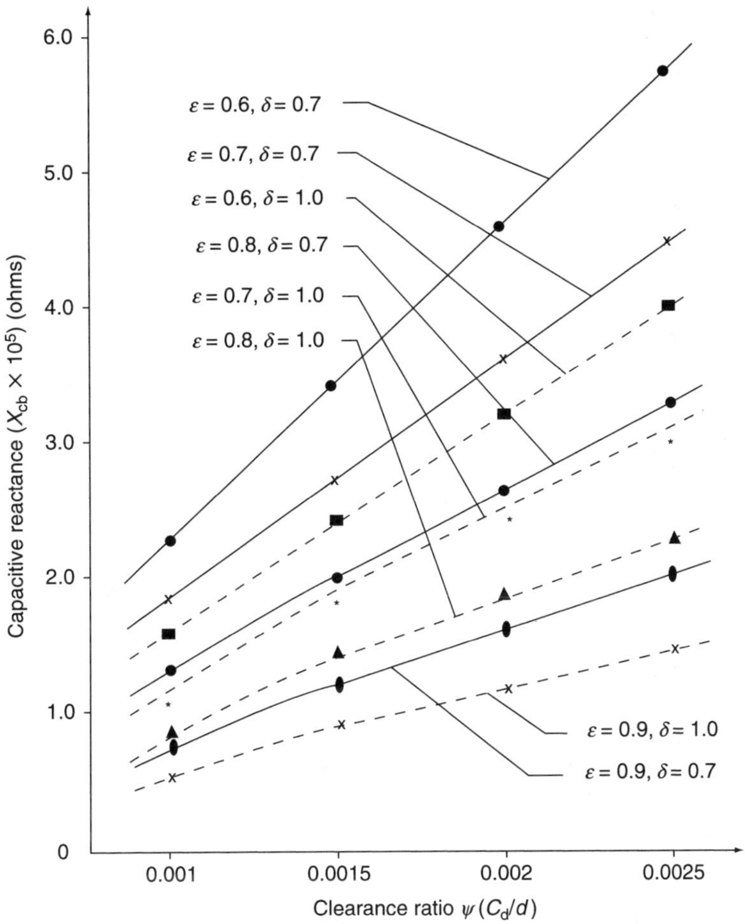

Fig. 15.6. Variation of bearing capacitive reactance (X_{cb}) with clearance ratio (ψ) at different eccentricity ratios (ε) for length/diameter (δ) ratio of 0.7 and 1 for bearing of diameter 300 mm

Impedance of a hydrodynamic journal bearing is governed by the resistance and is not considerably affected by the capacitive reactance. This is because the R_b is 10^4 times more than X_{cb}. In general, impedance of a bearing is of the same order as that of the bearing resistance (Figs. 15.4 and 15.6).

15.6.3 Variation of Bearing Capacitance with Speed

The bearing capacitance decreases with increase in speed. However, it increases with increase in L/d ratio. For $L/d = 1.5$, the bearing capacitance decreases from

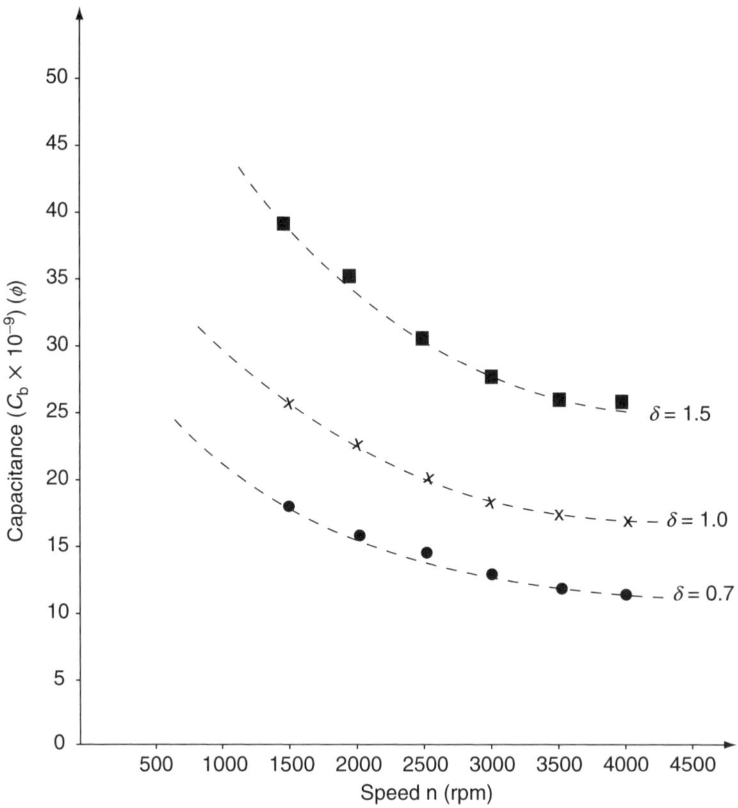

Fig. 15.7. Optimum ratio of bearing capacitance (C_b) with speed at different l/d (δ) ratios

$38.5\,n\phi$ to $24.5\,n\phi$ as against $25\,n\phi$ to $16\,n\phi$ for $L/d = 1$, and $17.5\,n\phi$ to $11\,n\phi$ for $L/d = 0.7$ as the speed changes from 1500 to 4000 rpm (Fig. 15.7). For the safe and reliable operation of a bearing the capacitance should be less than that depicted, as shown in Fig. 15.7, at any operating speed.

15.6.4 Ratio of Capacitive Reactance and Active Resistance

With increase in eccentricity ratio, the bearing capacitance increases but resistance and capacitive reactance reduce (Figs. 15.3 and 15.5). However, the ratio of capacitive reactance to the active resistance is constant and is less than unity (1.64×10^{-4}) up to the first threshold voltage for a bearing using oil having resistivity 10^{14} ohm cm and a relative permittivity of 2.2. It shows that bearing accumulates the electric charges up to the first threshold voltage. On the contrary, a bearing using low-resistivity lubricant (10^7 ohm cm) passes current without indicating threshold voltage phenomenon. Furthermore, passage of current disintegrates the low-resistivity lubricant and bearing fails in due course. This suggests that the X_{cb}/R_b should be less than unity for a bearing to show capacitive response similar to that of the rolling-element bearings (Chapter 4).

15.6.5 Width of Contact Between the Journal and Liner in the Zone of Load-Carrying Oil Film

The width of contact between the journal and liner in the zone of load-carrying oil film depends on clearance ratio (C_d/d), load per unit area (p) and liner material properties. Width of contact (W_b) increases with increase in load on a bearing. However, it is independent of L/d ratio [Eqn. (15.23)]. For bearing of 300 mm diameter having $C_d/d = 0.0015$ operating at 1500 rpm with $p = 0.2\,\text{Kg}\,\text{mm}^{-2}$, W_b is 117.84 mm and duration of contact in the zone of load-carrying oil film is 5×10^{-3} s (Table 15.1).

15.6.6 Temperature Rise in the Zone of Load-Carrying Oil Film Between the Journal and Bearing

The temperature rise in the zone of load-carrying oil film between the journal and a bearing depends on the journal diameter, the clearance ratio and the summation of length of high 'points' contact in each shaft rotation (L_b). The rate of temperature rise under the influence of different levels of shaft voltages decreases as the frequency of rotation increases, but increases as the square of the ratio of charge accumulation to bearing capacitance as shown in Eqn. (15.27).

For 300 mm diameter bearing having capacitance of $25 \times 10^{-9}(\phi)$, and operating at 1500 rpm with $\varepsilon = 0.82$, the rate of temperature rise at the high 'points' on the bearing surfaces increases as the charge accumulation/shaft voltage increases. As the shaft voltage increases from 0.25 to 2 V, the rate of temperature rise at the high 'points' in each shaft rotation (T_n) increases from 0.67 to 42.88 °C (Table 9.1). Up to a shaft voltage of 0.5 V, T_n is 2.68 °C, which increases steeply to 4.01, 6.25, 9.0 and 16 times as the shaft voltage rises to 1, 1.25, 1.5 and 2 Volts respectively (Table 15.1). However, below 0.5 V, temperature rise as well as rate of increase is much less as compared to that of above 0.5 V. The similar trend is found for various sizes of bearings under different conditions of operation.

15.6.7 Ratio of Thermal Stresses and Percentage Reduction in Bearing Life

The ratio of original stresses without the effect of instantaneous charge leakage/shaft voltage to that of the bearings operating under different levels of shaft voltages

(σ_o/σ_n) decreases from 0.99 to 0.54 as the shaft voltage increases from 0.25 to 2 V. Subsequently, $(\sigma_n - \sigma_o)/\sigma_o$ per cent, i.e. the minimum reduction in bearing life before appearance of craters on liner increases from 1 to 46 per cent. However, up to 0.5 V shaft voltage, the reduction in bearing liner life is only 5 per cent. When shaft voltage exceeds 0.5 V liner life reduces very quickly. The bearing life decreases from 2.2 to 9.2 times at 0.75 and 2 V as compared to 0.5 V shaft voltage. Below 0.5 V shaft voltage, reduction in bearing life is very much less as compared to that above 0.5 V (Table 15.1). It is therefore recommended that shaft voltage should not be allowed to reach more than 0.5 V, to guarantee safe and reliable operation of a hydrodynamic journal bearing. This closely matches with the recommendation of leading manufacturers of the rotating machines.

15.7 Damages Resulting from Shaft Voltages

Shaft voltages and resulting bearing currents lead to damages caused by arcing across bearing and shaft surfaces. The main consequence of arcing is the wear of soft bearing (Babbitt) material and rapid accumulation of wear debris in the lubricating oil. The electrical wear of the bearing and the deterioration of the lubricant accelerate the mechanical wear and hence results in further destruction of the bearings. This excessive wear results in the loss of their load-carrying capacity [7].

It is observed that oil films in hydrodynamic bearings cease to be insulators and behave like perfect conductors when subjected to high voltages or fast changing voltages [7]. The voltage at which this happens is similar to threshold voltage phenomenon occurring in rolling-element bearings lubricated with high-resistivity lubricants. Also this depends on oil film thickness, the rate of change of voltage and is accompanied by an arc. The current flows in the arc when high points on the asperities of the journal come close to or contact similar points on the bearing, or when the fused metal particles pass through the oil film [8]. The current flow is intermittently broken as the asperities are separated by the motion of the shaft. This interruption takes place in a microsecond or less, which causes high-induced voltages and accumulation of charges across the oil film due to the self-inductance. This is, however, a running-in effect and once running-in conditions is complete, this effect is removed under running-in conditions. Charge accumulation depends on bearing capacitance and may cause sustained arcing at the asperities contacts between journal and the bearing resulting in metal fusing and bearing damage. However, the shaft is less sensitive to arcing effects due to its higher melting point.

15.8 Conclusions

The following conclusions are drawn based on the analysis carried out in this chapter [9, 10]:

(1) Bearing capacitance increases with increase in eccentricity and *L*/*d* ratios, but reduces with increase in bearing clearance and operating speed.

(2) Bearing resistance and capacitive reactance increase with clearance ratio, but decrease with increase in eccentricity ratio.

(3) Bearing resistance and capacitive reactance decrease with increase in *L*/*d* ratio under the constant clearance and eccentricity ratios.

(4) The resistance of a bearing using oil with resistivity 10^{14} ohm cm and a relative permittivity of 2.2 is 10^5 times higher than the capacitive reactance.

(5) The impedance of a bearing is mainly governed by its resistance than its capacitive reactance.

(6) The ratio of capacitive reactance to active resistance depends on permittivity and resistivity of the lubricant.

(7) For the capacitive response of a bearing the ratio of capacitive reactance to active resistance should be less than unity.

(8) The width of circumferential contact between the journal and bearing depends on clearance ratio, the load per unit area and the liner material properties. It increases with load and is independent of the *L*/*d* ratio.

(9) The instantaneous temperature rise due to charge leakage in an oil film depends on the journal diameter, the clearance ratio and the summation of the length of high 'points' contact in each shaft rotation.

(10) Temperature rise under the influence of shaft voltage decreases with increase in frequency of rotation. However, it increases as the square of the ratio of charge accumulation to the bearing capacitance increases.

(11) Instantaneous thermal stresses on the liner depend on instantaneous temperature rise due to charge leakage.

(12) Temperature rise of liner below 0.5 V shaft voltage is much less and does not have a significant effect on the bearing life. Reduction in bearing life is 5 per cent at 0.5 V as against 46 per cent at 2 volts shaft voltage.

(13) At 2 V shaft voltage temperature rise of liner increases 16 times and its life decreases by a factor of 9.2 as compared to 0.5 V shaft voltage.

(14) For safe, reliable operation and adequate life of a bearing, shaft voltage of 0.5 V must not be exceeded.

The developed technique may have a potential as an effective tool to assess performance of hydrodynamic journal bearings and may act as a guide to select the design parameters for safe and reliable operation of a cylindrical journal bearing, and also, to assess reduction of life of a journal bearing under the effect of shaft voltages.

REFERENCES

1. Yakhin, Z. A. (1970). Determination of oil temperature in load-carrying oil film in journal bearing. *Russian Eng. J.*, YDK 621.436.233, 21.00, 1.5, 21–23.
2. Heshmat, H. and Pinkus, O. (1986). Mixing inlet temperatures in hydrodynamic bearing. Transactions of ASME, *J. Tribol.*, **108**, 231–248.
3. Prashad, H. (1988). The effects of viscosity and clearance on the performance of hydrodynamic journal bearings. *STLE Trans.*, **31**(1), 113–119.
4. Rodstein, L. (1974). *Electrical Control Equipment*, Mir Publishers, Moscow.
5. Prashad, H. (1988). Theoretical evaluation of impedance, capacitance and charge accumulation on roller bearings operated under electrical fields. *Wear*, **125**, 223–239.
6. Warnock, F. V. and Benham, P. P. (1965). *Mechanism of Solids and Strength of Materials*, Sir Isaac Pitman and Sons, London.
7. Verma, S. P., Girgis, R. S. and Fleming, R. J. (1975). The problems and failures caused by shaft-potentials and bearing-currents in turbo-generators. Method and Prevention Report, Power System Research Group. Department of Electrical Engineering., University of Saskatchewan (Canada).
8. Prashad, H. (1990). Theoretical analysis of the effects of instantaneous charge leakage on roller tracks of roller bearings lubricated with high-resistivity lubricants under the influence of electrical current. ASME, *J. Tribol.*, **112**, 37–43.
9. Prashad, H. (1991). Theoretical evaluation of capacitance, capacitive reactance, resistance and their effects on performance of hydrodynamic journal bearings. Transactions of ASME, *J. Tribol.*, **113**, 762–767.
10. Prashad, H. (1991). Theoretical evaluation of reduction in life of hydrodynamic journal bearings operating under the influence of different levels of shaft voltages. *Tribol. Trans.*, **44**(4), 623–627.

Nomenclature

c	specific heat of bearing liner
C_b	bearing capacitance
C_d	bearing diameter clearance
d	shaft diameter
d_1	diameter of roller 1
d_2	diameter of roller 2

E	Young's modulus of elasticity
f	frequency of applied voltage
f_s	shaft rotational frequency
F_b	bearing conductance (ohm^{-1})
h	oil film thickness
h_o	minimum oil film thickness
H	depth of craters on bearing liner
I_b	bearing current
l	peripheral length of load-carrying oil film
L	bearing length
L_b	summation of length of high 'points' contact in the zone of load-carrying oil film between the journal and bearing in each shaft rotation
n	rpm
p	load per unit area (P/Ld)
P	bearing load
q	electrical energy due to charge leakage between the journal and bearing liner
Q	stored electric charge
R_b	bearing resistance in operating conditions
R_s	bearing resistance in static conditions
S	loading coefficient
t_{in}	duration of contact between the journal and bearing to cross zone of load-carrying oil film during operation
T_o	rise of bearing temperature without the effect of charge leakage
T_{in}	instantaneous rise of bearing contact temperature due to charge leakage during passage of the journal in the zone of load-carrying oil film
T_n	bearing temperature rise due to charge leakage in each shaft rotation
V	shaft voltage
V_n	circumferential shaft speed ($\pi d f_s$)
w	$2\pi f$
W	$2\pi n/60$
W_b	width of contact between the journal and bearing in the zone of load-carrying oil film
W_r	width of contact between rollers
X_{cb}	capacitive reactance of bearing
Z_b	bearing impedance
α	coefficient of thermal expansion
$\delta = L/d$	length to diameter ratio
ε	eccentricity ratio

η	dynamic oil viscosity
μ	Poisson's ratio
θ	attitude angle
ρ	oil resistivity
ρ_1	density of bearing material
σ_{in}	instantaneous rise in contact stresses due to charge leakage on a ring liner in the zone of load-carrying oil film
σ_n	contact stresses due to charge leakage on part of the bearing liner in each shaft rotation
σ_o	contact stress on bearing liner in the zone of load-carrying oil film without the effect of charge leakage
ξ	permittivity/dielectric constant of oil (ϕm^{-1})
ξ_r	relative permittivity of oil
$\psi = C_d/d$	diameteral clearance ratio

Chapter 16

ANALYSIS OF CAPACITIVE EFFECT OF JOURNAL BEARINGS ON REPEATED STARTS AND STOPS OF A MACHINE AND MINIMUM NUMBER OF CYCLES FOR THE FORMATION OF CRATERS ON THE LINER SURFACE UNDER THE INFLUENCE OF SHAFT VOLTAGES

16.1 Introduction

A number of surveys have indicated that large numbers of bearing failures are due to bearing current. The origin of shaft voltage phenomenon in electric machines and flow of current through bearings and lubricated contacts have been dealt in different chapters besides electrical parameters of various bearings [1–8]. Various authors have reported that the passage of current through a hydrodynamic journal bearing in a zone of load-carrying oil film, which causes craters to develop over time and damages the bearing surface [9]. Also, it causes loss of load-carrying capacity of the bearing over time [10]. Theoretical analysis on the effects of instantaneous charge leakage on roller tracks of roller bearings and reduction in life of hydrodynamic journal bearings under the influence of electric current has been analyzed in the Chapters 7, 8 and 15 [11–12].

In this chapter, a study is reported on the capacitive effect and life estimation of hydrodynamic journal bearings on repeated starts and stops of a machine operating under the influence of shaft voltages. The purpose is to determine the increase in charge accumulation on the bearing liner with time when the machine is started and the gradual leakage of the accumulated charges on the liner as the shaft voltage falls when the power supply to the machine is switched-off.

Under these conditions, the variation of shaft revolutions to accumulate charges and discharge to the accumulated charges on the liner surface of hydrodynamic journal bearings at various levels of bearing-to-shaft voltage is analyzed. Also, the variation of safe limits of starts and stops with the ratio of bearing-to-shaft voltage is studied.

Furthermore, it has been established that during rotation, the oil film thickness of a hydrodynamic journal bearing depends on clearance and eccentricity ratios, which in turn depend on load, oil viscosity and operating speed [13, 14]. A passage of current due to various reasons through a bearing in a zone of load-carrying oil film develops craters on the liner surface in due course and damages it. That is why bearings are insulated to restrict the flow of current through them. However, in the event of insulation damage or in non-insulated bearings or by other unforeseen causes (deposition of dirt, foreign particles on the side planes of the insulation separating the conducting surfaces) or when the journal and the bearing liner come in very close contact due to various instabilities of bearings and the system, flow of localized current through the bearing and journal becomes unavoidable. This damages the bearing liner and causes loss of load-carrying capacity of the bearings in due course [10].

The theoretical model developed in this chapter brings out the effects of instantaneous leakage of charge energy between the journal and a bearing lubricated with high-resistivity oil and operating under the influence of shaft voltages of different levels, so as to determine the minimum number of shaft revolutions/cycles for the formation of craters of different sizes on the bearing liner. The gradual rise in temperature at the 'high' points, because of the leakage of charge energy, causes softening of the bearing liner at 'high' points contact and reduction in its toughness, leading to the formation of craters of different dimensions in due course. This temperature rise has been considered in the model. Besides this, the volume of craters can be assessed by the developed model, provided the minimum number of shaft revolutions/cycles for the appearance of craters, under the known parameters of bearing operation and shaft voltage has been established experimentally. Also, by the assessment of dimensions and number of craters on the bearing liner after the identified span and known operating parameters of the bearing, the shaft voltage, which has caused the craters formation, can be estimated.

16.2 Theoretical Analysis Pertaining to Repeated Starts and Stops of a Machine

16.2.1 Theoretical Background

Magnetic flux develops in electric machines due to asymmetry of the magnetic circuits, which closes in the circumference over the yoke and induces the voltage on the shaft. Shaft voltage and flux can occur in the electric rotating machines due to various reasons, which usually result in localized currents at each bearing rather than a potential difference between shaft ends as discussed in Chapters 2 and 4.

It is evident that in a hydrodynamic journal bearing, the zone of minimum film thickness i.e. the load-carrying oil film varies along the circumference of a bearing through the bearing length. This forms a capacitor of varying capacitance between the journal and the bearing, depending on the permittivity of the lubricant, circumferential length of load-carrying oil film, bearing length, eccentricity ratio, and the clearance in a bearing. Besides this, the load-carrying oil film offers resistance depending on the operating parameters and resistivity of the lubricant. Thus, the load-carrying oil film offers impedance to current flow in a hydrodynamic journal bearing. This has been analyzed in Chapter 15.

16.2.2 Time Required to Accumulate Charges on Bearing Liner After Start of a Machine

At the instant when the machine is started, the potential difference (V) across the journal and bearing liner surface is zero. But this gradually increases and approaches the shaft voltage (E). While the shaft voltage increases, the charge (Q_b) on the bearing liner (bearing lubricated with high-resistivity lubricant (ρ) 10^{14} ohm cm) builds up. Until then V is changing, and transient current is delivered from the shaft voltage.

To maintain continuity throughout the whole circuit, there is a rate of change of flux (dQ_b/dT) within the dielectric (lubricant), which is given as [15, 16]:

$$I = \frac{dQ_b}{dT} \tag{16.1}$$

And, the stored charge on the bearing surfaces is determined as [16, 14]:

$$Q_b = V \times C_b \tag{16.2}$$

And if the current is varying then [15]:

$$I = C_b \times \frac{dV}{dT} \tag{16.3}$$

Also, it can be expressed as:

$$I = \frac{(E - V)}{R_b} \tag{16.4}$$

So from Eqns. (16.1)–(16.4), it is evident that:

$$V = E - R_b \times C_b \times \frac{dV}{dT} \tag{16.5}$$

On integrating Eqn. (16.5) and applying the initial conditions as $T = 0$, $V = 0$, the solution of Eqn. (16.5) is obtained as:

$$V = E \times \left(1 - e^{-T/C_b R_b}\right) \tag{16.6}$$

and so

$$Q_b = E \times C_b(1 - e^{-T/C_b R_b}) \tag{16.7}$$

The time taken (T_{cb}) to develop charge Q_b on liner of a bearing having capacitance C_b and resistance R_b is expressed as [using Eqn. (16.6)]:

$$T_{cb} = -C_b \times R_b \times \log_e (1 - a) \tag{16.8}$$

16.2.3 Current Passing through Bearing

From Eqns. (16.2) and (16.4), shaft voltage is given as:

$$E = I \times R_b + \frac{Q_b}{C_b} \tag{16.9}$$

Using Eqn. (16.1) and applying initial conditions as $T = 0$, $I = E/R_b$, the solution of Eqn. (16.9) is:

$$I = \left(\frac{E}{R_b}\right) \times e^{-T/R_b C_b} \tag{16.10}$$

and potential drop across a bearing is given as:

$$V = I \times R_b = E \times e^{-T/R_b C_b} \tag{16.11}$$

16.2.4 Time Required to Discharge the Accumulated Charges From Bearing Liner During Stop

As soon as the power supply to a machine is switched-off, magnetic flux disappears and the shaft voltage becomes zero very quickly. The rate of discharge of bearing capacitor (formed during bearing operation) is determined by differentiating Eqn. (16.11), and given as:

$$\frac{-dV}{dT} = \left(\frac{E}{C_b \times R_b}\right) e^{-T/C_b R_b} \tag{16.12}$$

At the instant, when machine is stopped ($T = 0$), the rate of change of bearing potential drop is determined as:

$$\frac{dV}{dT} = \frac{-E}{C_b \times R_b} \tag{16.13}$$

In time constant ($C_b \times R_b$), the potential drop of bearing capacitor falls and is determined as [by Eqn. (16.11)]:

$$V = E \times e^{-1} = 0.368E \tag{16.14}$$

The time required (T_{db}) to decrease potential drop V to the value of 'a' times the shaft voltage (E) ($V = a \times E$) is given as [by Eqn. (16.11)]:

$$T_{db} = -R_b \times C_b \log_e a \tag{16.15}$$

16.2.5 Duration/Time Taken by Journal to Cross the Zone of Load-Carrying Oil Film

The duration/time taken by the journal to cross the zone of load-carrying oil film is determined as W_b/V_n, and is given as [12]:

$$t_{bn} = \left[\frac{0.68}{F_s}\right] \times \left[\frac{p(1 + 1/\psi)}{Y}\right]^{1/2} \tag{16.16}$$

The number of time the journal passes over region W_b, the circumferential width of load-carrying oil film, in each shaft rotation is determined as β_{bn}, and is determined as [12]:

$$\beta_{bn} = \frac{\pi d}{W_b} \tag{16.17}$$

where

$$W_b = 2.15 \times d \times \left[\frac{p^{\times}(1 + 1/\psi)}{Y}\right]^{1/2} \text{[12]} \tag{16.18}$$

16.2.6 Number of Shaft Rotations for the Charge Accumulation on Bearing Liner After Start of a Machine

The number of times the journal crosses the zone of the load-carrying oil film of a bearing (N_{bc}) for the charge accumulation Q_b after start of a machine can be determined as the ratio of time taken (T_{cb}) to accumulate charges to the time taken for the journal to cross the zone of load-carrying oil film during operation (t_{bn}), and can be determined as:

$$N_{bc} = \frac{T_{cb}}{t_{bn}} \tag{16.19}$$

By using Eqns. (16.8) and (16.16), the N_{bc} is expressed as:

$$N_{bc} = -1.47 \times F_s \times \left[\frac{P(1 + 1/\psi)}{Y}\right]^{-1/2} \times C_b \times R_b \times \log_e(1 - a) \tag{16.20}$$

The number of shaft rotations (N_{bcn}) to accumulate the charges (Q_b) is determined as the ratio of N_{bc} to β_{bn}, i.e. the number of times part of the circumference of journal crosses the zone of load-carrying oil film to the number of such respective crosses of the journal in each shaft rotation. Thus, N_{bcn} is expressed as follows using Eqns. (16.17)–(16.20):

$$N_{bcn} = -F_s \times C_b \times R_b \times \log_e(1 - a) \tag{16.21}$$

16.2.7 Number of Shaft Rotations for the Discharge of the Accumulated Charges After the Stop of a Machine

The number of times that a part of the circumference of the journal crosses the zone of load-carrying oil film to discharge of the accumulated charges from the bearing liner (N_{bd}) after the power supply to a machine is switched-off, is determined similar to the Eqn. (16.19) on using the time required for the discharge of the accumulated charges [Eqn. (16.15)], and is given as:

$$N_{bd} = \frac{T_{db}}{t_{bn}} \tag{16.22}$$

In a manner similar to that discussed above, relations for N_{bd} and N_{bdn} (number of shaft rotations to discharge of the accumulated charges on the liner surface) are developed using Equations (16.15) to (16.17), and is given as:

$$N_{bd} = \left[\frac{R_b \times C_b \times \log_e a}{0.68}\right] \times F_s \times \left[\frac{p(1 + 1/\psi)}{Y}\right]^{-1/2} \tag{16.23}$$

and

$$N_{bdn} = -F_s \times R_b \times C_b \times \log_e a \tag{16.24}$$

16.2.8 Determination of the Ratio of Number of Shaft Rotations for Charge Accumulation to Discharge of the Accumulated Charges

The ratio (N_{bc}/N_{bd}) indicating the number of times, a part of the circumference of the journal crosses the zone of load-carrying oil film for the charge accumulation to discharge the accumulated charges from the liner of a hydrodynamic journal bearing, and the ratio of the number of shaft revolutions (N_{bcn}/N_{bdn}) for the same is determined using Eqns. (16.20)–(16.24), and given as:

$$\frac{N_{bcn}}{N_{bdn}} = \frac{N_{bc}}{N_{bd}} = \frac{\log_e (1 - a)}{\log_e a} \tag{16.25}$$

16.2.9 Number of Starts and Stops before the Initiation of Craters on the Liner Surface of a Journal Bearing

The number of starts and stops before the initiation of craters on the liner surface of a hydrodynamic journal bearing (N_{ssb}) can be determined as the ratio of the net time required to initiate craters (C_{sb}/F_s) to the time lapse for charge and discharge of the accumulated charges (T_{cb} and T_{db}) in each start and stop of a machine to the number of cycles (C_{sp}) required before the machine come to standstill condition after the power supply to the machine is off. Thus N_{ssb} can be determined as:

$$N_{ssb} = \frac{-C_{sb}}{F_s \times (T_{cb} + T_{db}) \times C_{sp}} \tag{16.26}$$

The number of cycles C_{sp}, before the machine comes to standstill condition, depends on the machine inertia, and friction in bearings, apart from the other factors. Various experimental investigations reveal that, in general, C_{sp} varies between 20 to $90F_s$. Using Eqns. (16.8) and (16.15), the N_{ssb} is determined as:

$$N_{ssb} = \frac{-C_{sb}}{F_s \times C_b \times R_b \times \log_e a(1 - a) \times C_{sp}} \tag{16.27}$$

16.3 Theoretical Model to Determine Number of Cycles for the Appearance of Craters on the Liner Surface

16.3.1 Determination of Equivalent Bearing Capacitance and Stored Charges

The equivalent capacitance of a hydrodynamic journal bearing depends on its dimensions and operating parameters and has been determined in Chapter 15, and is given as [7]:

$$C_{\rm b} = \frac{\left\{ 4\delta\xi d \tan^{-1} 0.41 \left[\frac{(1-\varepsilon)}{(1-\varepsilon)} \right]^{1/2} \right\}}{\psi(1-\varepsilon^2)^{1/2}} \tag{16.28}$$

The equivalent capacitance of a bearing lubricated with high-resistivity oil (10^{12} ohm m) opposes any change in the existing voltage, and causes storing of charge. The instantaneous stored charge depends on the bearing capacitance ($C_{\rm b}$) and voltage across the oil film (V), and is given by Eqn. (16.2).

16.3.2 Wear in Sliding Electric Contacts and Its Effect on Bearing Performance

The major differences between sliding contacts and most other contacts are the problems related to wear during extended period of operation. Good conduction of electric current between opposing contact members under relative motion initiates essentially metal-to-metal contact, and such sliding characteristics produce surface changes like loss of surface finish, transfer of material from softer interacting metal surface, fretting corrosion, abrasion, adhesion, etc. at the contact interface.

The friction between lubricated contacts depends on the effectiveness of a lubricant (its resistivity, volatility and chemical stability) and other contact conditions, which contribute to the load support by the hydrodynamic effect. Furthermore, under the effect of shaft voltage exceeding threshold voltage, instantaneous flow of current takes place, leading to the softening of the high 'points' or prows of the bearing surface. This would tend to accentuate the adhesion between the interacting members and prolong the phase of reforming of the prow. In case of instantaneous actual softening of the contact member, the interaction may result in sliding seizure.

16.3.3 Determination of Minimum Number of Cycles/Revolutions for the Appearance of Craters on the Bearing Liner

When a bearing lubricated with high-resistivity oil (10^{12} ohm m) is operated under the influence of shaft voltage, current does not flow through the bearing and charge gets accumulated on the bearing surface till the shaft voltage is below the threshold voltage as discussed in Chapter 4 [4]. The flow of current takes place by itself only when the shaft voltage reaches or exceeds the value of threshold voltage. When the voltage is below the threshold voltage, the charge is induced on the inner peripheral surface of the bearing (i.e. liner) due to self-inductance, and an equal and opposite charge is induced on the outer peripheral surface of the journal bearing, which is earthed in case of non-insulated bearing. During rotation of the journal, primarily, slow continuous passage of charge takes place through the bearing when the 'high' points on the journal and the bearing in the zone of load-carrying oil film come close due to vibration effects resulting from various instabilities of hydrodynamic journal bearings. Also, when the high 'points' of the interacting surfaces are separated by oil film of a relatively higher thickness, because of bearing operating parameters (in case of lightly loaded bearings), arcing results due to instantaneous leakage of stored charge energy between the journal and the bearing liner under the effect of higher amplitude of vibrations due to instabilities of various sources, which causes the interacting surfaces to come in proximity instantaneously. For insulated bearings, this phenomenon leads to the instantaneous flow of localized current.

The instantaneous charge leakage between the journal and the high 'points' of a bearing generates an instant heat. The discharge of stored electrical energy, q, between journal and the hydrodynamic journal bearing of capacitance C_b, due to charge leakage Q_b at the bearing voltage V is given as [using Eqn. (16.2)]:

$$q = C_b \frac{V^2}{2} = 0.5 Q_b V \tag{16.29}$$

A part of the stored energy of charge, q_n, is dissipated to K 'high' points contact between the journal and the bearing after N cycles of rotation. The energy dissipated to K high 'points' after these N cycles raises their temperature to the softening limits, which leads to crater formation at each of these points. The field experience with journal bearings and literature survey indicate that instability phenomenon in hydrodynamic journal bearings causes vibrations of frequency $0.5F_s$ or so. However, these vibrations are of very low energy/amplitude in stable bearings. But, even then, this causes proximity/closeness of the high 'points' of the journal and liner surface [17]. This means that each 'high' point contact between the journal and bearing occurs $0.5F_s$ times per second. Thus, the number of asperity high 'point' contacts in N_a cycles is $0.5N_a$ i.e. N_p. This indicates that, in general, $0.5N_a$ number of contacts lead to the dissipation of stored energy of charge. So, the minimum energy (q_n) dissipated after N_a cycles, resulting in crater formation, is given as:

$$q_n = N_p q = 0.5 N_a q$$

which is [using Eqn. (16.29)]:

$$q_n = 0.25 N_a C_b V^2 \tag{16.30}$$

This dissipated energy creates K craters (of radius r and depth H) in the zone of load-carrying oil film, and increases temperature by T_n after N_a cycles of operation. Furthermore, each 'high' point of a bearing contacts the journal $(\pi d/W_b)\,\Delta$ times in each shaft rotation. So, using Eqn. (16.31) and neglecting loss of heat by oil flow and other means, the following is evident:

$$q_n = 0.25 N_a C_b V^2 = \pi r^2 H K \rho_1 c T_n \left(\frac{\pi d}{W_b}\right)\Delta \tag{16.31}$$

The number of cycles for accumulating charges (N_{bcn}) and discharge of the accumulated charges (N_{bdn}) play an important role in determining the minimum cycles to create the craters on a liner (N). So considering the ratio of N_{bcn} to N_{bdn}, the minimum number of cycles for the formation of craters of specified dimensions is determined as N_{bcn}/N_{bdn} times of N_a and is given as [using Eqn. (16.18)]:

$$N = \left\{\frac{5.8 V_v \rho_1 c T_n}{C_b V^2}\left[\frac{p(1 + 1/\psi)}{Y}\right]^{1/2}\right\}\Delta\,\frac{N_{bcn}}{N_{bdn}} \tag{16.32}$$

The number of shaft revolutions N_{bcn} to accumulate charges Q_b and number of shaft revolutions N_{bdn} to discharge of the accumulated charges have been worked out theoretically in the Eqns. (16.21)–(16.25), and can be used in the Eqn. (16.32) to determine the minimum number of cycles for the appearance of crates on bearing liner surface.

In general, formation of craters occurs in the zone of minimum film thickness along the bearing length. To determine the minimum revolutions for initiation of craters on the liner (N), localized 'high' point temperature of 140 °C – the softening temperature of the liner at which the material toughness drastically reduces – has been considered for the analysis. Initial bearing temperature has been taken as 20 °C. The softening temperature may vary between 120 and 150 °C, depending on the properties of the liner material. Furthermore, the ratio of N_{bcn} to N_{bdn} depends on shaft voltage and varies from 1 to 25 [18, 19]. This ratio has been taken as 12.5 for the data deduction, considering the bearing-to-shaft voltage (a) approximately as 0.84.

Also, in the Eqn. (16.32), if N is known, V is measured and C_b is determined by Eqn. (16.28) by known parameters of operation, then the net volume of craters V_v can be estimated. Similarly, shaft voltage can be determined if the values of N, V_v, C_b have been ascertained.

16.4 Data Deduction

16.4.1 Repeated Starts and Stops of a Machine

The ratios N_{bcn}/N_{bdn} and N_{bc}/N_{bd} are determined for different values of bearing-to-shaft voltage, $(a = V/E)$, varying from 0.1 to 0.9 using Eqn. (16.25), and the variation of N_{bcn}/N_{bdn} and N_{bc}/N_{bd} against bearing-to-shaft voltage (a) is shown in Fig. 16.1.

The number of starts and stops of a machine before the initiation of craters on the bearing liner surface is determined for $a = 0.1\text{–}0.9$ using Eqn. (16.27). Figure 16.2 indicates the variation of number of starts and stops (N_{ssb}) before initiation of craters on the liner at various levels of bearing-to-shaft voltage (V/E) of the hydrodynamic journal bearing of 300 mm diameter operating at 1500 rpm ($F_s = 25\,\mathrm{sec}^{-1}$), having C_b, and R_b as 25×10^{-9} (ϕ), and 7×10^8 ohm, respectively [7, 12]. The C_{sp} for the machine has been experimentally determined as approximately $20F_s$. And C_{sb} based on the temperature rise and thermal stress calculations [12] under frequent start and stop regimes is approximately established as 10^9 cycles.

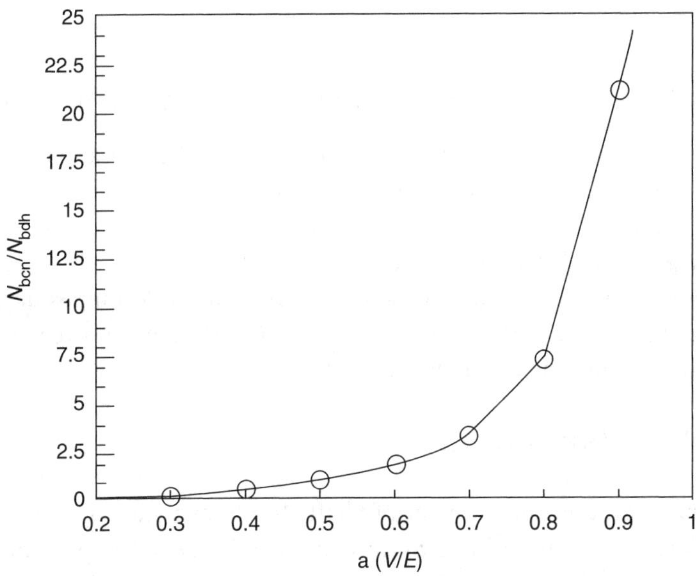

Fig. 16.1. Variation of the ratio of shaft revolution to accumulate and discharge of accumulated charges (N_{bcn}/N_{bdn}) at various levels of bearing-to-shaft voltages a (V/E) on the liner surface of a hydrodynamic journal bearing under the influence of shaft voltages

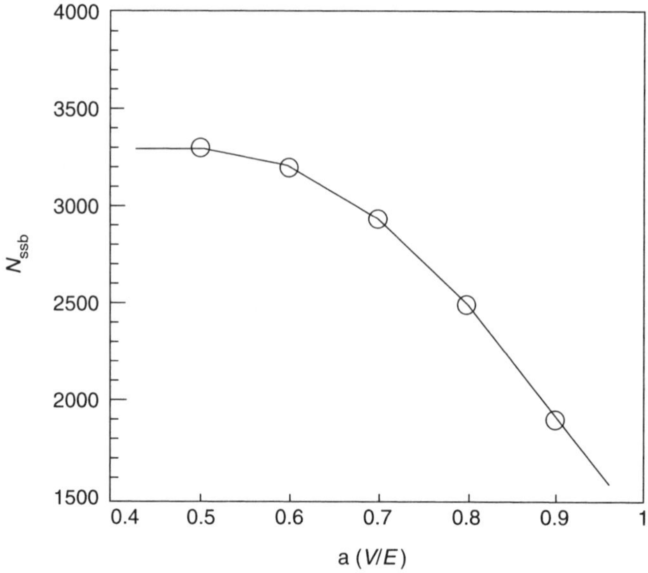

Fig. 16.2. Variation of number of starts and stops (N_{ssb}) of a machine before formation of craters on liner surface of the hydrodynamic journal bearing operating under the influence of electric currents at various levels of bearing-to-shaft voltage a (V/E)

16.4.2 Minimum Number of Shaft Revolutions for the Appearance of Craters

Using Eqn. (16.32), minimum number of cycles for the appearance of craters on bearing liner surface (N) has been determined for different values of shaft voltages varying from 0.25 to 2 V, for a bearing operating under stabilized regime and having parameters given in Table 16.1. The shaft voltages taken are less than the first threshold voltage (<3 V), to avoid the effect of chemical decomposition of the oil [4]. The various bearing parameters $L/d = 1$ and $d = 300$ mm, as given in Table 16.1, have been taken from the References [7, 13].

16.5 Results and Discussion

16.5.1 Time Duration and Number of Shaft Rotations required to Accumulate Charges on Bearing Liner Surface and Discharge of the Accumulated Charges during Repeated Starts and Stops of a Machine

Time required to accumulate the charges on the liner of a hydrodynamic journal bearing depends on bearing capacitance C_b and resistance R_b. Also, this is a function

Table 16.1. Minimum number of cycles for the formation of craters of specified dimensions on the liner surface of a hydrodynamic journal bearing due to leakage of stored charge energy between the journal and the bearing under the influence of shaft voltages of different levels

	(1)	(2)	(3)	(4)	(5)	(6)	(7)	(8)
V (V)	0.25	0.5	0.75	1.0	1.25	1.5	1.75	2.0
$(VC_b)10^{-9}$ (C)	6.25	12.5	18.8	25.0	31.25	37.5	43.7	50.0
$N \times 10^9$ (cycles)	38.4	9.6	4.27	2.4	1.54	1.07	0.78	0.6
Years (minimum)	48.6	12.16	5.4	3.04	1.95	1.36	0.99	0.76
Ratio of increase in reduction in life as against 0.5 V	–	–	2.25	4.0	6.25	8.97	12.3	16.0

Bearing dimensions, operating parameters and analytical values $d = 300\,\text{mm}$, $L = 300\,\text{mm}$, $L/d(\delta) = 1$, $n = 1500\,\text{rpm}$, $F_s = 25\,\text{s}^{-1}$, $C_d/d(\psi) = 0.0015$, $\varepsilon = 0.82$, $p = 2 \times 10^6\,\text{N m}^{-2}$, $\rho = 10^{12}\,\text{ohm m}$, $H = 0.75\,\text{mm}$, $r = 0.5\,\text{mm}$, $V_v = 14.92 \times 10^{-9}\,\text{m}^3$, $\xi = 2.5 \times 8.85419 \times 10^{-12}\,\phi/m$, $\xi_r = 2.5$, $C_b = 25 \times 10^{-9}(\phi)$ [Eqn. (16.28)], $K = 25$, $\rho_l = 7.46 \times 10^3\,\text{kg m}^{-3}$, $c = 2.26 \times 10^2\,\text{ws kg}^{-1}\,{}^\circ\text{C}^{-1}$, $Y = 40 \times 10^9\,\text{N m}^{-2}$, $\Delta = 0.05$ (0.01– 0.5), $N_{bcn}/N_{bdn} = 12.5$

of the natural logarithm of the difference of shaft and bearing voltage to the shaft voltage [Eqn. (16.8)]. Similarly, time required to discharge off the accumulated charges depends on C_b and R_b, and is a function of the natural logarithm of the ratio of bearing-to-shaft voltage [Eqn. (16.15)].

The number of times a journal circumferential location passes through the zone of load-carrying oil film to accumulate and discharge the accumulated charges depends on the shaft rotational frequency, clearance-to-diameter ratio C_d/d, load per unit area (p) and the natural logarithm of the shaft-to-bearing voltage, besides C_b, R_b and properties of the liner material [Eqns. (16.20) and (16.23)]. However, the shaft revolutions to accumulate charges and to discharge the accumulated charges depend on C_b, R_b, a and F_s [Eqns. (16.21) and (16.24)]. And the ratio of shaft revolutions to accumulate and discharge the accumulated charges on the liner surface is independent of all these parameters and depends only on the ratio of difference of natural logarithm of shaft and bearing voltage to the voltage across bearing, and on the ratio of the natural logarithm of the shaft and bearing voltage [Eqn. (16.25)].

The ratio of shaft rotations to accumulate and discharge of the accumulated charges (N_{bcn}/N_{bdn}) increases with V/E. As (V/E) increases from 0.1 to 0.9, the (N_{bcn}/N_{bdn}) increases from 0.05 to 20.91 (Fig. 16.1). It is evident that for the successively higher value of 'a', the ratio of N_{bcn}/N_{bdn} is higher than for the immediate lower value of V/E (Fig. 16.1).

16.5.2 Number of Start and Stop Cycles Before the Initiation of Craters on the Bearing Liner due to Discharge of the Accumulated Charges

The number of starts and stops before the initiation of craters on the liner surface of a hydrodynamic journal bearing depends on the number of cycles required to initiate craters, shaft rotational frequency, capacitance and resistance of the bearing, number of cycles before machine comes to standstill condition after the power supply to the machine is switched-off (C_{sp}), and natural logarithm of a (V/E) [Eqn. (16.27)].

The number of starts and stops to initiate craters on the bearing liner decreases as the ratio of bearing-to-shaft voltage increases. For a journal bearing of 300 mm diameter, number of starts and stops to initiate craters on the liner decreases from 3300 to 1905 as the bearing-to-shaft voltage (a) increases from 0.5 to 0.9 (Fig. 16.2).

16.5.3 Minimum Number of Cycles for the Formation of Craters of Various Sizes on the Liner Surface at Different Shaft Voltages

The minimum number of cycles for formation of craters of specified sizes/net volume on the liner surface of a hydrodynamic journal bearing depends on the clearance ratio, load per unit area and properties of liner material, and varies inversely with the square of shaft voltage, and capacitance of a bearing [Eqn. (16.32)]. Bearing capacitance, in turn, depends on L/d, eccentricity and C_d/d ratios [Eqn. (16.28)].

For 300 mm diameter bearing having capacitance of 25×10^{-9} (ϕ) operating at 1500 rpm with $\varepsilon = 0.82$, the value of N for the same volume of craters decreases as the shaft voltage increases from 0.25 to 2 V. At 0.5 V shaft voltage, N is 9.6×10^9 cycles, which decreases steeply to 0.6×10^9 cycles as the shaft voltage increases to 2 V (Table 16.1). However, at 0.25 V (below 0.5 V), N is 38.4×10^9 cycles. Similar trend is found for bearings of other sizes after analysis under different conditions of operation.

16.5.4 Effect of Shaft Voltage on Life Span of a Bearing

The minimum life span and operating cycles of operation of bearing decreases as the shaft voltage increases. As seen from Table 16.1, at 1500 rpm, the minimum years of operation for the formation of craters of different sizes of net volume (V_v)14.92×10^{-9} m^3 decreases from 12.16 to 0.76 years as the shaft voltage increases from 0.5 to 2 V. However, under ideal hydrodynamic conditions, the life of a bearing is infinite. In general, as compared to 0.5 V shaft voltage, ratio of increase in reduction of bearing life changes from 2.25 to 16 times as the shaft voltage increases from 0.75 to 2 V. Whereas, at 0.25 V, the effect on bearing life is insignificant (Table 16.1). It is, therefore, recommended that for safe and reliable operation of a hydrodynamic journal bearing, shaft voltage should not be allowed to exceed 0.5 V. This matches closely with the recommendation of leading manufacturers of the rotating machines. The experimental and theoretical investigation pertaining to diagnosis of failure of a hydrodynamic journal bearing under the effect of shaft voltage of a synchronous condenser has been reported, which indicates the formation of craters in 2 h after 12×10^4 cycles with volume of craters as $3.77 \times 10^{-12}\,\mathrm{m}^3$ at the shaft voltage of 2.5 V [20]. Also, laboratory tests conducted by different authors show that shaft voltage of 10 to 20 V may cause severe electro-erosion in bearings and shaft seals. With increasing shaft voltage the volume of erosion increases sharply and damages the bearings within no time [21].

It may be noted that under the influence of lower shaft voltage, the life expectancy of a journal bearing is much higher before the craters are initiated on the liner, because of less rise in contact temperature at the 'high' points. In addition, decrease in charge accumulation ($C_b V = Q_b$) and increase in bearing capacitance also result in less rise in contact temperature in the load-carrying zone of a bearing [Eqn. (16.31)], since $C_b V^2 = Q_b^2 / C_b$. This subsequently increases the number of minimum cycles for the appearance of craters on the liner surface of a bearing [Eqn. (16.32)]. This is because the square of fractional value of charge in the denominator of Eqn. (16.32) plays a significant role as compared to T_n (since $C_b V^2 = Q_b^2 / C_b$) and increases N significantly.

16.6 Conclusions

From the developed theoretical model and analysis, the following conclusions are brought out [19, 22]:

(1) Time required to accumulate and discharge the accumulated charges on the bearing liner depends on the bearing capacitance and resistance, and is a function of the natural logarithm of the ratio of bearing-to-shaft voltage.

(2) The number of shaft rotations to accumulate and discharge the accumulated charges on the liner surface depends on shaft rotational frequency besides the capacitance and resistance of the bearing, and the ratio of bearing-to-shaft voltage.

(3) As the ratio of bearing-to-shaft voltage increases, the ratio of shaft rotation to accumulate and discharge the accumulated charges increases.

(4) With increase of bearing-to-shaft voltage, the number of starts and stops to initiate craters on the liner surface of a hydrodynamic bearing decreases.

(5) The minimum number of cycles for the formation of craters of different sizes and of specified volume on the bearing liner depends on the clearance ratio, the load per unit area and the liner properties, and varies inversely with the square of shaft voltage, and bearing capacitance.

(6) For a hydrodynamic journal bearing operating under the influence of shaft voltage, the minimum number of cycles for the appearance of craters is much higher at shaft voltage below 0.5 V as compared to that at values above 0.5 V. Also shaft voltage of less than 0.5 V does not affect the life of the bearing. In contrast, at shaft voltage of 2 V, reduction in bearing life increases to 16 times of that at 0.5 V, for the identical volume of craters formed.

(7) For safe and reliable operation and adequate life of a bearing, shaft voltage of 0.5 V must not be exceeded.

This analysis, besides establishing the effect of capacitive response of the bearings, is useful for transient performance analysis and life assessment of bearings under the effect of shaft voltages.

REFERENCES

1. Verma, S. P. and Gupta, Vikas (1980). Shaft potential in modern turbo-generators with particular reference to static excitation. Report, Department of Electrical Engineering, University of Saskatchewan, Saskatoon, Canada S 7 N OWO.
2. Anderson, S. (1968). Passage of electric current through rolling bearings. *Ball Bearing J.*, **153**, 6–12.
3. Chu, P. S. Y. and Cameron, A. (1967). Flow of electric current through lubricated contacts. *ASLE Trans.*, **10**, 226–234.
4. Prashad, H. (1987). Effects of operating parameters on the threshold voltages and impedance response on non-insulated rolling-element bearings under the action of electrical currents. *Wear*, **117**, 223–240.
5. Prashad, H. (1991). Theoretical and experimental investigations on the pitch and width of corrugations on the surfaces of ball bearings operated. *Wear*, **143**, 1–14.
6. Prashad, H. (1988). Theoretical evaluation of impedance, capacitance and charge accumulation of roller bearings operated under electrical fields. *Wear*, **125**, 223–239.
7. Prashad, H. (1991). Theoretical evaluation of capacitance, capacitive reactance, resistance and their effects on performance of hydrodynamic journal bearings. ASME, *J. Tribol.*, **113**, 762–767.

8. Prashad, H. (1992). An approach to evaluate capacitance, capacitive reactance and resistance of pivoted pads of a thrust bearing. STLE, *Tribol. Trans.*, **36**(3).
9. Jones-Conway, J. M. and Leopard, A. J. (1976). Plain bearing damage. Proc. of the 4th Turbo machinery Symp., 55–63.
10. Verma, S. P., Girgis, P. S. and Fleming, R. J. (1975). The Problems and Failure Caused by Shaft Potentials and Bearing Currents in Turbo generators, method of prevention. Report, Power System Research Group, Department of Electrical Engineering, University of Saskatchawavan.
11. Prashad, H. (1990). Theoretical analysis of the effects of instantaneous charge leakage of roller tracks of roller bearings lubricated with high-resistivity lubricants under the influence of electric current. ASME, *J. Tribol.*, **112**, 37–43.
12. Prashad, H. (1991). Theoretical evaluation of reduction in life of hydrodynamic journal bearings operating under the influence of different levels of shaft voltages. STLE, *Tribol. Trans.*, **34**(4), 623–627.
13. Prashad, H. (1988). The effects of viscosity and clearance on the performance of hydrodynamic journal bearings. *STLE Trans.*, **31**(1), 113–119.
14. Martin, F. A. and Garner, D. R. (1973). Plain journal bearings under steady loads, design guidance and safe operation. Paper no. 313/73, 1st European Tribology Conference, 449–463.
15. Kaehne, Von P. (1964). Shaft voltage and bearing current. A survey of published work, Electrical Research Association, Report no. 5030, London.
16. Cotton, H. (1983). *Advanced Electrical Technology*, A. H. Wheeler and Company, London.
17. Maxwell, J. H. (1981). Diagnosing induction motor vibration. *Hydro Carbon Processing*, 117–120.
18. Prashad, H. (1992). Theoretical analysis of capacitive effect of roller bearings on repeated starts and stops of a machine operating under the influence of shaft voltages. Transactions of ASME, *J. Tribol.*, **114**, 818–822.
19. Prashad, H. and Rao, K. N. (1994). Analysis of capacitive effect and life estimation of hydrodynamic journal bearings on repeated starts and stops of a machine operating under the influence of shaft voltages. *Tribol. Trans.*, **37**(3), 641–645.
20. Prashad, H. (1995). Diagnosis of bearing problem of synchronous condenser – An experimental and theoretical investigation. *Wear*, **188**, 95–101.
21. Ammann, C., Reichert, K., Joho, R. and Posedel, Z. (1987). Shaft voltage in generators with static excitation system – Problems and solution. IEEE, Paper 87 SM 403–9, 1–8.
22. Prashad, H. (1995). A theoretical model to determine the minimum number of shaft revolutions/cycles for appearance of craters of various sizes on the liner surface of a hydrodynamic journal bearing operating under the influence of shaft voltages of different levels. *BHEL J.*, **16**(2), 27–33.

Nomenclature

a	ratio of potential difference across bearing to shaft versus casing voltage ($a = V/E < 1$)
c	specific heat of bearing liner, ws kg^{-1} °C
C_b	bearing capacitance
C_d	bearing diametral clearance
C_{sb}	number of cycles before initiation of craters on liner surface of journal bearing
C_{sp}	number of cycles before the machine comes to stand still condition after the power supply is cut-off
d	shaft diameter
E	shaft versus casing voltage
F_s	shaft rotational frequency
H	depth of craters on bearing liner, m
K	number of craters
I	current passing through bearing
L	bearing length
n	rpm
N	minimum number of shaft revolutions/cycles for appearance of craters on bearing liner surface
N_a	number of cycles for appearance of craters on bearing liner
N_p	number of 'high' point contacts in N_a cycles
N_{bc}	number of times a part of circumference of journal crosses the load-carrying oil film during operation, during which charges (Q_b) will accumulate on the liner surface of a bearing
N_{bcn}	number of shaft rotations to accumulate charges (Q_b)
N_{bd}	number of times a part of circumference of journal crosses the load-carrying oil film providing opportunity to discharge of the accumulated charges (Q_b)
N_{bdn}	number of shaft revolutions to discharge of the accumulated charge (Q_b)
N_{ssb}	number of starts and stops before the formation of craters on the bearing liner
P	bearing load
p	load per unit area (P/Ld)
Q_b	electric charge accumulation on bearing
q_n	electrical energy dissipated to K high points due to charge leakage between journal and bearing liner after N cycles, ws

q	stored electrical energy between journal and hydrodynamic journal bearing, ws
r	radius of crater, m
R_b	bearing resistance
t_{bn}	duration/time taken by the journal to cross the zone of load-carrying oil film during operation
T	time
T_{cb}	time required to accumulate charge (Q_b) on bearing liner
T_{db}	time required to discharge of the accumulated charges from bearing liner
V	potential drop across bearing
V_n	circumferential shaft speed (πdF_s)
V_v	net volume of craters of different sizes (πr^2KH), m^3
W_b	width of contact between journal and bearing in the zone of load-carrying oil film
Y	Young's modulus of elasticity
β_{bn}	number of times journal passes in the zone of load-carrying oil film in each shaft rotation
$\delta = L/d$	length to diameter ratio
ε	eccentricity ratio
μ	dynamic oil viscosity
θ	attitude angle
ρ	resistivity of lubricant
ρ_1	density of bearing material, $\mathrm{kg\,m^{-3}}$
ξ	permittivity/dielectric constant of lubricant ($\phi\,\mathrm{m}^{-1}$)
ξ_r	relative permittivity of lubricant
ψ	C_d/d, diametral clearance ratio
Δ	ratio of asperity to the net contact surface area of bearing liner and journal in each shaft revolution

Chapter 17

ELECTRICAL PARAMETERS AND DYNAMIC COEFFICIENTS OF DIFFERENT MULTI-LOBES AND TILTING-PAD JOURNAL BEARINGS BY ELECTRICAL ANALOGY

17.1 General

A theoretical approach to evaluate capacitance, resistance, capacitive reactance and impedance of different multi-lobes and tilting-pad hydrodynamic journal bearings under different operating conditions has been developed. It has been established that the change in capacitance and resistance with the change in eccentricity ratio is non-linear. The capacitance and resistance, thus determined, are correlated with the flexibility/stiffness^{-1} and damping coefficients of bearings, respectively, using electrical analogy approach and it is found that the stiffness and damping coefficients are higher for multi-lobe bearings as compared to those of the cylindrical bearings having identical dimensional parameters operating under similar operating conditions. Also, tilting-pad journal bearings are found to have better dynamic characteristics than the equivalent multi-lobe bearings.

The analysis may have a potential to diagnose the stability regime of a bearing through the bearings electrical parameters. The electrical analogy may be a useful alternative to the conventional techniques.

17.2 Introduction

Cylindrical journal bearings are not common in use in high-speed machines because of their higher temperature increase and bearing instability. Two-lobe, three-lobe and four-lobe, tilting-pad journal bearings are used extensively in high speed rotating machines because of their superior dynamic characteristics. Multi-lobe bearings can be chosen so that the rotor support system can be tuned to minimize the rotor response. Dynamic coefficients data for two-lobe, three-lobe, four-lobe and tilting-pad bearings are available in the literature. Perturbation techniques have been used, which require solving the Reynolds equation both for stiffness and damping coefficients [1–3].

In most of the cases, it is possible to represent a mechanical vibrating system by an electric circuit. This is because of analogy between electrical and mechanical systems exist because of the similarity of differential equations, which represent them [4]. The assembly and measurement of electrical system is much easy as compared to that of the mechanical system.

It is evident that in hydrodynamic journal bearings, the zone of minimum film thickness, i.e. the load-carrying oil films, varies along the circumference of a bearing through out the bearing length. This forms a capacitor of varying capacitance between the journal and bearing, depending on permittivity of the lubricant, circumferential length of load-carrying oil film, bearing length, eccentricity and clearance of the bearing. In addition, the load-carrying oil film offers resistance depending on operating parameters and resistivity of the lubricant. Thus, the load-carrying oil film forms a resistor–capacitor (RC) circuit and offers impedance to a current flow in a hydrodynamic journal bearing. Thus, by analyzing the RC Circuit the behavior of a journal bearing can be predicted. Similar studies have been carried out for cylindrical journal and rolling-element bearings in Chapters 15 and 5, respectively [5, 6].

This chapter discusses the theoretical studies undertaken to determine the capacitance and active resistance of the interacting surface of a journal with two-lobe, three-lobe, four-lobe and tilting-pad bearings in the load-carrying oil film under different operating parameters. Also, the equivalent capacitance and active resistance of these bearings have been determined and compared with that of a cylindrical bearing of the same size operating under identical parameters [5]. Based on the capacitance and resistance, dynamic coefficients of different multi-lobe, tilting-pad and cylindrical bearings have been computed, and the values have been compared.

17.3 Two-Lobe Journal Bearing

17.3.1 Geometry and Geometrical Characteristics of Two-Lobe Elliptical Bearings

The elliptical bearing is made up of two circular arcs, with their centers of curvature, O_L, and O_U, each displaced by a distance, e_p, from the geometric center, O_b, of the bearing as shown in Fig. 17.1. For any given journal position O_j, the eccentricities e_1, e_2 and attitude angles ϕ_1, ϕ_2 of the two lobes will be different as

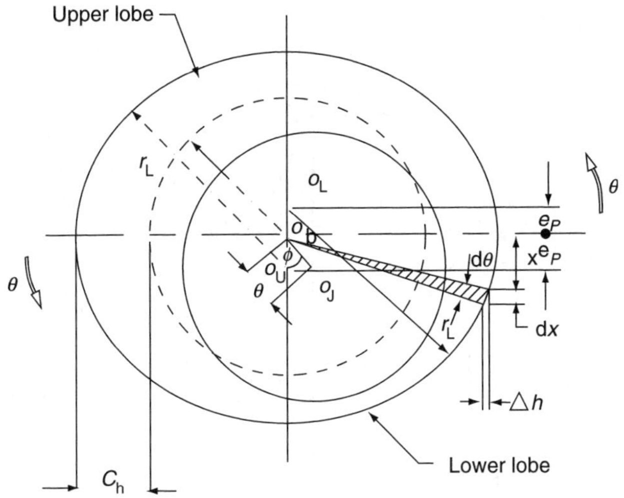

Fig. 17.1. Journal configuration in a hydrodynamic elliptical/two-lobe journal bearing under rotation

shown in Fig. 17.2. The various mathematical relations can be developed for an elliptical bearing and given for the lower lobe as [7]

$$\varepsilon_1 = \left(\varepsilon_p^2 + \varepsilon^2\ 2\varepsilon\ \varepsilon_p \cos\phi\right)^{1/2} \tag{17.1}$$

$$\phi_1 = \frac{\tan^{-1} \varepsilon \sin\phi}{(\varepsilon_p + \varepsilon\cos\phi)} \tag{17.2}$$

and

$$\lambda_1 = \frac{\pi}{2} - \phi_1$$

Likewise, for the upper lobe:

$$\varepsilon_2 = (\varepsilon_p^2 + \varepsilon^2 - 2\varepsilon\varepsilon_p \cos\phi)^{1/2} \tag{17.3}$$

$$\phi_2 = \frac{\pi - \tan^{-1} \varepsilon \sin\phi}{\varepsilon_p - \varepsilon\cos\phi} \tag{17.4}$$

and

$$\lambda_2 = 3\frac{\pi}{2} - \phi_2$$

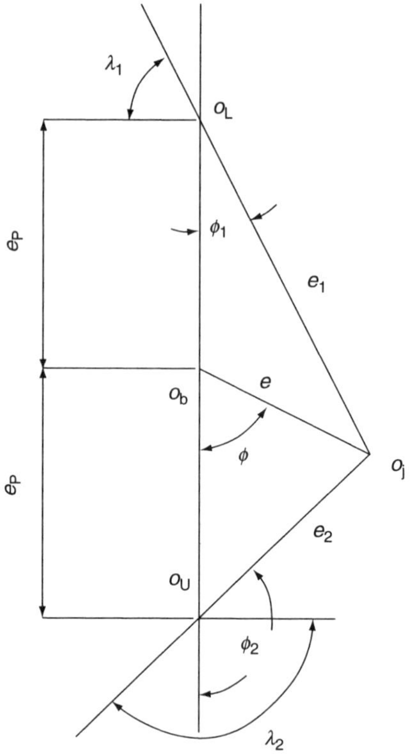

Fig. 17.2. Geometric parameters of elliptical/two-lobe journal bearing

The film thickness for the lobes, in non-dimensional form, can be expressed as [7]

$$H = 1 + \varepsilon_1 \cos(\theta + \lambda_1) \quad (17.5)$$

$$H = 1 + \varepsilon_2 \cos(\theta + \lambda_2)$$

where

$$H = \frac{h}{C_h}$$

and

$$\begin{aligned} h &= C_h[1 + \varepsilon_1 \cos(\theta + \lambda_1)] \\ h &= C_h[1 + \varepsilon_2 \cos(\theta + \lambda_2)] \end{aligned} \quad (17.6)$$

17.3.2 Theoretical Analysis

17.3.2.1 Capacitance and Active Resistance of Cylindrical Hydrodynamic Journal Bearings

For each revolution of the shaft, part of the circumference of the journal passes through a zone of load-carrying oil film. This forms a parallel plate capacitor of varying thickness between the interacting surfaces of the journal and the bearing. The equivalent capacitance of a hydrodynamic journal bearing has been determined in Chapter 15 and is given as [5]

$$C_{\text{cb}} = \frac{4\delta\xi d \tan^{-1} 0.41[(1+\varepsilon_{\text{c}})/(1-\varepsilon_{\text{c}})]^{1/2}}{\psi(1-\varepsilon_{\text{c}}^2)^{1/2}} \tag{17.7}$$

The equivalent resistance of a cylindrical bearing is determined as [5]

$$R_{\text{cb}} = \frac{L\rho(1-\varepsilon_{\text{c}}^2)^{1/2}}{4\delta d \tan^{-1} 0.41[(1+\varepsilon_{\text{c}})/(1-\varepsilon_{\text{c}})]^{1/2}} \tag{17.8}$$

17.3.2.2 Capacitance and Active Resistance of Upper and Lower Lobe of a Two-Lobe Elliptical Journal Bearings

17.3.2.2.1 Capacitance

In an elliptical bearing contrary to a cylindrical journal bearing, positive pressure builds up in both lower and upper lobes. The analysis is performed in a unitary manner, for both lobes, by starting the analysis with an isolated lobe placed in any position with respect to the shaft or to the bearing center. The characteristics of a bearing as a whole are found by summing the results from a number of lobes.

During rotation of the shaft, part of the circumference of the journal passes through lower lobe and part passes through the upper lobe in a zone of active load-carrying oil film. This forms a parallel plate capacitor of varying film thickness between the interacting surface of the journal and an individual lobe, whereas capacitance of both the lobes are in parallel to each other with respect to the journal. However, the capacitance of a parallel plate capacitor is greatest at the minimum film thickness of an individual lobe, and reduces on both sides as the film thickness increases.

17.3.2.2.1.1 Capacitance of Lower Lobe

The change in capacitance ΔC, between two rectangular plates of width dX and length L, located at the angular coordinate differential $d\theta$ (as shown in Fig. 17.1) and separated by a film thickness of Δh, is given as

$$\Delta C = \frac{2\xi[(X + \mathrm{d}X)L - XL]}{2h + \Delta h} \tag{17.9}$$

The change in film thickness (Δh) is very small in comparison to $\mathrm{d}X$ and hence can be neglected. Thus, the change in capacitance is given as

$$\Delta C = \frac{\xi \mathrm{d}XL}{h} \tag{17.10}$$

From Fig. 17.1, it is evident that

$$\mathrm{d}X = \mathrm{d}\theta(r + C_\mathrm{h}) \tag{17.11}$$

On substituting Eqns. (17.6) and (17.11) into Eqn. (17.10), the change in capacitance is given as

$$\Delta C = \frac{\xi L(r + C_\mathrm{h})\mathrm{d}\theta}{C_\mathrm{h}[1 + \varepsilon_1 \cos(\theta + \lambda_1)]} \tag{17.12}$$

The maximum change in capacitance depends on angle θ – the angular coordinate measured from leading edge – and occurs along the film thickness. Thus, integrating between the inlet angle θ_{pi} and the outlet angle θ_{pe}, the capacitance of the lower lobe of a two-lobe bearing is determined as

$$C_\mathrm{L} = \xi L C_\mathrm{h}^{-1}(r + C_\mathrm{h}) \int_{-\theta_{\mathrm{pi}}}^{\theta_{\mathrm{pe}}} [1 + \varepsilon_1 \cos(\theta + \lambda_1)]^{-1} \mathrm{d}\theta \tag{17.13}$$

Equation (17.13) can be integrated by Sommerfeld substitution [8] and capacitance is worked out using equation $\lambda_1 = \pi/2 - \phi_1$:

$$C_\mathrm{L} = \xi L C_\mathrm{h}^{-1}(r + C_\mathrm{h})(1 - \varepsilon_1^2)^{-1/2} \times \left\{ \cos^{-1}\left[\frac{\varepsilon_1 - \sin(\theta_{\mathrm{pe}} - \phi_1)}{1 - \varepsilon_1 \sin(\theta_{\mathrm{pe}} - \phi_1)}\right] - \cos^{-1}\left[\frac{\varepsilon_1 + \sin(\theta_{\mathrm{pi}} + \phi_1)}{1 + \varepsilon_1 \sin(\theta_{\mathrm{pi}} + \phi_1)}\right] \right\} \tag{17.14}$$

17.3.2.2.1.2 Capacitance of the Upper Lobe

Similar to Eqn. (17.14), the capacitance of the upper lobe of an elliptical bearing is given as

$$C_\mathrm{U} = \xi L C_\mathrm{h}^{-1}(r + C_\mathrm{h}) \int_{-\theta_{\mathrm{pi}}}^{\theta_{\mathrm{pe}}} [1 + \varepsilon_2 \cos(\theta + \lambda_2)]^{-1} \mathrm{d}\theta \tag{17.15}$$

On integrating and using equation $\lambda_2 = 3\pi/2 - \phi_2$, the capacitance of upper lobe is worked out similar to Eqn. (17.14) and given as

$$C_{\mathrm{U}} = \xi L C_{\mathrm{h}}^{-1}(r + C_{\mathrm{h}})(1 - \varepsilon_2^2)^{-1/2} \times \left\{ \cos^{-1}\left[\frac{\varepsilon_2 + \sin(\theta_{\mathrm{pe}} - \phi_2)}{1 + \varepsilon_2 \sin(\theta_{\mathrm{pe}} - \phi_2)}\right] - \cos^{-1}\left[\frac{\varepsilon_2 - \sin(\theta_{\mathrm{pi}} + \phi_2)}{1 - \varepsilon_2 \sin(\theta_{\mathrm{pi}} + \phi_2)}\right] \right\} \tag{17.16}$$

17.3.2.2.2 Resistance

The change in conductance/resistance^{-1}, ΔF, between two rectangular plates of width dX of bearing length L, located at an angular coordinate differential dθ and separated by a film of thickness Δh, is given as

$$\Delta F = \frac{\mathrm{d}X\,L}{\rho h} \quad (\Delta h \text{ being small and hence negligible}) \tag{17.17}$$

The maximum change in conductance/resistance^{-1} depends on an angle θ – the angular coordinate measured from the leading edge and along the film thickness. So, on integrating between the inlet angle θ_{pi} and the outlet angle θ_{pe}, the resistance of the lower lobe [using Eqns. (17.11) and (17.6)] is determined as

$$F_{\mathrm{L}} = R_{\mathrm{L}}^{-1} = \rho^{-1} L (r + C_{\mathrm{h}}) C_{\mathrm{h}}^{-1} \int_{-\theta_{\mathrm{pi}}}^{\theta_{\mathrm{pe}}} [1 + \varepsilon_1 \cos(\theta + \lambda_1)]^{-1} \mathrm{d}\theta \tag{17.18}$$

The above equation is integrated using Sommerfeld substituting [8], and resistance of lower lobe is worked out using equation $\lambda_1 = \pi/2 - \phi_1$, which is given as

$$R_{\mathrm{L}} = \rho C_{\mathrm{h}} L^{-1}(r + C_{\mathrm{h}})^{-1}(1 - \varepsilon_1^2)^{1/2} \times \left\{ \cos^{-1}\left[\frac{\varepsilon_1 - \sin(\theta_{\mathrm{pe}} - \phi_1)}{1 - \varepsilon_1 \sin(\theta_{\mathrm{pe}} - \phi_1)}\right] - \cos^{-1}\left[\frac{\varepsilon_1 + \sin(\theta_{\mathrm{pi}} + \phi_1)}{1 + \varepsilon_1 \sin(\theta_{\mathrm{pi}} + \phi_1)}\right] \right\}^{-1} \tag{17.19}$$

Similarly, resistance for upper lobe ($F_{\mathrm{U}}^{-1}/R_{\mathrm{U}}$) is worked out using equation $\lambda_2 = 3\pi/2 - \phi_2$ and given as

$$R_{\mathrm{U}} = \rho C_{\mathrm{h}} L^{-1}(r + C_{\mathrm{h}})^{-1}(1 - \varepsilon_2^2)^{1/2} \times \left\{ \cos^{-1}\left[\frac{\varepsilon_2 + \sin(\theta_{\mathrm{pe}} - \phi_2)}{1 + \varepsilon_2 \sin(\theta_{\mathrm{pe}} - \phi_2)}\right] - \cos^{-1}\left[\frac{\varepsilon_2 - \sin(\theta_{\mathrm{pi}} + \phi_2)}{1 - \varepsilon_2 \sin(\theta_{\mathrm{pi}} + \phi_2)}\right] \right\}^{-1} \tag{17.20}$$

17.3.2.3 Equivalent Capacitance and Active Resistance of Two-Lobe Bearing

17.3.2.3.1 Capacitance

As per the electrical analogy, both the lobes of an elliptical bearing are in parallel to each other with respect to the journal. So the equivalent capacitance of the two-lobe bearing is determined as

$$C_{\text{eq}} = C_{\text{L}} + C_{\text{U}} \tag{17.21}$$

17.3.2.3.2 Resistance

Since both the lobes are in parallel to each other with respect to the journal, the equivalent resistance of a two-lobe bearing is given as

$$R_{\text{eq}}^{-1} = R_{\text{L}}^{-1} + R_{\text{U}}^{-1}$$

and hence

$$R_{\text{eq}} = \frac{R_{\text{L}} R_{\text{U}}}{R_{\text{L}} + R_{\text{U}}} \tag{17.22}$$

17.3.2.4 Capacitive Reactance and Impedance of Two-Lobe Bearings

Capacitive reactance between the interacting surfaces of the journal and the bearing in the oil film is determined and given as [5]

$$X_{\text{eq}}^{-1} = wC_{\text{eq}} = w(C_{\text{L}} + C_{\text{U}}), \quad X_{\text{L}}^{-1} = wC_{L},$$
$$X_{\text{U}}^{-1} = wC_{\text{U}}$$

where $w = 2\pi f$

and

$$X_{\text{eq}}^{-1} = X_{\text{L}}^{-1} + X_{\text{U}}^{-1} \tag{17.23}$$

The impedance of bearing circuit is determined as [5]

$$Z_{\text{U}} = R_{\text{U}} - jX_{\text{U}}$$
$$Z_{\text{L}} = R_{\text{L}} - jX_{\text{L}}$$

and

$$Z_{\text{eq}} = R_{\text{eq}} - jX_{\text{eq}} \tag{17.24}$$

The ratio of capacitive reactance to active resistance is given as [Eqn. (17.23) and Eqns. (17.13)–(17.20)]

$$\frac{X_{\mathrm{eq}}}{R_{\mathrm{eq}}} = \frac{R_{\mathrm{L}}^{-1} + R_{\mathrm{U}}^{-1}}{X_{\mathrm{U}}^{-1} + X_{\mathrm{U}}^{-1}}, \frac{X_{\mathrm{L}}}{R_{\mathrm{L}}} = \frac{1}{wC_{\mathrm{L}}R_{\mathrm{L}}}, \frac{X_{\mathrm{U}}}{R_{\mathrm{U}}} = \frac{1}{wC_{\mathrm{U}}R_{\mathrm{U}}}$$
$$\frac{X_{\mathrm{eq}}}{R_{\mathrm{eq}}} = \frac{R_{\mathrm{L}} + R_{\mathrm{U}}}{wR_{\mathrm{L}}R_{\mathrm{U}}(C_{\mathrm{U}} + C_{\mathrm{U}})} = \frac{1}{w\rho\xi} = \frac{X_{\mathrm{L}}}{R_{\mathrm{L}}} = \frac{X_{\mathrm{U}}}{R_{\mathrm{U}}} \tag{17.25}$$

17.3.2.5 Determination of Stiffness and Damping Coefficients by Electrical Analogy

For an electrical circuit with components including inductance (L_{j}), resistance (R), and capacitance (C), under the influence of voltage (E), the following equation can be written as [4]

$$L_{\mathrm{j}}\frac{\mathrm{d}i}{\mathrm{d}t} + R_{\mathrm{i}} + C^{-1}\int i\,\mathrm{d}t = E \tag{17.26}$$

where t is the time and is current flowing through the circuit.

Since

$$Q \text{ (the charge)} = \int i\,\mathrm{d}t \tag{17.27}$$

Equation (17.26) is modified as

$$L_{\mathrm{j}}\ddot{Q} + R\dot{Q} + C^{-1}Q = E \tag{17.28}$$

The inductance L_{j} of a journal bearing is worked out by Prashad [9] and is expressed as [Chapter 21]

$$L_{\mathrm{j}} = 2 \times 10^{-9} L \log_{\mathrm{e}}\left(\frac{C_{\mathrm{d}} + d}{d}\right) \tag{17.29}$$

The equation representing a mechanical system with damping and a single degree of freedom is written as

$$M\ddot{X} + C_{\mathrm{c}}\dot{X} + KX = F \tag{17.30}$$

Equations (17.29) and (17.30) imply voltage–force analogy. Voltage (E) in an electrical circuit corresponds to force in a mechanical system. Charge (Q) in columns, current (i) in amperes, inductance (L_{j}) in Henry, resistance (R) in ohms, capacitance (C) in Farad of Eqns. (17.28) and (17.26) correspond to displacement (X), velocity ($\dot{X}$), mass (M), damping coefficients (C_{c}) in $\mathrm{N\,s\,m^{-1}}$ and flexibility/stiffness^{-1}($1/K$) in m N^{-1}, respectively of Eqn. (17.30). Stiffness (K_{eq} and K_{cb}) and damping (C_{eq} and C_{cb}) coefficients of the two-lobe and cylindrical bearings have been evaluated by this analogy.

17.3.2.6 Comparison of the Electrical Analogy Approach With Conventional Analysis

It may be noted that the electrical analogy is geometrically based on solutions for charge distribution along potential and isostatic surfaces, with an assumed analogy between force and voltage providing coupling to the mechanical hydrodynamic problem. The solution satisfies continuity. Ratios of solutions eliminate the need for heuristic data such as resistivity, dielectric constant and equivalent inductance of the fluid film; thus any value will suffice, as these values are not required for comparisons provided the fluid conditions remains constant. Solutions to the electrical equations provide potential lines and iso-surfaces.

Similarly, a direct solution to the Reynolds equation for a fluid with constant properties would evolve into a solution of the continuity equation and would only be geometry-dependent. The dynamic would depend on solutions to the steady and perturbed equations, and again for constant properties would only be geometry-dependent. Also, ratios of solutions would not require knowing the fluid properties. Hence, the electrical analogy can be used as an alternative to conventional analysis, as a potential tool for relative comparison in addition to the absolute.

17.3.3 Data Derived and Discussion

17.3.3.1 Data Derived from Analysis

The capacitance and resistance of cylindrical and two-lobe bearings (lower lobe and upper lobe), along with the equivalent dynamic coefficients, are computed under different eccentricity ratios for the bearing with $L/d = 0.5$, $C_h = 0.4$ mm, $e_p = 0.2$ mm, $C_d/d = 0.001$ as per the developed methodology and data is computed as shown in Tables 17.1 and 17.2. Figures 17.3 and 17.4 show the variation of stiffness and damping coefficients with eccentricity ratio, respectively. Attitude angle, eccentricity ratio and other relevant details for a two-lobe bearing have been taken from the published data [7] for an elliptical bearing. Based on the electrical analogy, stiffness and damping coefficients are worked out under different operating conditions for two-lobe and cylindrical bearings (Tables 17.1 and 17.2). Table 17.3 gives the ratio of damping and stiffness coefficients C_{dL}, C_{dU}, C_{deq}, K_L, K_U and K_{eq} of the two-lobe bearing to that of the respective coefficients of cylindrical bearing (C_{dc}, K_{cb}).

17.3.3.2 Capacitance of Two-Lobe and Cylindrical Bearings

The capacitance of a two-lobe bearing depends on the capacitance of the lower and upper lobes. This in turn is a function of lubricant permittivity, horizontal clearance (C_h), bearing length (L) and shaft radius (r), and depends on the eccentricity ratio and attitude angle of the respective lobe along with the angles θ_{pi} and θ_{pe}. With an increase in the eccentricity ratio, the capacitance of a lobe increases. For the lower lobe, as the eccentricity ratio (ε_1) increases from 0.7 to 0.96, C_L increases from $(2.48 \times 10^{-9})\phi$ to $(4.51 \times 10^{-9})\phi$. Similarly, for the upper lobe as (ε_2) increases from 0.253 to 0.54, C_U increases from $(0.6 \times 10^{-9})\phi$ to $(1.66 \times 10^{-9})\phi$ as shown in Table 17.1. For a lower eccentricity ratio, the capacitance of the lobe has a different pattern. This is because of the fact the eccentricity ratio (ε_1, ε_2) and attitude angle (ϕ_1, ϕ_2) of a lobe depends on the cosine of the attitude angle (ϕ), in addition to the global eccentricity ratio (ε) of the two-lobe bearing [Eqns. (17.1)–(17.4)]. The change in capacitance with eccentricity ratio is non-linear (Table 17.1).

The capacitance of the lower lobe (C_L) of a two-lobe bearing for ratio (ε) is greater than that of the upper lobe (C_U) under identical eccentricity as shown in Table 17.1. However the variation is non-linear. It is evident from the fact that the equivalent capacitance is high [$(5.11 \times 10^{-9})\phi$ and $(4.76 \times 10^{-9})\phi$] both at the high and low eccentricity ratios (0.494 and 0.273) of the two-lobe bearing as compared to the other values of ε varying between 0.372 and 0.46, for which the C_{eq} has lower value, and it is around $(4.1 \times 10^{-9})\phi$ (Table 17.1).

The capacitance of a cylindrical bearing depends on ξ, C_d/d, δ, and ε_c. It increases with the eccentricity ratio, and varies between $(6.85 \times 10^{-9})\phi$ and $(9.3 \times 10^{-9})\phi$ for the bearing having $L/d = 0.5$, $C_d/d = 0.001$ as the eccentricity ratio changes from 0.273 to 0.494 (Table 17.2). There is no change observed in pattern of capacitance with a change in the eccentricity ratio as in the case of a two-lobe bearing (Table 17.2). The change in capacitance with a change in eccentricity ratio is non-linear for a cylindrical bearing, similar to that of the two-lobe bearing.

17.3.3.3 Resistance of Two-Lobe and Cylindrical Bearings

The resistance of a two-lobe bearing depends on the individual resistance of the lower and upper lobes. This in turn is a function of lubricant resistivity (ρ), C_h, L, r, and depend on ϕ_1, ε_1 for the lower lobe, and ϕ_2, ε_2 for the upper lobe along with the angles θ_{pi} and θ_{pe}. With an increase in the eccentricity ratio, the resistance of a lobe decreases. For the lower lobe as the eccentricity ratio (ε_1) increases from 0.70 to 0.96, R_L decreases from 8.62×10^6 to 4.92×10^6 ohm. Similarly, for the upper lobe, as ε_2 increases from 0.253 to 0.54, R_U is reduced from 37×10^6 to 11.2×10^6 ohm. The

Table 17.1. Capacitance, resistance and dynamic coefficients of two-lobe bearing under different operating parameters

ε	$\phi^*(°)$	ε_1	$\phi_1^*(°)$	ε_2	$\phi_2^*(°)$	C_L $\times 10^{-9}(\phi)$	C_U $\times 10^{-9}$ (ϕ)	C_{eq} $\times 10^{-9}$ (ϕ)	R_L $\times 10^6$ (ohm)	R_U $\times 10^6$ (ohm)	R_{eq} $\times 10^6$ (ohm)	C_{dL} $\times 10^6$ $(N s m^{-1})$	C_{dU} $\times 10^6$ $(N s m^{-1})$	C_{deq} $\times 10^6$ $(N s m^{-1})$	K_L $\times 10^9$ $(N m^{-1})$	K_U $\times 10^9$ $(N m^{-1})$	K_{eq} $\times 10^9$ $(N m^{-1})$
0.494	29.5	0.96	14.7	0.253	106.10	4.51	0.60	5.11	4.92	37.0	4.34	4.92	37	4.34	0.22	1.67	0.20
0.460	57.0	0.84	27.47	0.460	122.66	3.02	1.10	4.12	7.34	20.2	5.38	7.34	20.2	5.38	0.33	0.91	0.24
0.410	69.0	0.75	30.31	0.520	132.65	2.49	1.57	4.06	8.87	14.1	5.50	8.87	14.1	5.50	0.40	0.64	40.25
0.372	74.43	0.70	30.96	0.540	138.01	2.48	1.66	4.14	8.62	13.4	5.25	8.62	13.4	5.25	0.40	0.60	0.24
0.273	79.54	0.61	26.16	0.524	149.04	2.75	2.01	4.76	8.14	11.2	4.71	8.14	11.2	4.71	0.36	0.50	0.21

*$L/d = 0.5$, $L = 150\,mm$, $d = 300\,mm$, $C_h = 0.4\,mm$, $e_p = 0.2\,mm$, $\varepsilon_p = 0.5$, $\xi = 2.5 \times 8.854 \times 10^{-12}\,\phi\,m^{-1}$, $\rho = 10^{11}$ ohm cm.

Table 17.2. Capacitance, resistance and dynamic coefficients of cylindrical hydrodynamic journal bearing under different operating parameters

S. No.	ε	$C_{cb} \times 10^{-9}(\phi)$	$R_{cb} \times 10^{6}$(ohm)	$C_{dc} \times 10^{6}(\mathrm{N\,s\,m^{-1}})$	$K_{cb} \times 10^{9}(\mathrm{N\,m^{-1}})$
1	0.494	9.3	2.38	2.38	0.107
2	0.460	8.95	2.48	2.48	0.112
3	0.410	8.24	2.71	2.71	0.121
4	0.372	7.76	2.88	2.88	0.130
5	0.273	6.85	3.21	3.21	0.146

$\delta = L/d = 0.5$, $L = 150$ mm, $d = 300$ mm, $\psi = C_d/d = 0.001$, $\rho = 10^{11}$ ohm cm, $\xi = 2.5 \times 8.854 \times 10^{-12}\ \phi\,\mathrm{m}^{-1}$.

change in resistance of lobes and bearing with a change in eccentricity ratio is not linear in all the ranges of eccentricity ratio, and nor the change in equivalent resistance with global eccentricity ratio (ε) is linear. The equivalent resistance is 4.34×10^6, 5.5×10^6 and 4.71×10^6 ohm at the eccentricity ratio (ε) of 0.494, 0.41 and 0.273, respectively (Table 17.1).

Similarly, for a cylindrical bearing, resistance decreases with an increase in eccentricity ratio and the change of resistance is non-linear for all ranges of eccentricity ratios. The resistance decreases from 3.21×10^6 to 2.38×10^6 ohm as the eccentricity increases from 0.273 to 0.494.

17.3.3.4 Stiffness and Damping Coefficients of Two-Lobe and Cylindrical Bearings

17.3.3.4.1 Damping Coefficient

The damping coefficient of the lower lobe (C_{dL}), of a two-lobe bearing varies non-linearly from 4.92×10^6 to 8.87×10^6 to $8.14 \times 10^6\ \mathrm{N\,s\,m^{-1}}$ as ε_1 decreases from 0.96 to 0.75 to 0.61, respectively. Also, C_{dU} follow the non-linear pattern and decreases from 37×10^6 to $11.20 \times 10^6\ \mathrm{N\,s\,m^{-1}}$ as ε_2 increases from 0.253 to 0.524 (Table 17.1). The equivalent damping coefficient (C_{deq}) follow a non-linear pattern similar to that of C_{dL}, C_{dU} and varies from 4.34×10^6 to 5.50×10^6 to $4.71 \times 10^6\ \mathrm{N\,s\,m^{-1}}$ as ε decreases from 0.494 to 0.41 to 0.273, respectively (Table 17.1) as shown in Fig. 17.3.

Also, for the cylindrical bearing, the damping coefficient (C_{dc}) increases non-linearly from 2.38×10^6 to $3.21 \times 10^6\ \mathrm{N\,s\,m^{-1}}$ with a variation in the eccentricity ratio of 0.494–0.273 (Table 17.2). The ratio of C_{dL}/C_{dc} and C_{dU}/C_{dc}, C_{deq}/C_{dc} under different eccentricity ratios vary between 2.07 and 3.27, 3.49 and 15.55, and 1.47 and 2.17, respectively as shown in Table 17.3 (Fig. 17.3). This indicates that a two-lobe

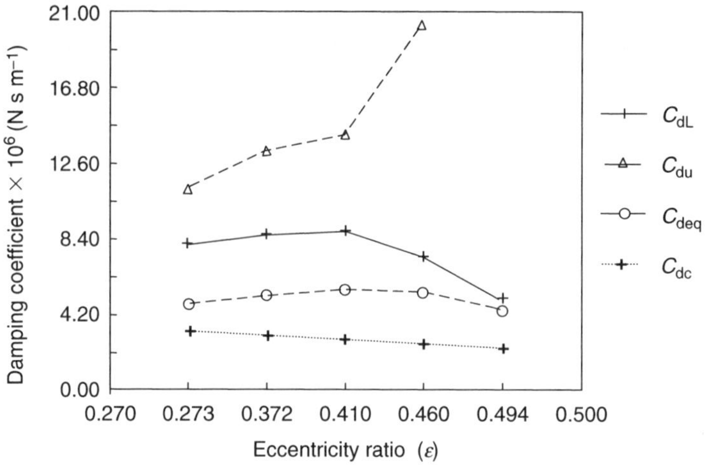

Fig. 17.3. Variation of damping coefficients with eccentricity ratio of lower lobe (C_{dL}), upper lobe (C_{dU}) and equivalent damping C_{deq} of two-lobe and C_{dc} of cylindrical bearings

bearing has better damping characteristics compared to a cylindrical bearing under the identical eccentricity ratios having the identical or comparable dimensional parameters. A similar trend exists in the published work by different authors [1, 2, 10].

17.3.3.4.2 Stiffness Coefficient

The stiffness coefficient of the lower lobe (K_{L}) varies non-linearly from 0.22×10^9 to 0.40×10^9 to $0.36 \times 10^9\,\mathrm{N\,m^{-1}}$ as ε_1 decreases from 0.96 to 0.75 and 0.70 to 0.61, respectively. Similarly, K_{U} follows non-linear pattern and decreases from 1.67×10^9 to $0.50 \times 10^9\,\mathrm{N\ m^{-1}}$ as ε_2 increases from 0.253 to 0.524 (Table 17.1) as shown in Fig. 17.4. The equivalent stiffness coefficient (K_{eq}) also

Table 17.3. Comparison of dynamic coefficients of two-lobe and cylindrical bearings

S. No.	ε_{c}	$C_{\mathrm{dL}}/C_{\mathrm{dc}}$	$C_{\mathrm{dU}}/C_{\mathrm{dc}}$	$C_{\mathrm{eq}}/C_{\mathrm{dc}}$	$K_{\mathrm{L}}/K_{\mathrm{cb}}$	$K_{\mathrm{U}}/K_{\mathrm{cb}}$	$K_{\mathrm{eq}}/K_{\mathrm{cb}}$
1	0.494	2.07	15.55	1.82	2.06	15.61	1.87
2	0.464	2.96	8.15	2.17	2.95	8.13	2.14
3	0.410	3.27	5.21	2.03	3.31	5.29	2.07
4	0.372	2.99	4.65	1.82	3.08	4.62	1.85
5	0.273	2.54	3.49	1.47	2.47	3.42	1.44

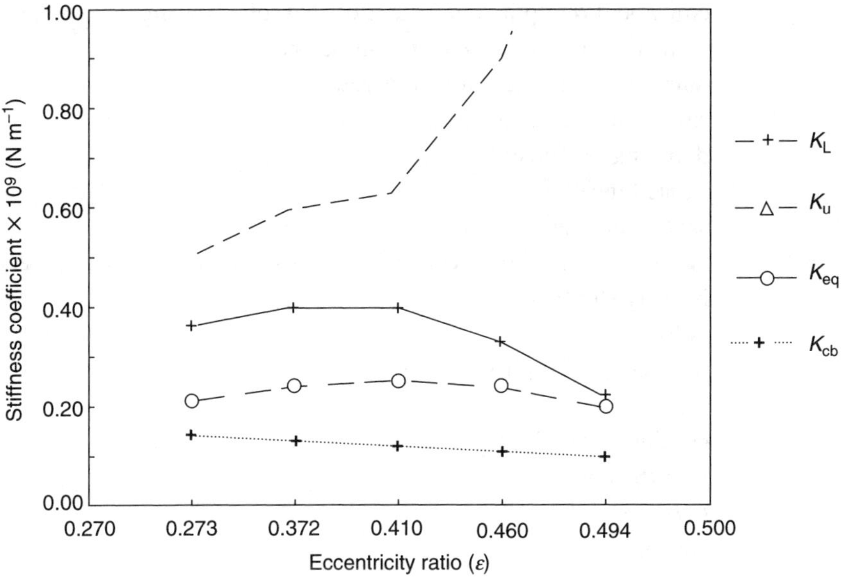

Fig. 17.4. Variation of stiffness coefficients with eccentricity ratio of lower lobe (K_L), upper lobe (K_U) and equivalent stiffness K_{eq} of two-lobe and K_{cb} of cylindrical bearings

follows the non-linear pattern, similar to K_U, K_L, and varies from 0.20×10^9 to 0.25×10^9 to 0.21×10^9 N m^{-1} as ε decreases from 0.494 to 0.41 to 0.273, respectively (Table 17.1), as shown in Fig. 17.4.

For a cylindrical bearing, stiffness coefficient (K_{cb}) increases non-linearly from 0.107×10^9 to 0.145×10^9 N m^{-1} as the eccentricity ratio (ε_c) changes from 0.494 to 0.273 (Table 17.2). The ratios of K_L/K_{cb}, K_U/K_{cb} and K_{eq}/K_{cb} under different eccentricity ratios vary between 2.06 and 3.31, 3.42 and 15.61, and 1.44 and 2.07, respectively, as shown in Table 17.3. This shows that a two-lobe bearing has better stiffness characteristics as compared to a cylindrical bearing under the identical eccentricity ratios having the identical or comparable dimensional parameters. A similar trend exists in the published work [1, 2, 10].

Nomenclature

C_d	diametral clearance of cylindrical bearing
C_h	horizontal radial clearance of two-lobe bearing
C_L, C_U	capacitance of lower and upper lobe, respectively
C_{dL}	damping coefficient of lower lobe
C_{dU}	damping coefficient of upper lobe

C_{deq}	equivalent damping coefficient of two-lobe bearing
C_{eq}	equivalent capacitance of two-lobe bearing
C_{dc}	damping coefficient of cylindrical bearing
C_{cb}	capacitance of cylindrical bearing
C_c	damping coefficient
C	capacitance
d	shaft diameter
e, e_1, e_2	bearing eccentricity, eccentricity of lower and upper lobe, respectively
e_p	bearing ellipticity
E	voltage
f	frequency of applied voltage
F	force
F_L, F_U	conductance of lower and upper lobe, respectively
h	film thickness
H	h/c_h
i	current
K	stiffness
K_L	stiffness of lower lobe
K_U	stiffness of upper lobe
K_{eq}	equivalent stiffness of two-lobe bearing
K_{cb}	stiffness of cylindrical bearing
L	bearing length
L_j	inductance
M	mass
Q	charge
r	radius of journal
r_L	radius of lobe
R	resistance
R_L, R_U	resistance of lower and upper lobe, respectively
R_{cb}	resistance of cylindrical journal bearing
R_{eq}	equivalent resistance of two-lobe bearing
t	time
w	$2\pi f$
X_L, X_U, X_{eq}	capacitive reactance of lower lobe, upper lobe and equivalent capacitive reactance of two-lobe bearing, respectively
Z_L, Z_U, Z_{eq}	impedance of lower lobe, upper lobe and equivalent impedance of two-lobe bearing, respectively
$\delta = L/d$	length to diameter ratio
ε_c	eccentricity ratio of cylindrical bearing

$\varepsilon, \varepsilon_p, \varepsilon_1, \varepsilon_2$	$(e, e_p, e_1, e_2)/C_h$
ϕ	attitude angle of two-lobe bearing
ϕ_1, ϕ_2	attitude angle for lower and upper lobe, respectively
ϕ_c	attitude angle of cylindrical bearing
λ_1, λ_2	location of loading edge of the positive pressure fluid-film with reference to line of centers for lower and upper lobes
θ	angular coordinate measured from leading edge for either lower or upper lobe
θ_{pi}	inlet angle from line of centers to leading edge film in each lobe
θ_{pe}	outlet angle from line of centers to trailing edge film in each lobe
ρ	resistivity of lubricant
ξ	permittivity/dielectric constant of lubricant $[\xi_o \xi_r (\phi\,m^{-1})]$
ξ_r	relative permittivity of lubricant
ξ_o	premittivity of free space $(8.854 \times 10^{-12}\,\phi\,m^{-1})$
ψ	diametral clearance ratio of cylindrical bearing (C_d/d)

17.4 Three-Lobe Bearing

17.4.1 Geometry and Geometrical Characteristics of Three-Lobe Bearings

The three-lobe bearing is made up of three equal non-concentric arcs, with their centers of curvatures O_1, O_2 and O_3, each displaced by a distance, e_p, from the geometric center O_b, of the bearing, as shown in Fig. 17.5. For any given journal positions O_j, the eccentricities e_1, e_2, and e_3 and attitude angles ϕ_1, ϕ_2, ϕ_3 for each lobe will be different, as shown in Fig. 17.6. The angle ϕ, and distance between bearing center and journal center give the global attitude angle and eccentricity e, respectively (Fig. 17.6). Nica [7] has developed various mathematical relationships for a three-lobe bearing, which are given as

$$\varepsilon_1 = (\varepsilon_p^2 + \varepsilon^2 + 2\varepsilon_p \varepsilon \cos\phi)^{1/2} \tag{17.31}$$

$$\phi_1 = \tan^{-1}\left(\frac{\varepsilon \sin\phi}{\varepsilon_p + \varepsilon\cos\phi}\right) \tag{17.32}$$

$$\varepsilon_2 = \left[\varepsilon_p^2 + \varepsilon^2 - 2\varepsilon_p\varepsilon\cos\left(\frac{\pi}{3} + \phi\right)\right]^{1/2} \tag{17.33}$$

$$\phi_2 = \frac{2\pi}{3} - \tan^{-1}\left[\frac{\varepsilon\sin(\pi/3 + \phi)}{\varepsilon_p - \varepsilon\cos(\pi/3 + \phi)}\right] \tag{17.34}$$

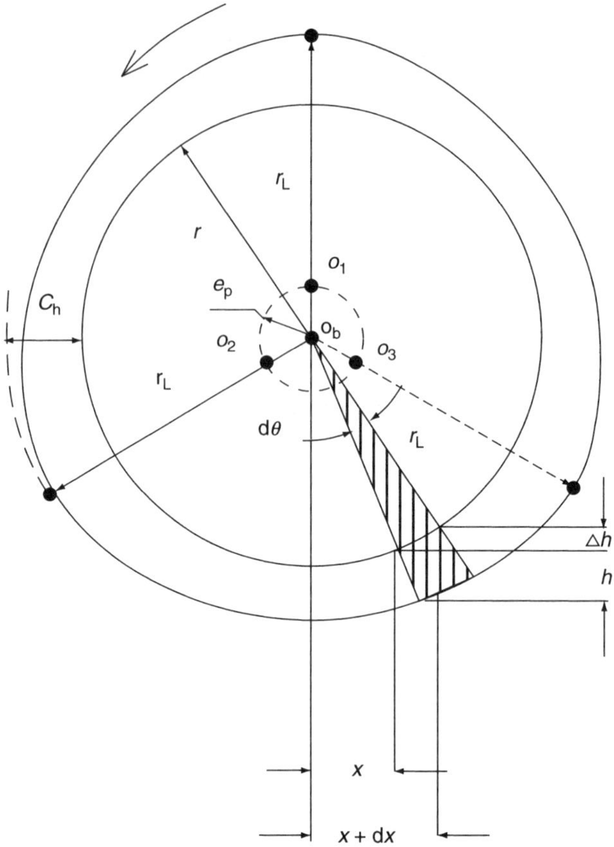

Fig. 17.5. Journal configuration in a hydrodynamic three-lobe journal bearing under rotation

$$\varepsilon_3 = \left[\varepsilon_p^2 + \varepsilon^2 - 2\varepsilon\varepsilon_p \cos\left(\frac{\pi}{3} - \phi\right)\right]^{1/2} \tag{17.35}$$

$$\phi_3 = \frac{4\pi}{3} + \tan^{-1}\left[\frac{\varepsilon \sin(\pi/3 - \phi)}{\varepsilon_p - \varepsilon \cos(\pi/3 - \phi)}\right] \tag{17.36}$$

and

$$\begin{aligned} \lambda_1 &= \frac{2\pi}{3} - \phi_1 \\ \lambda_2 &= \frac{4\pi}{3} - \phi_2 \\ \lambda_3 &= 2\pi - \phi_3 \end{aligned} \tag{17.37}$$

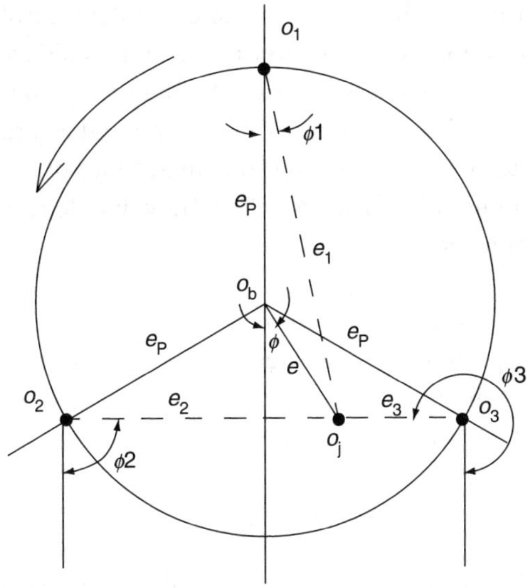

Fig. 17.6. Geometrical parameters of a three-lobe journal bearing

The film thickness for first, second and third lobe can be expressed as [7]

$$h = C_{\rm h}[1 + \varepsilon_1 \cos(\theta + \lambda_1)] \tag{17.38}$$

or

$$h = C_{\rm h}[1 + \varepsilon_2 \cos(\theta + \lambda_2)]$$

or

$$h = C_{\rm h}[1 + \varepsilon_3 \cos(\theta + \lambda_3)]$$

17.4.2 Theoretical Analysis of Capacitance and Active Resistance of the Individual Lobes of Three-Lobe Bearings

17.4.2.1 Capacitance

In a three-lobe bearing, contrary to an elliptical or a cylindrical bearing, positive pressure builds in all the three lobes. The analysis is performed in a unitary manner, for three lobes, by starting the analysis with an isolated lobe placed in any position with respect to the shaft or to the bearing center. The characteristics of the bearing as a whole are found by summing up the results from the individual lobes similar to that of a two-lobe bearing.

During rotation of the shaft, part of the circumference of the journal passes through each lobe in a zone of load-carrying oil film. This forms a parallel plate capacitor of varying film thickness between the interacting surfaces of the journal and an individual lobe. The capacitances of all three lobes are in parallel to each other with respect to the journal. However, the capacitance of parallel plate capacitor is greatest at the minimum film thickness of an individual lobe, and reduces on both the sides as the film thickness increases.

17.4.2.1.1 Capacitance of the Individual Lobes

The change in capacitance ΔC between two rectangular plates of width $\mathrm{d}X$ of length L, located at the angular coordinate differential $\mathrm{d}\theta$ (as shown in Fig. 17.5) and separated by a film thickness Δh, is given by

$$\Delta C = 2\xi \frac{(X + \mathrm{d}X)L - XL}{2h + \Delta h} \tag{17.39}$$

The change in film thickness (Δh) is very small in comparison to $\mathrm{d}X$ and hence can be neglected. Thus the change in capacitance is given as

$$\Delta C = \frac{\xi \mathrm{d}XL}{h} \tag{17.40}$$

From Fig. 17.5, it is evident that

$$dX = d\theta(r + C_\mathrm{h}) \tag{17.41}$$

On substituting the values from Eqns. (17.38) and (17.41) into Eqn. (17.40), the change in capacitance of the first lobe is expressed as

$$\Delta C = \frac{\xi L(r + C_\mathrm{h})d\theta}{C_\mathrm{h}[1 + \varepsilon_1 \cos(\theta + \lambda_1)]} \tag{17.42}$$

The maximum change in capacitance depends on an angle θ – the angle coordinate measured from leading edge to trailing edge – and it is along the film thickness. Thus, integrating between the inlet angle θ_{pi} and outlet angle θ_{pe}, the capacitance of the first lobe of a three-lobe bearing is determined as

$$C_1 = \xi L C_\mathrm{h}^{-1}(r + C_\mathrm{h}) \int_{\theta_{\mathrm{pi}}}^{\theta_{\mathrm{pe}}} [1 + \varepsilon_1 \cos(\theta + \lambda_1)]^{-1} d\theta \tag{17.43}$$

Equation (17.43) can be integrated by Sommerfeld substitution [8]. Capacitance is worked out by using Eqn. (17.37) ($\lambda_1 = 2\pi/3 - \phi_1$) and by taking θ_{pe} and θ_{pi} equal to 110° and 10° for the first lobe on considering 20° as the angle for the oil supply for each lobe. Capacitance of the first lobe is given as

$$C_1 = \xi L C_{\mathrm{h}}^{-1}(r + C_{\mathrm{h}})(1 - \varepsilon_1^2)^{-1/2}$$
$$\times \left\{ \cos^{-1}\left[\frac{\varepsilon_1 + \cos(230^\circ - \phi_1)}{1 + \varepsilon_1 \cos(230^\circ - \phi_1)}\right] - \cos^{-1}\left[\frac{\varepsilon_1 + \cos(130^\circ - \phi_1)}{1 + \varepsilon_1 \cos(130^\circ - \phi_1)}\right]\right\} \quad (17.44)$$

Similar to Eqn. (17.43), capacitance of the lobes 2 and 3 (C_2 and C_3) can be determined using ε_1, ε_2 and ε_3 for the respective lobes and integrating by taking θ_{pe} and θ_{pi} between 230° and 130° for lobe 2, and between 350° and 250° for lobe 3, respectively. C_2 and C_3 are determined by the following equations:

$$C_2 = \xi L C_{\mathrm{h}}^{-1}(r + C_{\mathrm{h}})(1 - \varepsilon_2^2)^{-1/2}$$
$$\times \left\{ \cos^{-1}\left[\frac{\varepsilon_2 + \cos(110^\circ - \phi_2)}{1 + \varepsilon_2 \cos(110^\circ - \phi_2)}\right] - \cos^{-1}\left[\frac{\varepsilon_2 + \cos(10^\circ - \phi_2)}{1 + \varepsilon_2 \cos(10^\circ - \phi_2)}\right]\right\} \quad (17.45)$$

and

$$C_3 = \xi L C_{\mathrm{h}}^{-1}(r + C_{\mathrm{h}})(1 - \varepsilon_3^2)^{-1/2}$$
$$\times \left\{ \cos^{-1}\left[\frac{\varepsilon_3 + \cos(350^\circ - \phi_3)}{1 + \varepsilon_3 \cos(350^\circ - \phi_3)}\right] - \cos^{-1}\left[\frac{\varepsilon_3 + \cos(250^\circ - \phi_3)}{1 + \varepsilon_3 \cos(250^\circ - \phi_3)}\right]\right\} \quad (17.46)$$

17.4.2.2 Resistance

The change in conductance/resistance^{-1}, ΔF, between two rectangular plates of width $\mathrm{d}X$ of bearing length L, located at an angular coordinate differential $\mathrm{d}\theta$ and separated by a film thickness Δh, is given as

$$\Delta F = \frac{\mathrm{d}XL}{\rho h} \quad (\Delta h \text{ being small and hence negligible}) \quad (17.47)$$

The maximum change in conductance/resistance^{-1} depends on an angle θ – the angular coordinate is measured from the leading edge and is along the oil film thickness. So, on integrating between the inlet angle θ_{pi} and the outlet angle θ_{pe}, the resistance of the first lobe of a three-lobe bearing is determined using Eqns. (17.41) and (17.38), and is given as

$$F = R_1^{-1} = \rho^{-1} L(r + C_{\mathrm{h}}) C_{\mathrm{h}}^{-1} \int_{\theta_{\mathrm{pi}}}^{\theta_{\mathrm{pe}}} [1 + \varepsilon_1 \cos(\theta + \lambda_1)]^{-1} \mathrm{d}\theta \quad (17.48)$$

The above equation is integrated using Sommerfeld substitution [8]. The resistance of each lobe is worked out using similar integration limits as for the capacitance of the respective lobe and is given as

$$R_1 = \rho C_h L^{-1}(r + C_h)^{-1}(1 - \varepsilon_1^2)^{1/2} \times \left\{\cos^{-1}\left[\frac{\varepsilon_1 + \cos(230^\circ - \phi_1)}{1 + \varepsilon_1 \cos(230^\circ - \phi_1)}\right] - \cos^{-1}\left[\frac{\varepsilon_1 + \cos(130^\circ - \phi_1)}{1 + \varepsilon_1 \cos(130^\circ - \phi_1)}\right]\right\}^{-1} \tag{17.49}$$

$$R_2 = \rho C_h L^{-1}(r + C_h)^{-1}(1 - \varepsilon_2^2)^{1/2} \times \left\{\cos^{-1}\left[\frac{\varepsilon_2 + \cos(110^\circ - \phi_2)}{1 + \varepsilon_2 \cos(110^\circ - \phi_2)}\right] - \cos^{-1}\left[\frac{\varepsilon_2 + \cos(10^\circ - \phi_2)}{1 + \varepsilon_2 \cos(10^\circ - \phi_2)}\right]\right\}^{-1} \tag{17.50}$$

$$R_3 = \rho C_h L^{-1}(r + C_h)^{-1}(1 - \varepsilon_3^2)^{1/2} \times \left\{\cos^{-1}\left[\frac{\varepsilon_3 + \cos(350^\circ - \phi_3)}{1 + \varepsilon_3 \cos(350^\circ - \phi_3)}\right] - \cos^{-1}\left[\frac{\varepsilon_3 + \cos(250^\circ - \phi_3)}{1 + \varepsilon_3 \cos(350^\circ - \phi_3)}\right]\right\}^{-1} \tag{17.51}$$

17.4.3 Equivalent Capacitance, Active Resistance, Capacitive Reactance and Impendance of Three-Lobe Bearings

17.4.3.1 Capacitance

As per the electrical analogy, all the lobes of a three-lobe bearing are in parallel to each other with respect to the journal. So the equivalent capacitance of the three-lobe bearing is determined as

$$C_{eq} = C_1 + C_2 + C_3 \tag{17.52}$$

17.4.3.2 Resistance

Since all the lobes are in parallel to each other with respect to the journal, the equivalent resistance of the three-lobe bearing is given as

$$R_{eq} = \frac{R_1 R_2 R_3}{R_1 R_2 + R_2 R_3 + R_1 R_3} \tag{17.53}$$

17.4.3.3 Capacitive Reactance and Impedance of Three-Lobe Bearings

Capacitive reactance between the interacting surfaces of the journal and each lobe in the oil film, as well as the equivalent capacitive reactance of the bearing, is determined by Prashad [5] and is expressed as

$$X_1 = \frac{1}{wC_1},\ X_2 = \frac{1}{wC_2},\ X_3 = \frac{1}{wC_3}$$

and

$$X_{eq} = \frac{1}{wC_{eq}} \tag{17.54}$$

The impedance of bearing circuit is determined as

$$\begin{aligned} &Z_1 = R_1 - jX_1,\ Z_2 = R_2 - jX_2,\ Z_3 = R_3 - jX_3 \\ &Z_{eq} = R_{eq} - jX_{eq} \end{aligned} \tag{17.55}$$

The ratio of capacitive reactance to active resistance is given by [Eqns. (17.43)–(17.54)]

$$\frac{X_{eq}}{R_{eq}} = \frac{X_1}{R_1} = \frac{X_2}{R_2} = \frac{X_3}{R_3} = \frac{1}{\rho\xi w} \tag{17.56}$$

17.4.4 Data Deduction and Discussion

The capacitance, resistance and dynamic coefficients for an individual lobe and an equivalent value of these parameters for the three-lobe bearing having $L/d = 0.5$ ($L = 150\,\text{mm}$, $d = 300\,\text{mm}$), $C_h = 0.4\,\text{mm}$ and $e_p = 0.2\,\text{mm}$ have been computed as per the developed methodology and are shown in Table 17.4, and Figs.17.7 and 17.8. The global attitude angles, eccentricity ratios, and other relevant details for the three-lobe bearing have been taken from the published data by Nica [7]. The attitude angles as well as the eccentricity ratios for an individual lobe have been computed separately using Eqns. (17.31)–(17.36). Based on the electrical analogy, dynamic coefficients of each lobe have been worked out for the global eccentricity ratios of 0.293, 0.195, 0.141, and 0.077 with the corresponding attitude angles of 31.3°, 39.8°, 44.8° and 49.7°, respectively. Also the equivalent values of dynamic coefficients of the three-lobe bearing have been compared with those of two-lobe and cylindrical bearings, and are shown in Figs.17.7 and 17.8 [6, 11].

The effective dynamic coefficients of the three-lobe bearings determined by the approach developed have also been compared with those obtained through the conventional approaches [12]. The approach used for this comparison is similar to that reported in Section 17.3 for two-lobe and cylindrical bearings [13].

17.4.4.1 Capacitance

The capacitance of a three-lobe bearing depends on the capacitance of each lobe. This, in turn, is a function of lubricant permittivity, horizontal clearance (C_h), bearing length (L), and shaft radius (r), and depends on the eccentricity ratio and attitude angle of the respective lobe along with the angles θ_{pi} and θ_{pe} [Eqns. (17.43)–(17.46)].

With a decrease in the global eccentricity ratio from 0.293 to 0.077, the attitude angle increases from 31.3° to 49.7° (Table 17.4). ε_1, ε_2 decrease, as against ε_3 which increases, with decrease in global eccentricity ratio. Under these conditions, the capacitances of lobes 1, 2 and 3 change between $(0.975 \times 10^{-9})\phi$ and $(0.421 \times 10^{-9})\phi$, $(0.970 \times 10^{-9})\phi$ and $(1.735 \times 10^{-9})\phi$, and $(1.133 \times 10^{-9})\phi$ and $(1.960 \times 10^{-9})\phi$, respectively. The variation of capacitance with eccentricity ratio is non-linear. Under global eccentricity ratios of less than 0.293, the capacitance of lobe 3 is more than that of lobes 1 and 2. This is because of the relatively less eccentricity ratio (ε_3) of lobe 3 and the influence of film thickness and attitude angle on the capacitance as compared to that of the lobes 1 and 2 [Eqn. (17.38)]. Equivalent capacitance of the bearing with global eccentricity ratio varies between $(3.13 \times 10^{-9})\phi$ and $(4.21 \times 10^{-9})\phi$, and in general increases with decrease in global eccentricity ratio (Table 17.4). Also, the capacitance of lobes 2 and 3 increases, as against that of lobe 1, which decreases, with decrease in global eccentricity ratio (Table17.4).

17.4.4.2 Resistance

The resistance of a three-lobe bearing depends on the individual resistance of each lobe. This, in turn, is a function of lubricant resistivity, clearance, bearing length, shaft radius and depends on the attitude angle and eccentricity ratio of each lobe and varies with the angles θ_{pi} and θ_{pe} [Eqns. (17.48)–(17.51)].

Under the change of global eccentricity ratio from 0.293 to 0.077 and attitude angle from 31.3° to 49.7°, the resistances of lobes 1, 2 and 3 change from 22.7×10^6 to 52.5×10^6 ohm, 22.7×10^6 to 12.7×10^6 ohm, and 19.54×10^6 to 11.29×10^6 ohm, respectively (Table 17.4). The variation of resistance with eccentricity ratio is non-linear. The resistance of the lobe 3 is less than that of lobes 1 and 2, as shown in Table 17.4. This is attributed to the less eccentricity ratio of lobe 3 and the influence of relatively higher attitude angle as compared to that of the lobes 1 and 2 (as shown in Table 17.4). Resistance of lobes 2 and 3 decrease, as against that of lobe 1, which increases with decrease in global eccentricity ratio. However, the equivalent resistance of the three-lobe bearing in general decreases with decrease in global eccentricity ratios, and varies between 7.18×10^6 and 5.24×10^6 ohm, as shown in Table 17.4. This is because of the net effect of the variable film thickness in each lobe of a three-lobe bearing under the influence of different eccentricity ratios, and operating under variable conditions.

Table: 17.4. Capacitance and resistance of three-lobe bearing under different operating parameters

$\varepsilon \cos\phi$	ε	ϕ (°)	ε_1	ε_2	ε_3	ϕ_1 (°)	ϕ_2 (°)	ϕ_3 (°)	C_1 $\times 10^{-9}$ (ϕ)	C_2 $\times 10^{-9}$ (ϕ)	C_3 $\times 10^{-9}$ (ϕ)	C_{eq} $\times 10^{-9}$ (ϕ)	R_1 $\times 10^6$ (ohm)	R_2 $\times 10^6$ (ohm)	R_3 $\times 10^6$ (ohm)	R_{eq} $\times 10^6$ (ohm)
0.25	0.293	31.3	0.766	0.574	0.280	11.31	89.95	270.1	0.975	0.970	1.133	3.13	22.7	22.7	19.54	7.18
0.15	0.195	39.8	0.662	0.560	0.320	10.46	100.3	252.34	0.799	1.340	1.790	3.93	27.7	16.5	12.32	5.62
0.10	0.141	44.8	0.608	0.553	0.366	9.41	105.7	246.34	0.676	1.606	1.920	4.21	32.4	13.7	11.50	5.24
0.05	0.077	49.7	0.553	0.531	0.425	6.23	112.3	241.85	0.421	1.735	1.960	4.12	52.5	12.7	11.29	5.41

$L/d = 0.5$, $L = 150\,\text{mm}$, $d = 300\,\text{mm}$, $C_h = 0.4\,\text{mm}$, $e_p = 0.2\,\text{mm}$, $\varepsilon_p = 0.5$, $\xi = 2.5 \times 8.854 \times 10^{-12}\,\phi\,\text{m}^{-1}$, $\rho = 10^{11}$ ohm cm.

17.4.4.3 Damping and Stiffness Coefficients of Three-Lobe, Two-Lobe and Cylindrical Bearings

17.4.4.3.1 Damping Coefficient

The damping coefficient of each lobe varies differently with the eccentricity ratio. As ε changes from 0.293 to 0.077, the variations of C_{d1}, C_{d2} and C_{d3} are non-linear, changing from 22.7×10^6 to 52.5×10^6 N s m^{-1}, 22.7×10^6 to 12.7×10^6 N s m^{-1} and 19.54×10^6 to 11.29×10^6 N s m^{-1}, respectively as shown in Table 17.5 (Fig. 17.7). The damping coefficient of lobes 2 and 3 (C_{d2}, C_{d3}) increase, as against that of C_{d1}, which decreases, with increase in global eccentricity ratio. Also, the equivalent damping coefficient of the three-lobe bearing follows non-linear pattern and varies between 5.41×10^6 and 7.18×10^6 N s m^{-1} as the global eccentricity changes between 0.077 and 0.293 as shown in Table 17.5 (Fig. 17.7).

Under the identical conditions of operation, dynamic coefficient of a three-lobe bearing is found to be more than that of the equivalent two-lobe and cylindrical bearings, as shown in Fig. 17.7. The data for cylindrical and two-lobe bearings have been taken from the published work by Prashad [11]. A similar trend exists in published work by Glienicke et al. [1], Knight and Barrett [2], and Dargaiah et al. [10]. Also, the values of dynamic coefficients determined by the approach developed are found to be comparable to those obtained through the conventional approaches.

17.4.4.3.2 Stiffness Coefficient

The stiffness coefficient of an individual lobe of a three-lobe bearing depends on reciprocal of its capacitance [Eqns. (17.28) and (17.30)]. The equivalent stiffness of a three-lobe bearing is the resultant stiffness of all the lobes connected in parallel. The stiffness of an individual lobe shows non-linear behavior with the eccentricity ratio. K_1, K_2 and K_3 vary from 2.38×10^9 to 0.970×10^9 N m^{-1}, 0.58×10^9 to 1.031×10^9 N m^{-1}, and 0.51×10^9 to 0.833×10^9 N m^{-1} respectively, as ε changes from 0.077 to 0.293 as shown in Table 17.5 (Fig. 17.8). The stiffness of second and third lobes (K_2, K_3) increase with increase in global eccentricity ratio, as against that of K_1, which decreases. The equivalent stiffness of the three-lobe bearing varies between 0.24×10^9 and 0.32×10^9 N m^{-1} (Table 17.5). This is more than that of two-lobe and cylindrical bearings having comparable dimensional parameters and operating under identical conditions, as shown in Fig. 17.8. The data for cylindrical and two-lobe bearings have been taken from the published work by Prashad (1996) [11]. A similar trend exists in published work by Glienicke et al. [1], Knight and Barrett [2], and Dargaiah et al. [10]. The variation (0.24×10^9 to 0.32×10^9 N m^{-1}) of the equivalent stiffness coefficient as shown in Fig. 17.8, determined by this electrical analogy approach, matches closely with that obtained by Lund and Thomsen [12].

Table: 17.5. Dynamic coefficients of three-lobe journal bearing under different operating parameters

S. No.	$\varepsilon \cos \phi$	$C_{d1} \times 10^6$ $(N\,s\,m^{-1})$	$C_{d2} \times 10^6$ $(N\,s\,m^{-1})$	$C_{d3} \times 10^6$ $(N\,s\,m^{-1})$	$C_{deq} \times 10^6$ $(N\,m^{-1})$	$K_1 \times 10^9$ $(N\,m^{-1})$	$K_2 \times 10^9$ $(N\,m^{-1})$	$K_3 \times 10^9$ $(N\,m^{-1})$	$K_{eq} \times 10^9$ $(N\,m^{-1})$
1	0.45	23.60	−67.43	−206.46	−44.06	1.070	−3.110	−9.17	1.96
2	0.25	22.70	22.70	19.54	7.18	0.970	1.031	0.883	0.32
3	0.15	27.70	16.50	12.32	5.62	1.25	0.75	0.560	0.25
4	0.10	32.40	13.70	11.50	5.24	1.48	0.620	0.520	0.24
5	0.05	52.50	12.70	11.29	5.41	2.38	0.58	0.510	0.24

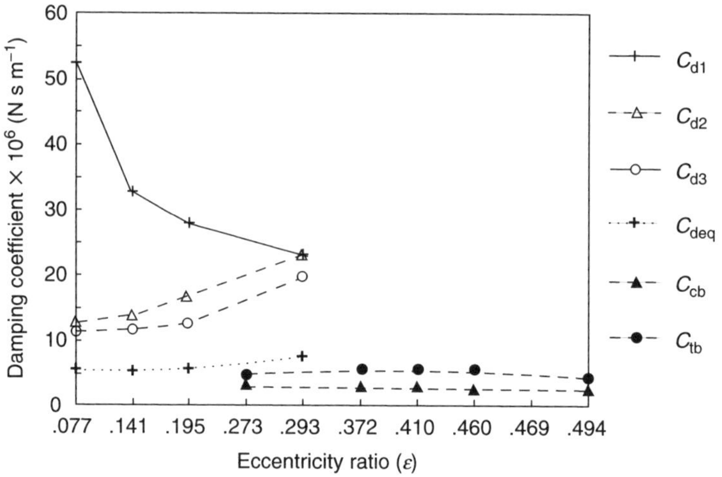

Fig. 17.7. Variation of damping coefficients with eccentricity ratio of first lobe (C_{d1}), second lobe (C_{d2}), third lobe (C_{d3}), and equivalent damping of three-lobe, and C_{tb} and C_{cb} of two-lobe and cylindrical bearings

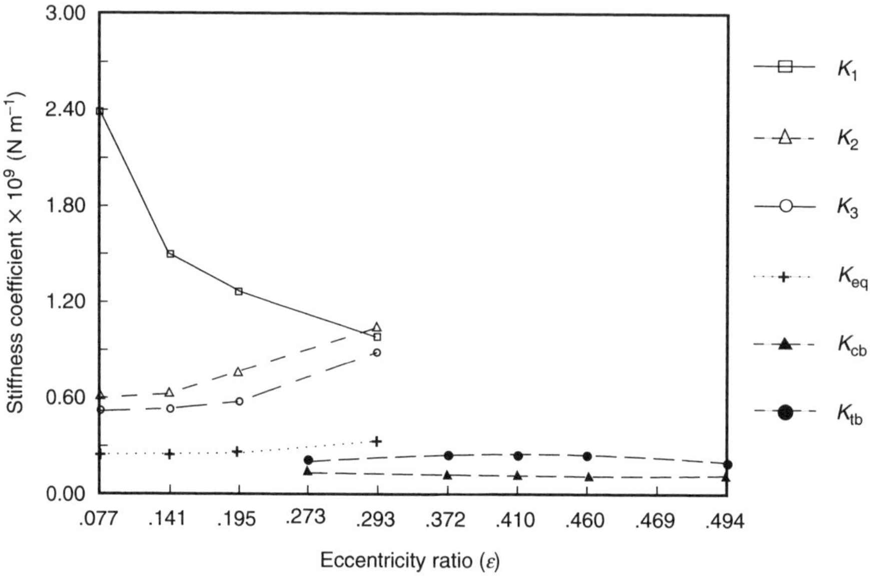

Fig. 17.8. Variation of stiffness coefficients with eccentricity ratio of first lobe (K_1), second lobe (K_2), third lobe (K_3), and equivalent stiffness (K_{eq}) of three-lobe, and K_{tb} and K_{cb} of two-lobe and cylindrical bearings

Nomenclature

C	capacitance
C_d	diametral clearance
C_h	horizontal radial clearance of three-lobe bearing
C_1, C_2, C_3	capacitance of first, second and third lobe, respectively
C_{eq}	equivalent capacitance of three-lobe bearing
C_{d1}, C_{d2}, C_{d3}	damping coefficients of first, second and third lobe, respectively
C_{deq}	equivalent/effective-damping coefficient of three-lobe bearing
C_c	damping coefficient
C_{tb}	equivalent/effective damping coefficient of two-lobe bearing
C_{cb}	equivalent/effective damping coefficient of cylindrical bearing
d	shaft diameter
e	bearing eccentricity
e_1, e_2, e_3	eccentricity of first, second and third lobe, respectively
e_p	bearing ellipticity
E	voltage
f	frequency of applied voltage
F	force
F_1	conductance
h	film thickness
i	current
K	stiffness
K_1, K_2, K_3	stiffness of first, second and third lobe, respectively
K_{eq}	equivalent/effective stiffness coefficient of three-lobe bearing
K_{tb}	equivalent/effective stiffness coefficient of two-lobe bearing
K_{cb}	equivalent/effective stiffness coefficient of cylindrical bearing
L	bearing length
L_j	inductance
M	mass
Q	charge
r	radius of journal
r_L	radius of lobe
R	resistance
R_1, R_2, R_3	resistance of first, second and third lobe, respectively
R_{eq}	equivalent resistance of three-lobe bearing
t	time
w	$2\pi f$
X_1, X_2, X_3	capacitive reactance of first, second and third lobe, respectively

X_{eq}, Z_{eq}	equivalent capactive reactance and impedance of three-lobe bearing, respectively
Z_1, Z_2, Z_3	impedance of first, second and third lobe, respectively
ε, ε_p, ε_1, ε_2, ε_3	$(e, e_p, e_1, e_2, e_3)/C_h$
ϕ	global attitude angle of three-lobe bearing
ϕ_1, ϕ_2, ϕ_3	attitude angle of first, second and third lobe, respectively
λ_1, λ_2, λ_3	location of leading edge of the positive pressure fluid film with reference to line of centers for first, second and third lobe, respectively
θ	angular coordinate measured from leading edge of a lobe
θ_{pi}	inlet angle from line of center to leading edge film in each lobe
θ_{pe}	outlet angle from line of center to trailing edge film in each lobe
ρ	resistivity of lubricant
ξ	permittivity/dielectric constant of lubricant [$\xi_o\ \xi_r(\phi\,\mathrm{m}^{-1})$]
ξ_r	relative permittivity of lubricant
ξ_o	permittivity of free space ($8.854 \times 10^{-12}\,\phi\,\mathrm{m}^{-1}$)

17.5 Four-Lobe Bearing

17.5.1 Geometry and Geometrical Characteristics of Four-Lobe Bearings

The four-lobe bearing is made up of four equal non-concentric arcs with their centers of curvatures O_1, O_2, O_3, and O_4 each displaced by a distance, e_p, from the geometric center O_b, of the bearing as shown in Fig. 17.9. For any given journal position O_j, the eccentricities e_1, e_2, e_3 and e_4 and attitude angles ϕ_1, ϕ_2, ϕ_3 and ϕ_4 for each lobe will be different as shown in Fig. 17.10. The angle ϕ and distance between bearing center and journal center give the global attitude angle and eccentricity e, respectively (Fig. 17.10). The various mathematical relations have been developed for a four-lobe bearing which are given as

$$\varepsilon_1 = \left(\varepsilon_p^2 + \varepsilon^2 + 2\varepsilon_p\varepsilon\cos\phi\right)^{1/2} \tag{17.57}$$

$$\phi_1 = \tan^{-1}\left(\frac{\varepsilon\sin\phi}{\varepsilon_p + \varepsilon\cos\phi}\right) \tag{17.58}$$

$$\varepsilon_2 = \left(\varepsilon_p^2 + \varepsilon^2 - 2\varepsilon_p\varepsilon\sin\phi\right)^{1/2} \tag{17.59}$$

$$\phi_2 = \frac{\pi}{2} - \tan^{-1}\left(\frac{\varepsilon\cos\phi}{\varepsilon_p - \varepsilon\sin\phi}\right) \tag{17.60}$$

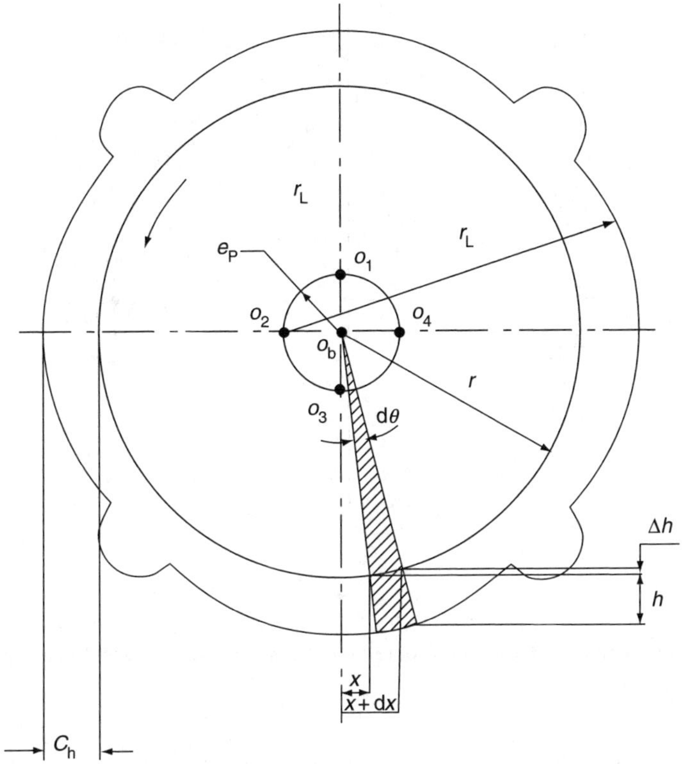

Fig. 17.9. Journal configuration in a hydrodynamic four-lobe journal bearing under rotation

$$\varepsilon_3 = \left(\varepsilon_\mathrm{p}^2 + \varepsilon_2 - 2\varepsilon\varepsilon_\mathrm{p}\cos\phi\right)^{1/2} \tag{17.61}$$

$$\phi_3 = \pi + \tan^{-1}\left(\frac{\varepsilon\sin\phi}{\varepsilon_\mathrm{p} - \varepsilon\cos\phi}\right) \tag{17.62}$$

$$\varepsilon_4 = \left(\varepsilon_\mathrm{p}^2 + \varepsilon_2 - 2\varepsilon\varepsilon_\mathrm{p}\sin\phi\right)^{1/2} \tag{17.63}$$

$$\phi_4 = \frac{3\pi}{2} + \tan^{-1}\left(\frac{\varepsilon\cos\phi}{\varepsilon_\mathrm{p} + \varepsilon\sin\phi}\right) \tag{17.64}$$

and

$$\left.\begin{aligned} \lambda_1 &= \frac{\pi}{2} - \phi_1 \\ \lambda_2 &= \pi - \phi_2 \\ \lambda_3 &= \frac{3\pi}{2} - \phi_3 \\ \lambda_4 &= 2\pi - \phi_4 \end{aligned}\right\} \tag{17.65}$$

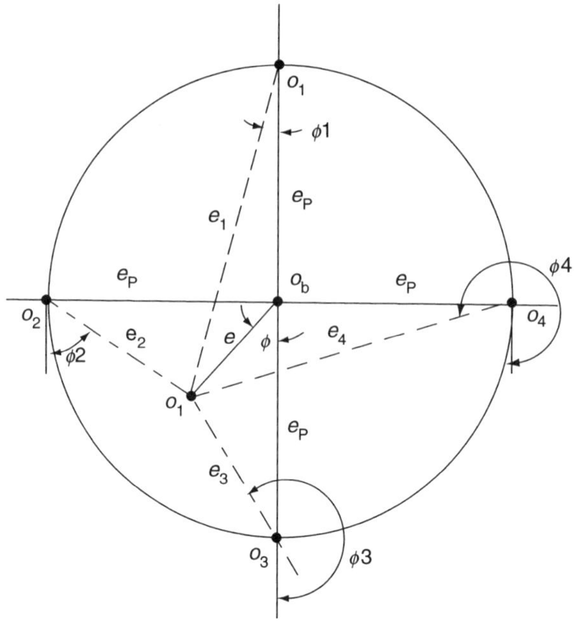

Fig. 17.10. Geometrical parameters of a four-lobe journal bearing

The film thickness for first, second, third and fourth lobe can be expressed as [8]

$$\begin{aligned} h &= C_{\mathrm{h}}[1 + \varepsilon_1 \cos(\theta + \lambda_1)] \\ h &= C_{\mathrm{h}}[1 + \varepsilon_2 \cos(\theta + \lambda_2)] \\ h &= C_{\mathrm{h}}[1 + \varepsilon_3 \cos(\theta + \lambda_3)] \end{aligned} \tag{17.66}$$

and

$$h = C_{\mathrm{h}}[1 + \varepsilon_4 \cos(\theta + \lambda_4)]$$

17.5.2 Theoretical Analysis of Capacitance and Active Resistance of Four-Lobe Bearings

In a four-lobe bearing, positive pressure is built up in all the four lobes. The analysis is performed in a unitary manner (for four lobes) by starting the analysis with an isolated lobe placed in any position with respect to the shaft or to the bearing center. The characteristics of the bearing as a whole are found by summing up the results from the individual lobes.

17.5.2.1 Capacitance

During rotation of the shaft, part of the circumference of the journal passes through each lobe in a zone of load-carrying oil film. This forms a parallel plate capacitor of varying film thickness between the interacting surface of the journal and an individual lobe. The capacitances of all the four lobes are in parallel to each other with respect to the journal. However, the capacitance of parallel plate capacitor is greatest at the minimum film thickness of an individual lobe, and reduces both sides as the film thickness increases.

17.5.2.1.1 Capacitance of the Individual Lobes

The change in capacitance ΔC between two rectangular plates of width dX of length L, located at the angular coordinate differential dθ (as shown in Fig. 17.9) and separated by a film thickness Δh, is given by

$$\Delta C = 2\xi \frac{(X + \mathrm{d}X)L - XL}{2h + \Delta h} \tag{17.67}$$

The change in film thickness (Δh) is very small in comparison to dX and, hence, can be neglected. Thus the change in capacitance is given as

$$\Delta C = \frac{\xi \; dXL}{h} \tag{17.68}$$

From Fig. 17.9, it is evident that

$$dX = d\theta(r + C_{\mathrm{h}}) \tag{17.69}$$

On substituting the values from Eqns. (17.66) and (17.69) into Eqn. (17.68), the change in capacitance of first lobe is expressed as

$$\Delta C = \frac{\xi L(r + C_{\mathrm{h}}) \; d\theta}{C_{\mathrm{h}}[1 + \varepsilon_1 \cos(\theta + \lambda_1)]} \tag{17.70}$$

The maximum change in capacitance depends on an angle θ, the angle coordinate measured from leading edge to trailing edge, and it is along the film thickness. So on integrating between inlet angle θ_{pi} and outlet angle θ_{pe}, the capacitance of the first lobe of a four-lobe bearing is determined as

$$C_1 = \xi L C_{\mathrm{h}}^{-1}(r + C_{\mathrm{h}}) \int_{\theta_{\mathrm{pi}}}^{\theta_{\mathrm{pe}}} [1 + \varepsilon_1 \cos(\theta + \lambda_1)]^{-1} \; \mathrm{d}\,\theta \tag{17.71}$$

Equation (17.71) can be integrated by Sommerfeld substitution [8], and capacitance is worked out using Eqn. (17.66) ($\lambda_1 = \pi/2 - \phi_1$) and taking θ_{pe} and θ_{pi} equal to 80° and 10° for lobe 1 on considering 20° as the angle for the oil supply for each lobe. So capacitance of the first lobe is given as

$$C_1 = \xi L C_h^{-1}(r + C_h)(1 - \varepsilon_1^2)^{-1/2} \times \left\{\cos^{-1}\left[\frac{\varepsilon_1 + \cos(170^\circ - \phi_1)}{1 + \varepsilon_1 \cos(170^\circ - \phi_1)}\right] - \cos^{-1}\left[\frac{\varepsilon_1 + \cos(100^\circ - \phi_1)}{1 + \varepsilon_1 \cos(100^\circ - \phi_1)}\right]\right\} \quad (17.72)$$

Similar to Eqn. (17.72), the capacitance of lobes 2, 3 and 4 (C_2, C_3 and C_4) can be determined using ε_2, λ_2; ε_3, λ_3; and ε_4, λ_4 for the respective lobes and integrating on taking θ_{pe} and θ_{pi} between 170° and 100° for the lobe 2, and between 260° and 190° for the lobe 3, and between 350° and 280° for the lobe 4, respectively. Thus, C_2, C_3, and C_4 are determined by the following equations:

$$C_2 = \xi L C_h^{-1}(r + C_h)(1 - \varepsilon_2^2)^{-1/2} \times \left\{\cos^{-1}\left[\frac{\varepsilon_2 + \cos(350^\circ - \phi_2)}{1 + \varepsilon_2 \cos(350^\circ - \phi_2)}\right] - \cos^{-1}\left[\frac{\varepsilon_2 + \cos(280^\circ - \phi_2)}{1 + \varepsilon_2 \cos(280^\circ - \phi_2)}\right]\right\} \quad (17.73)$$

$$C_3 = \xi L C_h^{-1}(r + C_h)(1 - \varepsilon_3^2)^{-1/2} \times \left\{\cos^{-1}\left[\frac{\varepsilon_3 + \cos(170^\circ - \phi_3)}{1 + \varepsilon_3 \cos(170^\circ - \phi_3)}\right] - \cos^{-1}\left[\frac{\varepsilon_3 + \cos(100^\circ - \phi_3)}{1 + \varepsilon_3 \cos(100^\circ - \phi_3)}\right]\right\} \quad (17.74)$$

and

$$C_4 = \xi L C_h^{-1}(r + C_h)(1 - \varepsilon_4^2)^{-1/2} \times \left\{\cos^{-1}\left[\frac{\varepsilon_4 + \cos(350^\circ - \phi_4)}{1 + \varepsilon_4 \cos(350^\circ - \phi_4)}\right] - \cos^{-1}\left[\frac{\varepsilon_4 + \cos(280^\circ - \phi_4)}{1 + \varepsilon_4 \cos(280^\circ - \phi_4)}\right]\right\} \quad (17.75)$$

17.5.2.1.2 Resistance

The change in conductance/resistance^{-1}[ΔF], between two rectangular plates of width dX of bearing length L located at an angular coordinate differential dθ and separated by a film thickness Δh can be expressed as

$$\Delta F = \frac{\mathrm{d}XL}{\rho h} \text{ (Δh being small and hence negligible)} \quad (17.76)$$

The maximum change in conductance/resistance^{-1} depends on an angle θ, the angular coordinate measured from leading edge, and it is along the oil film thickness. So, on integrating between inlet angle θ_{pi} and outlet angle θ_{pe}, the resistance of a first lobe of a four-lobe bearing is determined using Eqns. (17.69) and (17.66), and is given as

$$F = R_1^{-1} = \rho^{-1} L(r + C_h)C_h^{-1} \int_{\theta_{pi}}^{\theta_{pe}} [1 + \varepsilon_1 \cos(\theta + \lambda_1)]^{-1} \mathrm{d}\theta \quad (17.77)$$

The above equation is integrated using Sommerfeld substitution [8], and resistance of each lobe is worked out using similar integration limits as for the capacitance of the respective lobes and is given as

$$R_1 = \rho C_h L^{-1}(r + C_h)^{-1}(1 - \varepsilon_1^2)^{1/2} \times \left[\cos^{-1}\left\{\frac{\varepsilon_1 + \cos(170° - \phi_1)}{1 + \varepsilon_1 \cos(170° - \phi_1)}\right\}\right] - \left[\cos^{-1}\left\{\frac{\varepsilon_1 + \cos(100° - \phi_1)}{1 + \varepsilon_1 \cos(100° - \phi_1)}\right\}\right]^{-1} \quad (17.78)$$

$$R_2 = \rho C_h L^{-1}(r + C_h)^{-1}(1 - \varepsilon_2^2)^{1/2} \times \left[\cos^{-1}\left\{\frac{\varepsilon_2 + \cos(350° - \phi_2)}{1 + \varepsilon_2 \cos(350° - \phi_2)}\right\}\right] - \left[\cos^{-1}\left\{\frac{\varepsilon_2 + \cos(280° - \phi_2)}{1 + \varepsilon_2 \cos(280° - \phi_2)}\right\}\right]^{-1} \quad (17.79)$$

$$R_3 = \rho C_h L^{-1}(r + C_h)^{-1}(1 - \varepsilon_3^2)^{1/2} \times \left[\cos^{-1}\left\{\frac{\varepsilon_3 + \cos(170° - \phi_3)}{1 + \varepsilon_3 \cos(170° - \phi_3)}\right\}\right] - \left[\cos^{-1}\left\{\frac{\varepsilon_3 + \cos(100° - \phi_3)}{1 + \varepsilon_3 \cos(100° - \phi_3)}\right\}\right]^{-1} \quad (17.80)$$

$$R_4 = \rho C_h L^{-1}(r + C_h)^{-1}(1 - \varepsilon_4^2)^{1/2} \times \left[\cos^{-1}\left\{\frac{\varepsilon_4 + \cos(350° - \phi_4)}{1 + \varepsilon_4 \cos(350° - \phi_4)}\right\}\right] - \left[\cos^{-1}\left\{\frac{\varepsilon_4 + \cos(280° - \phi_4)}{1 + \varepsilon_4 \cos(280° - \phi_4)}\right\}\right]^{-1} \quad (17.81)$$

17.5.3 Equivalent Capacitance and Active Resistance of Four-Lobe Bearings

The calculation details of equivalent capacitance and active resistance of four-lobe bearings are given below.

17.5.3.1 Capacitance

As per the electrical analogy, all the lobes of a four-lobe bearing are in parallel to each other with respect to the journal. So, the equivalent capacitance of the four-lobe bearing is determined as

$$C_{eq} = C_1 + C_2 + C_3 + C_4 \quad (17.82)$$

17.5.3.2 Resistance

Since all the lobes are in parallel to each other with respect to the journal, the equivalent resistance of the four-lobe bearing is given as

$$R_{eq} = \frac{R_1 R_2 R_3 R_4}{R_1 R_2 R_3 + R_2 R_3 R_4 + R_1 R_3 R_4 + R_4 R_1 R_2} \quad (17.83)$$

17.5.4 Capacitive Reactance and Impedance of Four-Lobe Bearings

Capacitive reactance between the interacting surface of the journal and each lobe in the oil film as well as equivalent capacitive reactance of the bearing is expressed as [5, 6]

$$X_1 = \frac{1}{wC_1},\ X_2 = \frac{1}{wC_2},\ X_3 = \frac{1}{wC_3},\ X_4 = \frac{1}{wC_4},\quad \text{and}\quad X_{eq} = \frac{1}{wC_{eq}} \tag{17.84}$$

And impedance of bearing circuit is determined as

$$Z_1 = R_1 - jX_1,\ Z_2 = R_2 - jX_2,\ Z_3 = R_3 - jX_3,\ Z_4 = R_4 - jX_4,\quad \text{and}\quad Z_{eq} = R_{eq} - jX_{eq} \tag{17.85}$$

The ratio of capacitive reactance to active resistance is given by [Eqns. (17.71)–(17.83)]

$$\frac{X_{eq}}{R_{eq}} = \frac{X_1}{R_1} = \frac{X_2}{R_2} = \frac{X_3}{R_4} = \frac{X_4}{R_4} = \frac{1}{w\rho\xi} \tag{17.86}$$

17.5.5 Data Derived From Analysis

The capacitance, resistance and dynamic coefficients for an individual lobe and an equivalent value of these parameters for the four-lobe bearing having $L/d = 0.5(L = 150\,\text{mm},\ d = 300\,\text{mm})$, $C_h = 0.4\,\text{mm}$ and $e_p = 0.2\,\text{mm}$ have been computed as per the developed methodology and are given in Table 17.6. The global attitude angles, eccentricity ratios, and other relevant details for the four-lobe bearing have been taken from the published data [1, 7, 10]. The attitude angles as well as eccentricity ratios for an individual lobe have been computed separately using Eqns. (17.57)–(17.64). Based on the electrical analogy, dynamic coefficients of each lobe have been worked out for the global eccentricity ratios of 0.2, 0.3, 0.4, 0.45 and 0.5 with the corresponding attitude angles of 22.6°, 22.55°, 22.55°, 22.60° and 23.5°, respectively. Also the equivalent values of dynamic coefficients of the four-lobe bearing have been compared with those of three-lobe, two-lobe and cylindrical bearings and are shown in Figs. 17.11 and 17.12.

The effective dynamic coefficients of the four-lobe bearings established by the developed approach have also been compared with those of the conventional approaches [12]. The similar approach for this comparison as reported by Prashad for two-lobe and cylindrical bearings has been used [13].

17.5.6 Discussion on Analysis and Derived Data

17.5.6.1 Capacitance

The capacitance of a four-lobe bearing depends on the capacitance of each lobe. This, in turn, is a function of lubricant permittivity, horizontal clearance (C_h), bearing length (L) and shaft radius (r), and depends on the eccentricity ratio and attitude angle of the respective lobe along with the angles θ_{pi} and θ_{pe} [Eqns. (17.71)–(17.75)].

As the global eccentricity ratio of the bearing increases from 0.2 to 0.5, the attitude angle does not change significantly (Table 17.6). The values of ε_1, ε_2 and ε_4, increase as against ε_3, which decrease, when global eccentricity ratio increases from 0.2 to 0.5. Under these conditions, the capacitances of lobes 1, 2, 3 and 4 change between $(2.84 \times 10^{-9})\phi$ and $(4.90 \times 10^{-9})\phi$, $(1.86 \times 10^{-9})\phi$ and $(1.43 \times 10^{-9})\phi$, $(1.28 \times 10^{-9})\phi$ and $(1.48 \times 10^{-9})\phi$, and $(0.92 \times 10^{-9})\phi$ and $(0.24 \times 10^{-9})\phi$, respectively. The variation of capacitance with eccentricity ratio is non-linear. In general, the capacitance of lobe 1 is more than that of lobes 2, 3 and 4. This is because of the relatively higher eccentricity ratio (ε_1) of the lobe 1 and the influence of film thickness and attitude angle on the capacitance as compared to that of the lobes 2, 3 and 4 (Table 17.6). The equivalent capacitance of the bearing increases with global eccentricity ratio and varies between $(6.46 \times 10^{-9})\phi$ and $(8.02 \times 10^{-9})\phi$, (Table 17.6). This is because of the net effect of variable film thickness in each lobe under the influence of different eccentricity ratios and attitude angles for a four-lobe bearing operating under variable conditions.

17.5.6.2 Resistance

The resistance of a four-lobe bearing depends on the individual resistance of each lobe. This, in turn, is a function of lubricant resistivity, bearing clearance, bearing length and shaft radius, and depends on the attitude angle and eccentricity ratio of each lobe and varies with the angles θ_{pi} and θ_{pe} [Eqns. (17.78)–(17.81)].

Under the variation of global eccentricity ratio from 0.2 to 0.5, the resistances of lobes 1, 2, 3 and 4 change between 7.82×10^6 and 4.54×10^6 ohm; 11.93×10^6 and 15.49×10^6 ohm; $17.15 \times 10^6$and 13.55×10^6 ohm; and 24.11×10^6 and 95.58×10^6 ohm, respectively (Table 17.6). The variation of resistance with eccentricity ratio is non-linear. Contrary to the capacitance, the resistance of the lobe 1 is less than that of lobes 2, 3 and 4 and resistance of lobe 4 is higher than other lobes as shown in Table 17.6. This is attributed to the higher eccentricity ratio of lobe 1 and the influence of relatively less attitude angle as compared to that of the lobes 2, 3 and 4, and much higher attitude angle (303.31°) of lobe 4 (Table 17.6). Also, resistance of

Table 17.6. Capacitance and resistance of four-lobe bearing under different operating parameters

S. No.	ε	ε_1	ε_2	ε_3	ε_4	ϕ (°)	ϕ_1 (°)	ϕ_2 (°)	ϕ_3 (°)	ϕ_4 (°)	$C_1 \times 10^{-9}$ (ϕ)	$C_2 \times 10^{-9}$ (ϕ)	$C_3 \times 10^{-9}$ (ϕ)	$C_4 \times 10^{-9}$ (ϕ)	$C_{eq} \times 10^{-9}$ (ϕ)
1	0.20	0.69	0.46	0.33	0.61	22.60	6.66	66.80	184.04	287.24	2.84	1.86	1.28	0.92	6.46
2	0.30	0.78	0.48	0.24	0.67	22.55	8.07	54.32	206.57	294.66	3.20	1.69	1.45	0.57	6.91
3	0.40	0.88	0.51	0.21	0.75	22.55	9.78	43.41	229.10	299.65	3.72	1.57	1.56	0.56	7.41
4	0.45	0.93	0.53	0.19	0.79	22.60	10.54	40.49	242.1	301.46	4.31	1.52	1.63	0.30	7.76
5	0.50	0.98	0.55	0.20	0.84	23.50	11.8	33.1	258.69	303.31	4.90	1.43	1.48	0.24	8.02

S. No.	$R_1 \times 10^6$ (ohm)	$R_2 \times 10^6$ (ohm)	$R_3 \times 10^6$ (ohm)	$R_4 \times 10^6$ (ohm)
1	7.82	11.93	17.15	24.11
2	6.90	13.09	15.33	38.52
3	5.94	14.09	14.22	38.94
4	5.14	14.61	13.55	73.16
5	4.54	15.49	14.96	95.58

$L/d = 0.5$, $L = 150\,\text{mm}$, $d = 300\,\text{mm}$, $C_h = 0.4\,\text{mm}$, $e_p = 0.2\,\text{mm}$, $\varepsilon_p = 0.5$, $\xi = 2.5 \times 8.854 \times 10^{-12}\,\phi\,\text{m}^{-1}$, $\rho = 10^{11}$ ohm cm.

lobe 4 increases with increase in global eccentricity ratio from 0.2 to 0.5 as against that of lobe 1, which decreases, with increase in global eccentricity ratio. However, the equivalent resistance of the four-lobe bearing decreases with increase in global eccentricity ratio and varies between 3.21×10^6 and 2.76×10^6 ohm, as eccentricity ratio increases from 0.2 to 0.5 as shown in Table 17.6. This is because of the net effect of the variable film thickness in each lobe of a four-lobe bearing under the influence of different eccentricity ratios and operating under variable conditions.

17.5.6.3 Damping and Stiffness Coefficients of Four-Lobe Bearing

17.5.6.3.1 Damping Coefficient

The damping coefficient of each lobe varies differently with the eccentricity ratio. The damping coefficient in $\mathrm{N\,s\,m^{-1}}$ is equivalent to that of the resistance in ohm of each lobe [Eqns. (17.28)–(17.30)]. As ε changes from 0.2 to 0.5, the variation of C_{d1}, C_{d2}, C_{d3} and C_{d4} are non-linear, and vary from 7.82×10^6 to $4.54 \times 10^6\,\mathrm{N\,s\,m^{-1}}$, 11.93×10^6 to $15.49 \times 10^6\,\mathrm{N\,s\,m^{-1}}$, 17.15×10^6 to $13.55 \times 10^6\,\mathrm{N\,s\,m^{-1}}$, and 24.11×10^6 to $95.58 \times 10^6\,\mathrm{N\,s\,m^{-1}}$, respectively (Table 17.7). The damping coefficient of lobe 4 (C_{d4}) increases as against that of C_{d1}, which decreases with increase in global eccentricity ratio. However, the equivalent damping coefficient of the four-lobe bearing decreases with increase in global eccentricity ratio and varies between 3.21×10^6 and $2.76 \times 10^6\,\mathrm{N\,s\,m^{-1}}$ as the global eccentricity ratio increases from 0.2 to 0.5 (Table 17.7).

Under the identical conditions of operation, trend of variation of equivalent dynamic coefficient of a four-lobe bearing with eccentricity ratios is found to be similar to that of the three-lobe, two-lobe and cylindrical bearings as shown in Fig. 17.11. The data for cylindrical, two-lobe and three-lobe bearings have been taken from the published work [11, 14].

17.5.6.3.2 Stiffness Coefficient

The stiffness coefficient of an individual lobe of a four-lobe bearing depends on reciprocal of its capacitance [Eqns. (17.28)–(17.30)]. The equivalent stiffness of a four-lobe bearing is the resultant stiffness of all the lobes connected in parallel. The stiffness of an individual lobe shows non-linear behavior with the eccentricity ratio. The K_1, K_2, K_3, and K_4 vary from 0.35×10^9 to $0.20 \times 10^9\,\mathrm{N\,m^{-1}}$; 0.54×10^9 to $0.7 \times 10^9\,\mathrm{N\,m^{-1}}$; 0.78×10^9 to $0.61 \times 10^9\,\mathrm{N\,m^{-1}}$; and 1.09×10^9 to $4.17 \times 10^9\,\mathrm{N\,m^{-1}}$, respectively, as ε changes from 0.2 to 0.5 (Table 17.7). The stiffness of fourth lobe (K_4) increases as against that of K_1, which decreases with increase in global eccentricity ratios (Table 17.7). Furthermore, stiffness of second and third lobe varies with their respective eccentricity ratios (ε_2, ε_3). The equivalent stiffness of

Table 17.7. Dynamic coefficients of four-lobe journal bearing under different operating parameters

S. No.	ε	$C_{d1} \times 10^6$ (N s m^{-1})	$C_{d2} \times 10^6$ (N s m^{-1})	$C_{d3} \times 10^6$ (N s m^{-1})	$C_{d4} \times 10^6$ (N s m^{-1})	$C_{deq} \times 10^6$ (N s m^{-1})	$K_1 \times 10^9$ (N m^{-1})	$K_2 \times 10^9$ (N m^{-1})	$K_3 \times 10^9$ (N m^{-1})	$K_4 \times 10^9$ (N m^{-1})	$K_{eq} \times 10^9$ (N s m^{-1})
1	0.2	7.82	11.93	17.15	24.11	3.21	0.35	0.54	0.78	1.09	0.14
2	0.3	6.90	13.09	15.33	38.52	3.20	0.31	0.59	0.69	1.75	0.14
3	0.4	5.94	14.09	14.22	38.94	2.98	0.27	0.64	0.64	1.79	0.13
4	0.45	5.14	14.61	13.55	73.16	2.85	0.23	0.66	0.61	3.33	0.13
5	0.5	4.54	15.49	14.96	95.58	2.76	0.20	0.70	0.68	4.17	0.12

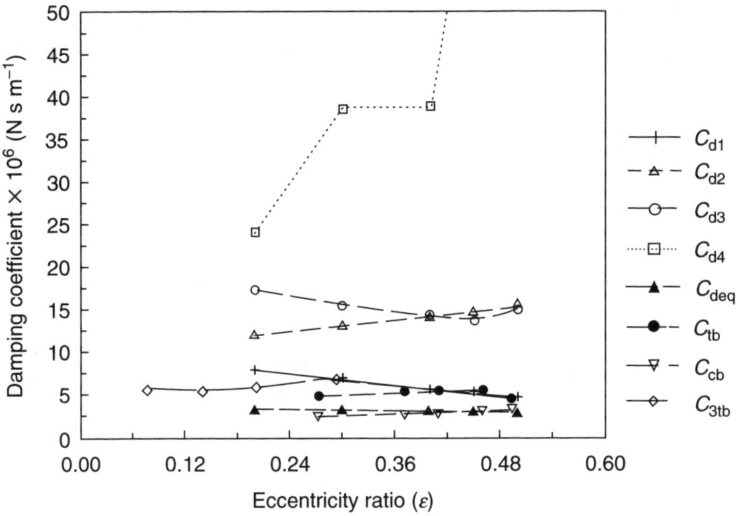

Fig. 17.11. Variation of damping coefficients with eccentricity ratio of first, second, third and fourth lobe (C_{d1}, C_{d2}, C_{d3} and C_{d4}), and equivalent damping coefficient of four-lobe (C_{deq}), three-lobe (C_{3lb}), two-lobe (C_{tb}) and cylindrical (C_{cb}) bearings

the four-lobe bearing decreases from 0.14×10^9 to 0.12×10^9 N m^{-1}, with increase in global eccentricity ratio (Table 17.7). However, the variation of K_{eq} of four-lobe bearing is similar to that of three-lobe, two-lobe and cylindrical bearings having comparable dimensional parameters and working under comparable global eccentricity ratios [11, 14]. The variation of equivalent stiffness coefficient as shown in Fig. 17.12, determined by electrical analogy approach, matches closely with those determined by the conventional approaches.

Nomenclature

C	capacitance
C_d	diametral clearance
C_h	horizontal radial clearance of four-lobe bearing
C_1, C_2, C_3, C_4	capacitance of first, second, third and fourth lobe, respectively
C_{eq}	equivalent capacitance of four-lobe bearing
C_{d1}, C_{d2}, C_{d3}, C_{d4}	damping coefficients of first, second, third and fourth lobe, respectively
C_{deq}	equivalent damping coefficient of four-lobe bearing
C_c	damping coefficient

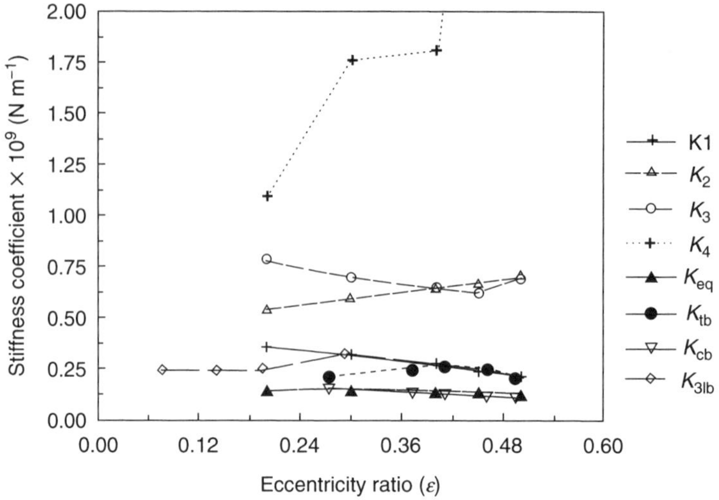

Fig. 17.12. Variation of stiffness coefficients with eccentricity ratio of first, second, third and fourth lobe (K_1, K_2, K_3 and K_4), and equivalent stiffness coefficient of four-lobe (K_{eq}), three-lobe (K_{3lb}), two-lobe (K_{tb}) and cylindrical (K_{cb}) bearings

C_{3lb}, C_{tb}, C_{cb}	equivelent/effective damping coefficient of three-lobe, two-lobe and cylindrical bearing, respectively
d	shaft diameter
e	bearing eccentricity
e_1, e_2, e_3, e_4	eccentricity of first, second, third and fourth lobe, respectively
e_p	bearing ellipticity
E	voltage
f	frequency of applied voltage
F_1/R_1^{-1}	conductance
F	force
h	film thickness
i	current
K	stiffness
K_1, K_2, K_3, K_4	stiffness of first, second, third, and fourth lobe, respectively
K_{eq}	equivalent stiffness of four-lobe bearing
K_{3lb}, K_{tb}, K_{cb}	equivalent/effective stiffness coefficient of three-lobe, two-lobe and cylindrical bearing, respectively
L	bearing length
L_j	inductance
M	mass

Q	charge
r	radius of journal
r_L	radius of lobe
R	resistance
R_1, R_2, R_3, R_4	resistance of first, second, third and fourth lobe, respectively
R_{eq}	equivalent resistance of four-lobe bearing
t	time
w	$2\pi f$
X_1, X_2, X_3, X_4	capacitive reactance of first, second, third and fourth lobe, respectively
X_{eq}, Z_{eq}	equivalent capacitive reactance and impedance of four-lobe bearing respectively
Z_1, Z_2, Z_3, Z_4	impedance of first, second, third and fourth lobe, respectively
$\varepsilon, \varepsilon_p, \varepsilon_1, \varepsilon_2, \varepsilon_3, \varepsilon_4$	$(e, e_p, e_1, e_2, e_3, e_4)/C_h$
ϕ	global attitude angle of four-lobe bearing
$\phi_1, \phi_2, \phi_3, \phi_4$	attitude angle of first, second, third and fourth lobe, respectively
$\lambda_1, \lambda_2, \lambda_3, \lambda_4$	location of leading edge of the positive pressure fluid film with reference to line of centers for first, second, third and fourth lobe, respectively
θ	angular coordinate measured from leading edge of a lobe
θ_{pi}	inlet angle from line of centers to leading edge film in each lobe
θ_{pe}	outlet-angle from line of centers to trailing edge film in each lobe
ρ	resistivity of lubricant
ξ	permittivity/dielectric constant of lubricant ($\phi\,m^{-1}$)
ξ_r	relative permittivity of lubricant

17.6 Tilting-Pad Journal Bearings

17.6.1 Introduction

Nicholas has used pad assembly method for determining the bearing coefficients of tilting-pad bearings [16]. The dynamic properties of a tilting pad bearing has been studied by Lund [17], and Hashimato et al. [18]. The approach used to drive the dynamic coefficients of such bearings follows basically the same reasoning as for the fixed lobe bearings to change the pad angular orientation according to the loading. It is this ability that gives the bearings such special characteristics as negligibly small cross-coupling terms [17, 18]. This chapter highlights the electrical analogy approach to determine effective dynamic coefficients of a tilting-pad journal with four pads besides the theoretical determination of electrical parameters of these bearings similar to those of multi lobe bearings as given in the previous sections of this chapter.

17.6.2 Geometry and Geometrical Characteristics of Tilting-Pad Journal Bearings

Tilting pad journal bearings are of more compact design as compared to those of the multi-lobe bearings and are less sensitive to deformations and thermal distortions. Tilting-pad bearings allow a more reliable design for heavy loaded, and as well for high-speed bearings. The clearance in the bearings can be precisely controlled by adjustable preload (δ^*/c_p). The small and medium size bearings are available with 3–5 pads, while in very large size bearings 12–16 pads are used. The pivot position is usually at the location of 0.55L, 0.6L from one end of the pad, and for both directions of operation, however, it requires a symmetrical pivot location. In tilting pad bearings two situations exists, namely load passing through the pivot and load midway between pivots. Figure 17.13 shows the geometrical configuration of a tilting-pad bearing with four pads. Figure 17.14 gives details of a single pad.

Various mathematical relations for a tilting pad journal bearing are given as [16]

$$\left.\begin{aligned} C_\mathrm{p} &= r_\mathrm{p} - r \\ \delta &= C_\mathrm{p} - C_\mathrm{b} \end{aligned}\right\} \tag{17.87}$$

$$\delta_x = \frac{\delta}{C_\mathrm{p}} = 1 - \frac{C_\mathrm{b}}{C_\mathrm{p}} \tag{17.88}$$

For a given preload and eccentricity ratio, the pivot film thickness is calculated for a given angle θ of the pivot location as

$$h = C_\mathrm{p}(1 - \delta_x)(1 + \varepsilon_\mathrm{b} \sin\theta) \tag{17.89}$$

$$\varepsilon_\mathrm{b} = -\frac{e_\mathrm{b}}{C_\mathrm{b}} \tag{17.90}$$

17.6.3 Theoretical Analysis of Capacitance and Active Resistance of Tilting-Pad Journal Bearings

In a tilting-pad journal bearing positive pressure is built up in all the pads depending on the pad location and operating parameters. The analysis is performed in a unitary manner, for four pads, by starting the analysis with an isolated pad placed in any position with respect to the shaft or to the bearing center. The

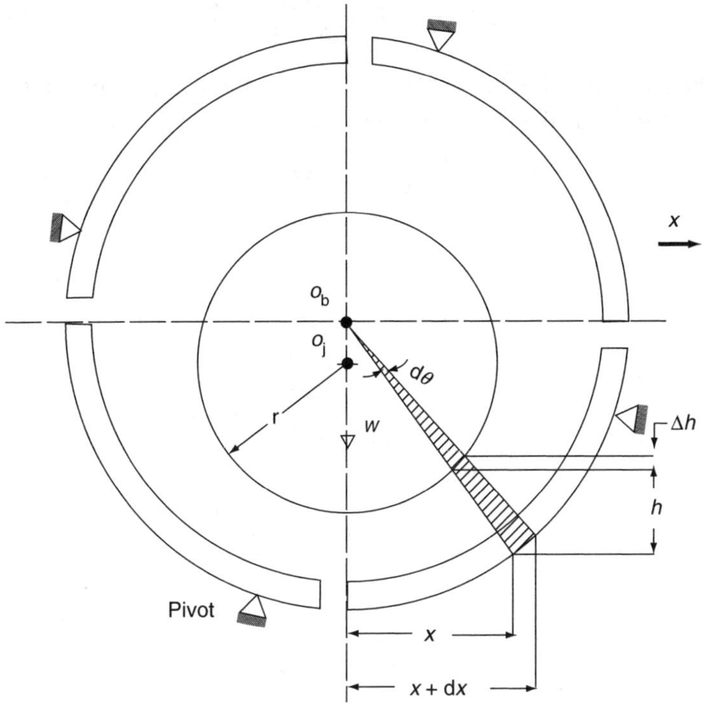

Fig. 17.13. Tilting-pad bearing with four pads

characteristics of the bearing as a whole are found by summing up the results from the individual pads.

17.6.3.1 Capacitance

During rotation of the shaft, part of the circumference of the journal passes through each pad in a zone of load-carrying oil film. This forms a parallel plate capacitor of varying film thickness between the interacting surface of the journal and an individual pad. The capacitances of all the four pads are in parallel to each other with respect to the journal. However, the capacitance of parallel plate capacitor is greatest at the minimum film thickness of an individual pad, and reduces both sides as the film thickness increases.

17.6.3.1.1 Capacitance of the Individual Pads

The change in capacitance ΔC between two rectangular plates of width $\mathrm{d}X$ of length L, located at the angular coordinate differential $\mathrm{d}\theta$ (as shown in Fig. 17.13) and separated by a film thickness Δh, is given by

$$\Delta C = 2\xi \frac{(X + \mathrm{d}X)L - XL}{2h + \Delta h} \tag{17.91}$$

The change in film thickness (Δh) is very small in comparison to dX and hence can be neglected. Thus the change in capacitance is given as

$$\Delta C = \frac{\xi \mathrm{d}XL}{h} \tag{17.92}$$

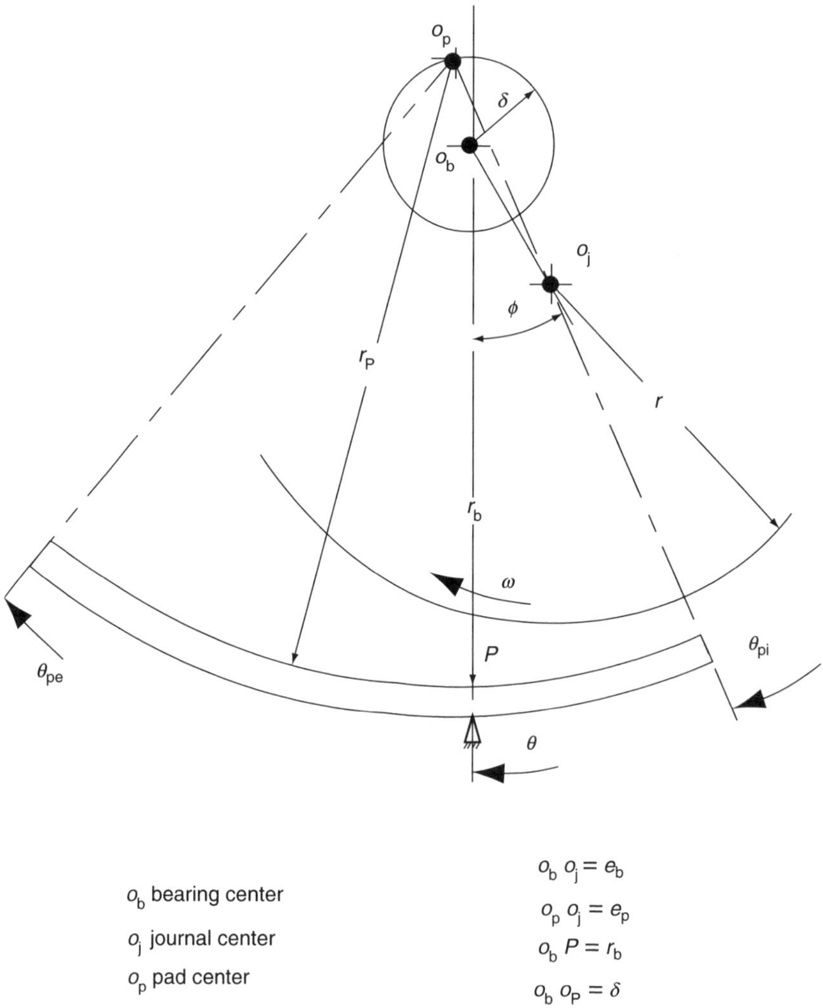

Fig. 17.14. Details of a single pad

From Fig. 17.13, it is evident that

$$dX = d\theta(r + C_p) \tag{17.93}$$

On substituting the values from Eqns. (17.89) and (17.93) into Eqn. (17.92), the change in capacitance of first pad is expressed as

$$\Delta C = \frac{\xi L(r + C_p)d\theta}{C_p(1 - \delta_x)(1 + \varepsilon_b \sin\theta)} \tag{17.94}$$

The maximum change in capacitance depends on an angle θ, the angle coordinate measured from leading edge to trailing edge, and it is along the film thickness. So on integrating between inlet angle θ_{pi} and outlet angle θ_{pe}, the capacitance of the first pad of a tilting-pad bearing is determined as

$$C = \xi L C_p^{-1}(r + C_p)(1 - \delta_x)^{-1} \int_{\theta_{pi}}^{\theta_{pe}} (1 + \varepsilon_b \sin\theta)^{-1} d\theta \tag{17.95}$$

Equation (17.95) can be integrated by Sommerfeld substitution [8] and capacitance is worked out as

$$C = \xi L C_p^{-1}(r + C_p)(1 - \delta_x)^{-1}(1 - \varepsilon_b^2)^{-1/2} \times \left[\cos^{-1}\left(\frac{\varepsilon_b + \sin\theta_{pe}}{1 + \varepsilon_b \sin\theta_{pe}}\right) - \cos^{-1}\left(\frac{\varepsilon_b + \sin\theta_{pi}}{1 + \varepsilon_b \sin\theta_{pi}}\right)\right] \tag{17.96}$$

On taking θ_{pe} and θ_{pi} equal to 90° and 10° for the first pad on considering the 10° as the gap between the pads (Fig. 17.13), the capacitance of the first pad is determined as

$$C_1 = -K_c(1 - \varepsilon_b^2)^{-1/2} \cos^{-1}\left(\frac{\varepsilon_b + \sin 10^\circ}{1 + \varepsilon_b \sin 10^\circ}\right) \tag{17.97}$$

where

$$K_c = \xi L C_p^{-1}(r + C_p)(1 - \delta_x)^{-1} \tag{17.98}$$

Similar to Eqn. (17.97), the capacitance of pads 2, 3 and 4 (C_2, C_3 and C_4) can be determined on taking θ_{pe} and θ_{pi} between 180° and 100° for the pad 2, and between 270° and 190° for the pad 3, and between 360° and 280° for the pad 4, respectively. The angle of pad tilt is not considered since it is relatively vary small as compared to θ and does not affect the capacitance of a tilting pad. Thus, C_2, C_3 and C_4 are determined by the following equations:

$$C_2 = K_c(1-\varepsilon_b^2)^{-1/2}\left[\cos^{-1}\varepsilon_b - \cos^{-1}\left(\frac{\varepsilon_b + \sin 100^\circ}{1+\varepsilon_b \sin 100^\circ}\right)\right] \tag{17.99}$$

$$C_3 = K_c(1-\varepsilon_b^2)^{-1/2}\left[\pi - \cos^{-1}\left(\frac{\varepsilon_b + \sin 190^\circ}{1+\varepsilon_b \sin 190^\circ}\right)\right] \tag{17.100}$$

and

$$C_4 = K_c(1-\varepsilon_b^2)^{-1/2}\left[\cos^{-1}\varepsilon_b - \cos^{-1}\left(\frac{\varepsilon_b + \sin 280^\circ}{1+\varepsilon_b \sin 280^\circ}\right)\right] \tag{17.101}$$

17.6.3.2 Resistance

The change in conductance/resistance^{-1}, $[\Delta F]$, between two rectangular plates of width $\mathrm{d}X$ of a pad length L located at an angular coordinate differential $\mathrm{d}\theta$ and separated by a film thickness Δh is given as

$$\Delta F = \frac{\mathrm{d}XL}{\rho h} \quad (\Delta h \text{ being small and hence negligible}) \tag{17.102}$$

The maximum change in conductance/resistance^{-1} depends on an angle θ – the angular coordinate measured from leading edge and it is along the oil film thickness. So, on integrating between inlet angle θ_{pi} and outlet angle θ_{pe}, the resistance of a first pad of a tilting pad journal bearing is determined using Eqns. (17.93) and (17.89), and is given as

$$F = R_1^{-1} = \rho^{-1}L(r + C_p)C_p^{-1}(1-\delta_x)^{-1}\int_{\theta_{pi}}^{\theta_{pe}}(1+\varepsilon_b \sin\theta)^{-1}\mathrm{d}\theta \tag{17.103}$$

The above equation is integrated using Sommerfeld substitution [8], and resistance of each pad is worked out using similar integration limits as for the capacitance of the respective pads and is given as

$$R_1 = -K_R(1-\varepsilon_b^2)^{1/2}\left[\cos^{-1}\left(\frac{\varepsilon_b + \sin 10^\circ}{1+\varepsilon_b \sin 10^\circ}\right)\right]^{-1} \tag{17.104}$$

where

$$K_R = \rho C_p L^{-1}(r + C_p)^{-1}(1-\delta_x) \tag{17.105}$$

$$R_2 = -K_R(1-\varepsilon_b^2)^{1/2}\left[\cos^{-1}\varepsilon_b - \cos^{-1}\left(\frac{\varepsilon_b + \sin 100^\circ}{1+\varepsilon_b \sin 100^\circ}\right)\right]^{-1} \tag{17.106}$$

$$R_3 = K_R(1-\varepsilon_b^2)^{1/2}\left[\pi - \cos^{-1}\left(\frac{\varepsilon_b + \sin 190°}{1+\varepsilon_b \sin 190°}\right)\right]^{-1} \tag{17.107}$$

$$R_4 = K_R(1-\varepsilon_b^2)^{1/2}\left[\cos^{-1}\varepsilon_b - \cos^{-1}\left(\frac{\varepsilon_b + \sin 280°}{1+\varepsilon_b \sin 280°}\right)\right]^{-1} \tag{17.108}$$

17.6.4 Equivalent Capacitance and Active Resistance of Tilting-Pad Journal Bearings

17.6.4.1 Capacitance

As per the electrical analogy, all the pads of a tilting-pad journal bearing are in parallel to each other with respect to the journal. So the equivalent capacitance of a tilting-pad journal bearing is determined as

$$C_{eq} = C_1 + C_2 + C_3 + C_4 \tag{17.109}$$

17.6.4.2 Resistance

Since all the pads are in parallel to each other with respect to the journal, the equivalent resistance of a tilting-pad journal bearing is given as

$$R_{eq} = \frac{R_1R_2R_3R_4}{R_1R_2R_3 + R_2R_3R_4 + R_1R_3R_4 + R_4R_1R_2} \tag{17.110}$$

17.6.5 Capacitive Reactance and Impedance of Tilting-Pad Journal Bearings

Capacitive reactance between the interacting surface of the journal and each pad in the oil film as well as equivalent capacitive reactance of the bearing is expressed as [5, 6]

$$X_1 = \frac{1}{wC_1},\ X_2 = \frac{1}{wC_2},\ X_3 = \frac{1}{wC_3},\ X_4 = \frac{1}{wC_4}$$
$$\text{and} \quad X_{eq} = \frac{1}{wC_{eq}} \tag{17.111}$$

And impedance of bearing circuit is determined as

$$Z_1 = R_1 - jX_1,\ Z_2 = R_2 - jX_2,\ Z_3 = R_3 - jX_3,\ Z_4 = R_4 - jX_4$$
$$\text{and} \quad Z_{eq} = R_{eq} - jX_{eq} \tag{17.112}$$

The ratio of capacitive reactance to active resistance is given by [Eqns. (17.95)–(17.111)]

$$\frac{X_{eq}}{R_{eq}} = \frac{X_1}{R_1} = \frac{X_2}{R_2} = \frac{X_3}{R_3} = \frac{X_4}{R_4} = \frac{1}{w\rho\xi} \tag{17.113}$$

17.6.6 Data Derived from Analysis and Discussion

The capacitance, resistance, stiffness and damping coefficients for an individual pad and an equivalent value of these parameters for the tilting-pad journal bearing having L/d as 0.5 ($L = 150\,\text{mm}$, $d = 300\,\text{mm}$), $C_p = 0.25\,\text{mm}$ and $\delta_x(\delta/C_p) = 0.5\,\text{mm}$ have been computed as per the developed methodology and shown in Table 17.8. The eccentricity ratios, and other relevant details for the tilting-pad bearing have been taken from the published data [18]. The pad and bearing radial clearance, preload ratio and pivot film thickness of a pad and bearing electricity ratio have been computed separately using Eqns. (17.87)–(17.90), respectively. Based on the electrical analogy, damping coefficients of each pad have been worked out for the bearing eccentricity ratios of 0.2, 0.4, 0.6, 0.8 and 0.9 with the corresponding angles of θ_{pi} and θ_{pe}, and data is shown in Table 17.9. Also, the equivalent values of damping coefficients of the tilting-pad journal bearing are worked out and have been compared with the equivalent size of two-lobe, three-lobe and four-lobe bearings [11–15].

17.6.6.1 Capacitance

The capacitance of a tilting-pad journal bearing depends on the capacitance of each pad. This, in turn, is a function of lubricant permittivity, pad radial clearance (C_p), bearing length (L), and shaft radius (r), and depends on the bearing eccentricity ratio and preload factor along with the angles θ_{pi} and θ_{pe} [Eqns. (17.95)–(17.101)].

As the eccentricity ratio of the bearing increases from 0.2 to 0.9, the capacitance of pads 1, 2, 3 and 4 changes between $(-4.90 \times 10^{-9})\phi$ and $(-3.40 \times 10^{-9})\phi$, $(4.89 \times 10^{-9})\phi$ and $(4.08 \times 10^{-9})\phi$, $(6.52 \times 10^{-9})\phi$ and $(23.61 \times 10^{-9})\phi$, and $(-6.24 \times 10^{-9})\phi$ and $(-16.77 \times 10^{-9})\phi$, respectively. The variation of capacitance with eccentricity ratio is non-linear. As per the configuration of the pads in the tilting bearing, the capacitance of pad 2 is more than that of pad 1, and capacitance of pad 3 is higher than of pad 4. However, capacitance of pad 3 is more than that of all other pads. This is because of the pad location and the loading condition of pad 3 (Fig. 17.13). The equivalent capacitance of the bearing increases with global eccentricity ratio and varies between $(0.27 \times 10^{-9})\phi$ and $(7.52 \times 10^{-9})\phi$ (Table 17.8). This is because of the net effect of variable film thickness of each pad under the influence

Table 17.8. Capacitance and resistance of the tilting-pad journal bearing under different operating parameters

ε_b	$C_1 \times 10^{-9}$ (ϕ)	$C_2 \times 10^{-9}$ (ϕ)	$C_3 \times 10^{-9}$ (ϕ)	$C_4 \times 10^{-9}$ (ϕ)	$C_{eq} \times 10^{-9}$ (ϕ)	$R_1 \times 10^6$ (ohm)	$R_2 \times 10^6$ (ohm)	$R_3 \times 10^6$ (ohm)	$R_4 \times 10^6$ (ohm)	$R_{eq} \times 10^6$ (ohm)
0.2	−4.90	4.89	6.52	−6.24	0.27	−4.53	4.53	3.40	−3.55	80.42
0.4	−4.37	4.51	7.90	−7.23	0.81	−5.07	4.91	2.81	−3.06	32.63
0.6	−3.99	4.10	10.32	−9.12	1.31	−5.55	5.40	2.14	−2.43	26.37
0.8	−3.68	4.26	16.48	−12.54	4.52	−6.05	5.20	1.34	−1.76	16.27
0.9	−3.40	4.08	23.61	−16.77	7.52	−6.43	5.43	0.94	−1.32	17.07

$L/d = 0.5$, $L = 150\,\text{mm}$, $d = 300\,\text{mm}$, $r = 150\,\text{mm}$, $C_p = 0.25\,\text{mm}$, $\delta_x = \delta/C_p = 0.5$, $\xi = 2.5 \times 8.854 \times 10^{-12}\,\phi\,\text{m}^{-1}$, $\rho = 10^{11}$ ohm cm.

Table 17.9. Dynamic coefficients of the tilting-pad journal bearing under different operating parameters

ε_b	$C_{d1} \times 10^6$ (N s m^{-1})	$C_{d2} \times 10^6$ (N s m^{-1})	$C_{d3} \times 10^6$ (N s m^{-1})	$C_{d4} \times 10^6$ (N s m^{-1})	$C_{deq} \times 10^6$ (N s m^{-1})	$K_1 \times 10^9$ (N m^{-1})	$K_2 \times 10^9$ (N m^{-1})	$K_3 \times 10^9$ (N m^{-1})	$K_4 \times 10^9$ (N m^{-1})	$K_{eq} \times 10^9$ (N m^{-1})
0.2	−4.53	4.53	3.40	−3.55	80.42	−0.20	0.20	0.15	−0.16	3.70
0.4	−5.07	4.91	2.81	−3.06	32.63	−0.23	0.22	0.13	−0.14	1.23
0.6	−5.55	5.40	2.14	−2.43	26.37	−0.25	0.24	0.10	−0.11	0.76
0.8	−6.05	5.20	1.34	−1.76	16.27	−0.27	0.23	0.06	−0.08	0.22
0.9	−6.43	5.43	0.94	−1.32	17.07	−0.29	0.25	0.04	−0.06	0.13

of different eccentricity ratios and bearing parameters for the tilting-pad bearing operating under variable conditions (Table 17.8).

17.6.6.2 Resistance

The resistance of a tilting-pad bearing depends on the individual resistance of each pad. This, in turn, is a function of lubricant resistivity, pad radial clearance, bearing length, shaft radius, preload factor and depends on the bearing eccentricity ratio and varies with the angles θ_{pi} and θ_{pe} [Eqns. (17.104)–(17.108)].

Under the change of eccentricity ratio from 0.2 to 0.9, the resistance of pads 1, 2, 3 and 4 changes between -4.53×10^6 and -6.43×10^6 ohm, 4.53×10^6 and 5.43×10^6 ohm, 3.40×10^6 and 0.94×10^6 ohm, and -3.55×10^6 and -1.32×10^6 ohm, respectively (Table 17.8). The variation of resistance with eccentricity ratio is non-linear. Contrary to the capacitance, the resistance of the pad 2 is less than that of pad 1, and resistance of the pad 4 is higher than pad 3 as shown in Table 17.8. However, resistance of pad 3 is less than all the pads under different eccentricity ratios. This is attributed to the location and the loading of condition of the pad 3 as compared to that of the pads 2, 3 and 4, as shown in Fig. 17.13. However, the equivalent resistance of the tilting-pad bearing decreases with increase in bearing eccentricity ratio and varies between 80.42×10^6 and 17.07×10^6 ohm, as eccentricity ratio increases from 0.2 to 0.9 as shown in Table 17.8. This is because of the net effect of the variable film thickness of each pad of a tilting pad bearing under the influence of different eccentricity ratios while operating under variable conditions.

17.6.6.3 Damping and Stiffness Coefficients of Tilting-Pad Bearing

17.6.6.3.1 Damping Coefficient

The damping coefficient of each pad varies differently with the eccentricity ratio. The damping coefficient in $\mathrm{N\,s\,m^{-1}}$ is equivalent to that of the resistance in ohm of each pad. As ε changes from 0.2 to 0.9, the variation of C_{d1}, C_{d2}, C_{d3} and C_{d4} are non-linear, and vary from -4.53×10^6 to $-6.43 \times 10^6\,\mathrm{N\,s\,m^{-1}}$, 4.53×10^6 to $5.43 \times 10^6\,\mathrm{N\,s\,m^{-1}}$, 3.40×10^6 to $0.94 \times 10^6\,\mathrm{N\,s\,m^{-1}}$ and -3.35×10^6 to $-1.32 \times 10^6\,\mathrm{N\,s\,m^{-1}}$, respectively (Table 17.9). The damping coefficient of the pads 3 and 4 (C_{d3} and C_{d4}) decreases with increase in eccentricity ratio varying from 0.2 to 0.9 as against that of C_{d1} and C_{d2}, which increases with increase in eccentricity ratio, and variation is reverse for the pad 2 and pad 3 as compared to that of the pad 1 and pad 4, which is in negative direction. However, the equivalent damping coefficient of the four-pad tilting bearing decreases with increase in eccentricity ratio and varies between 80.42×10^6 and $17.076 \times 10^6\,\mathrm{N\,s\,m^{-1}}$ as the global eccentricity ratio increases from 0.2 to 0.9 (Table 17.9).

Under the identical conditions of operation, trend of variation of equivalent dynamic coefficient of a four-pad tilting bearing with eccentricity ratios is found to be similar to that of the equivalent four-lobe bearing [15]. However, the values of equivalent dynamic coefficients are relatively more as compared to four-lobe, three-lobe and two-lobe bearings as determined in various sections of this chapter (Tables 17.1–17.9), and reported in references [15, 14, 11].

17.6.6.3.2 Stiffness Coefficient

The stiffness coefficient of an individual pad of a tilting-pad journal bearing depends on reciprocal of its capacitance. The equivalent stiffness of a tilting-pad bearing is the resultant stiffness of all the pads connected in parallel. The stiffness of an individual pad shows non-linear behavior with the eccentricity ratio. The K_1, K_2, K_3, and K_4 vary from -0.20×10^9 to $-0.29 \times 10^9\,\mathrm{N\,m^{-1}}$, 0.2×10^9 to $0.25 \times 10^9\,\mathrm{N\,m^{-1}}$, 0.15×10^9 to $0.04 \times 10^9\,\mathrm{N\,m^{-1}}$ and -1.16×10^9 to $-0.06 \times 10^9\,\mathrm{N\,m^{-1}}$, respectively, as ε changes from 0.2 to 0.9 (Table 17.9). The stiffness of the pads 1 and 2 (K_1 and K_2) increases as against that of K_3 and K_4, which decreases, with increase in eccentricity ratios (Table 17.9). Furthermore, variation of stiffness of second and third pads is positive as against that of the pads first and fourth, which is negative, but vary with their respective eccentricity ratios. The equivalent stiffness of the tilting-pad journal bearing decreases from 3.70×10^9 to $0.13 \times 10^9\,\mathrm{N\,m^{-1}}$, with increase in eccentricity ratio (Table 17.9). However, the value of K_{eq} of the tilting pad is more than that of two, three and four-lobe bearings having comparable dimensional parameters and working under identical global eccentricity ratios [11, 13–15]. The variation of equivalent stiffness coefficient determined by electrical analogy approach match closely with those determined by the conventional approach.

17.7 Conclusions

Based on the analysis pertaining to lobe-2, lobe-3, lobe-4 and tilting-pad journal bearings, the followings can be concluded [5, 11, 14, 15]:

1. The capacitance (stiffness^{-1}) and resistance/damping of two-lobe, three-lobe, four-lobe and tilting-pad journal bearings are function of the lubricant permittivity and resistivity, and depend on horizontal clearance, journal radius, eccentricity ratio, attitude angle and inlet/outlet angles from lines of centers from leading/trailing edge film of each lobe/pad.
2. The change in capacitance (stiffness^{-1}) and resistance/damping for an individual lobe/pad with eccentricity ratio and attitude angle is non-linear. However, equivalent capacitance of a bearing increases as against that of

equivalent resistance which decreases with the increase in global eccentricity ratio.

3. The capacitance (in ϕ) and resistance (in ohm) of each lobe/pad under various operating parameters can be taken as flexibility (stiffness^{-1}) (in m N^{-1}) and damping coefficient (in N s m^{-1}), respectively.
4. By electrical analogy, the damping coefficient and stiffness coefficients of each lobe/pad vary non-linearly with the eccentricity ratio.
5. The equivalent damping and stiffness coefficient for a four-lobe and tilting-pad journal bearing decrease with increase in global eccentricity ratio varying from 0.2 to 0.5, and from 0.2 to 0.9, respectively.
6. The ratio of capacitive reactance to active resistance depends on premittivity and resistivity of the lubricant.
7. By electrical analogy, the damping and stiffness coefficient of lower and upper lobe of a two-lobe and that of a cylindrical bearing are found to vary with the eccentricity ratio. The ratio of damping and stiffness coefficient of two-lobe, three-lobe and four-lobe versus cylindrical bearing is more than unity for different operating conditions. However, tilting pad bearings have superior dynamic characteristics than multi-lobe bearings. This matches closely with the published work by the conventional methods.

The developed technique and further development and extension of methodology of electrical analogy have the potential to determine dynamic coefficients more reliably and much faster under the complex situations as compared to that of the conventional techniques.

Nomenclature

C	capacitance
C_b	radial clearance of tilting pad becomes $(r_b - r)$
C_p	radial clearance of pad $(r_p - r)$
C_1, C_2, C_3, C_4	capacitance of first, second, third and fourth pad, respectively
C_{eq}	equivalent capacitance of tilting-pad journal bearing with four pads
C_{d1}, C_{d2}, C_{d3}, C_{d4}	damping coefficients of first, second, third and fourth pad, respectively
C_{deq}	equivalent damping coefficient of tilting pad journal bearing with four pads
C_c	damping coefficient

d	shaft diameter
e_b	bearing eccentricity (distance between shaft center and bearing center)
e_p	pad eccentricity
E	voltage
f	frequency of applied voltage
F	force
F_1/R_1^{-1}	conductance
h	film thickness
i	current
K	stiffness
K_c	capacitance constant
K_1, K_2, K_3, K_4	stiffness of first, second, third and fourth pad, respectively
K_R	resistance constant
K_{eq}	equivalent stiffness of tilting pad journal bearing with four pads
L	pad length
L_j	inductance
M	mass
Q	charge
r	radius of journal
r_p	pad radius
r_b	bearing radius
R	resistance
R_1, R_2, R_3, R_4	resistance of first, second, third and fourth pad, respectively
R_{eq}	equivalent resistance of tilting-pad bearing
t	time
w	$2\pi f$
X_1, X_2, X_3, X_4	capacitive reactance of first, second, third and fourth pad, respectively
X_{eq}, Z_{eq}	equivalent capacitive reactance and impedance of tilting-pad bearing with four pads, respectively
Z_1, Z_2, Z_3, Z_4	impedance of first, second, third and fourth pad, respectively
δ	distance between pad center to bearing center $(r_p - r_b)$ or $(c_p - c_b)$
δ_x	preload ratio (δ/c_p)
ε_b	e_b/C_b
ϕ	attitude angle
θ	angular coordinate indicating pivot position of a pad of a tilting journal pad bearing
θ_{pi}	inlet angle from line of centers to leading edge film in each pad

θ_{pe}	outlet-angle from line of centers to trailing edge film in each pad
ρ	resistivity of lubricant
ξ	permittivity/dielectric constant of lubricant ($\phi\,m^{-1}$)
ξ_r	relative permittivity of lubricant

REFERENCES

1. Glienicke, J., Han, D. C. and Leonhard, M. (1980). Practical determination and use of bearing dynamic coefficients. *Tribol. Int.*, **13**, 297–308.
2. Knight, J. D. and Barrett, L. E. (1983). An approximate solution technique for multi-lobe journal bearings including thermal effects with comparison to experiments. *ASME Trans.*, **26**(4), 501–508.
3. Shang, L. and Dien, I. K. (1989). A matrix method for computing the stiffness and damping coefficients of multi-arc journal bearings. *Tribol. Trans.*, **32**(3), 396–404.
4. Pujara, K. K. (1972). *Vibrations for Engineers*, Dhanpat Rai and Sons, Delhi.
5. Prashad, H. (1991). Theoretical evaluation of capacitance, capacitive reactance, resistance and their effects on performance of hydrodynamic journal bearings. ASME, *J. Tribol.*, **113**, 762–767.
6. Prashad, H. (1988). Theoretical evaluation of impedance, capacitance and charge accumulation of roller bearings operated under electrical fields. *Wear*, **125**, 223–239.
7. Nica, Al. (1985). *Sliding Bearings*, Allerton Press, Inc., New York.
8. Cameron, A. (1987). *Basic Lubrication Theory*, Wiley Eastern Limited.
9. Prashad, H. (1994). Analysis of inductive effects of bearings under the effect of shaft voltages. *BHEL J.*, **15**(1), 26–31.
10. Dargaiah, K., Parthasarthy, K. and Prabhu, B. S. (1993). Finite element method for computing dynamic coefficients of multi-lobe bearings. *Tribol. Trans.*, **36**(1), 73–83.
11. Prashad, H. (1996). Evaluation of dynamic coefficients of a two-lobe journal bearing using an electrical analogy approach. Transaction of ASME, *J. Tribol.*, **118**, 657–662.
12. Lund, J. W. and Thomsen K. K. (1978). A calculation method and data for the dynamic coefficients of oil-lubricated journal bearings. *Topics in Fluid Film Bearings and Rotor Bearing System Design and Optimization*, ASME, New York, 1–28.
13. Prashad, H. (1997). Determination of effective stiffness of cylindrical and two-lobe journal bearings by an electrical analogy and its comparison with the conventional approach. Proceeding of Second International Conference on Hydrodynamic Bearing Rotor System Dynamics (IC-HBRSD-97), held on March 20–22, 1997, Xi'an, P. R. China.
14. Prashad, H. (1997). Assessment of dynamic coefficients of three-lobe journal bearings through evaluation of electrical parameters – A new approach. *BHEL J.*, **17(1), 40–48**.
15. Prashad, H. (2000). Determination of capacitance, resistance and dynamic coefficients of four-lobe bearings through electrical analogy. *IE (I), J.-MC*, **81**, 30–36.
16. Nicholas J. C., Allaire, P. E. and Gunter, E. J. (1976). *A Manual for Use with Computer Program PADFEMI-A finite Element Analysis of Tilt Pad Bearings Coefficients Using the*

Pad Assembly Method. School of Engineering and Applied Science University of Virginia, Charlottesville, Virginia, 22901.No.VVA/643092/MAE81/116.

17. Lund, J. W. (1964). Spring and damping coefficients of the tilting pad journal bearings. *ASLE Trans.*, **17**(4), 342–352.
18. Hashimoto, H., Wada, S. and Maru Kawa, T. (1985). Performance characteristics of large scale tilting-pad journal bearings. *Bull. JSME*, **28**(242).

Chapter 18

EVALUATION OF STIFFNESS AND DAMPING COEFFICIENTS OF CYLINDRICAL JOURNAL BEARINGS BY ELECTRICAL ANALOGY

18.1 Introduction

In order to carry out vibration calculations for rotors using hydrodynamic journal bearings, the non-linear isotropic spring and damping properties of the lubricant film are each linearized in the vicinity of static working point. Thus four stiffness coefficients and four damping coefficients are obtained, which depend on the bearing design and the operational parameters. These coefficients can be determined by calculation by using the theory of an unsteadily loaded slide bearing and experimentally on the individual bearing or on real machines. Perturbation techniques have been used, which require solving the Reynolds equation for both stiffness and damping coefficients [1–3].

In most cases, it is possible to represent a mechanical vibrating system by an electric circuit. The analogy between electrical and mechanical system exists because of the similarity of the differential equations which represents them as discussed in Chapters 17 [4]. Theoretical studies have been undertaken to determine the capacitance and active resistance of the interacting surfaces of a journal and cylindrical, two-lobe, three-lobe and four-lobe journal bearings in the load-carrying oil film under different operating parameters in the Chapter 17 [5–7]. Based on the equivalent capacitance and resistance, the effective stiffness and damping coefficients of cylindrical, two-lobe, three-lobe and four lobe bearings have been computed (Chapter 17).

This chapter brings out the theoretical studies undertaken to derive the expressions to determine the stiffness and damping coefficients of cylindrical bearings in

XX, *YY*, *XY* and *YX* directions under different operating conditions by non-conventional electrical approach. The values of dimensionless stiffness and damping coefficients, thus determined, have been compared with those obtained by the conventional approach and potential of the developed non-conventional electrical approach has been established. Also, effective stiffness evaluated by electrical analogy has been compared with that of the existing technique.

18.2 Background

The capacitance and active resistance of cylindrical hydrodynamic journal bearing as well as comparison of electrical analogy approach with the conventional analysis have been discussed in details in this section.

18.2.1 Capacitance and Active Resistance of Cylindrical Hydrodynamic Journal Bearings

For each revolution of the shaft, part of the circumference of the journal passes through a zone of load-carrying oil film. This forms a parallel plate capacitor of varying thickness between the interacting surfaces of the journal and the bearing. The effective capacitance of a hydrodynamic journal bearing has been determined by [5] in the Chapter 15 and is given as:

$$C_{\text{cb}} = \frac{\left(4\delta\xi d \tan^{-1} 0.41 \left[\frac{(1+\varepsilon)}{(1-\varepsilon)}\right]^{1/2}\right)}{\psi(1-\varepsilon^2)^{1/2}} \tag{18.1}$$

The effective resistance of a cylindrical bearing is determined in Chapter 15 is given as [5]:

$$R_{cb} = \frac{\psi\rho\left[(1-\varepsilon^2)^{1/2}\right]}{4\,\delta d \tan^{-1} 0.41 \left[\frac{(1+\varepsilon)}{(1-\varepsilon)}\right]^{1/2}} \tag{18.2}$$

18.2.2 Electrical Analogy Approach

Electrical analogy is geometrically based on solutions for charge distribution along potential and isostatic surfaces, with an assumed analogy between force and voltage providing coupling to the mechanical hydrodynamic problem. The solution satisfies continuity. Ratios of solutions eliminate the need for heuristic data such as resistivity, dielectric constant and equivalent inductance of the fluid film, thus any value will suffice, as these values are not required for comparisons provided the fluid conditions remain constant. Solutions to the electrical equations provide potential lines and isosurfaces [6].

Similarly, a direct solution to the Reynolds equation for a fluid with constant properties would evolve into a solution of the continuity equation and would only be geometry dependent. The dynamic would depend on solutions to steady and perturbed equations, and again for constant properties would only be geometry dependent. Also, ratios of solutions would not require knowing the fluid properties. Hence, the electrical analogy can be used as an alternative to conventional analysis, as a potential tool for relative comparison in addition to the absolute.

18.3 Theoretical Derivations

The details of bearing and shaft voltage, stored charges on the bearing surfaces, forces acting on the bearings, procedure using electrical analogy for the determination of bearing stiffness and damping coefficients and effective stiffness are discussed here.

18.3.1 Bearing and Shaft Voltage

Magnetic flux develops in electric machines due to asymmetry of the magnetic circuits, which closes in the circumference over the yoke and induces the voltages on the shaft. Shaft voltage and flux can develop in the electric machines for various reasons namely, winding faults, unbalanced supplies, electrostatic effects, air gap fields, magnetized shafts or other machine members, or asymmetries of the magnetized fields [8]. This has been analyzed in Chapter 9. Also, shaft voltage depends on the operating parameters, namely, load, speed, eccentricity ratio apart from the variation in applied voltage, etc. [9–11]. Furthermore, shaft voltage vary in each revolution between minimum and maximum values and is a function of frequency. For lightly loaded bearings, the air-gap field tends to be lower in a zone of minimum film thickness as compared to that of the heavily loaded bearings. This is because of change in film thickness with eccentricity ratio and flux density variation apart from various other unforeseen reasons. That is why, the overall variation of shaft voltage in each revolution is higher for lightly loaded bearing as compared to that of the heavily loaded bearing. The effect of shaft voltage on bearing life has been theoretically estimated, and has been brought out in Chapter 16 [11].

It is evident that in a hydrodynamic journal bearing, the zone of minimum film thickness, i.e. the load-carrying oil film, varies along the circumference of a bearing through the bearing length. This forms a capacitor of varying capacitance between the journal and bearing, depending on the permittivity of the lubricant, circumferential length of load-carrying oil film, bearing length, eccentricity ratio and the clearance in a bearing. Besides this, the load-carrying oil film offers resistance depending on the operating parameters and resistivity of the lubricant. Thus, the load-carrying oil film acts like an impedance to current flow in a bearing. This has been analyzed in Chapter 16. Furthermore, during rotation, the minimum oil film thickness of a hydrodynamic journal bearing besides other parameters and operating conditions depends on clearance and eccentricity ratio, and varies approximately between 0.05 to 0.02 mm depending on the journal diameter [12].

18.3.2 Stored Charges on Bearing Surfaces

When a journal bearing lubricated with high-resistivity lubricant having resistivity more than 10^9 ohm m and relative permitivity (1.7–2.5), operates under the influence of shaft voltage, the current does not begin to flow until a first threshold voltage is reached. Journal bearing and shaft separated by an insulating high-resistivity lubricant form a capacitor of varying thickness. Under the effect of shaft voltage, bearing surface and shaft develop equal and opposite charges by induction and the net charge on the capacitor, thus formed, as a whole is zero. The electric field in the oil film region between the bearing and shaft is proportional to the charge and similarly the potential difference between the shaft and the bearing. Below the threshold voltage, the capacitance of a bearing opposes any change in the existing shaft voltage and causes an electric charge to be stored [10–11]. The stored charge on the surfaces of bearing and shaft is given as:

$$Q = EC_{\mathrm{cb}} \tag{18.3}$$

At the first threshold voltage (the increase in voltage from zero to a specific maximum value with negligible current), the increase in flow of current is momentary. On exceeding the first threshold voltage, even a further slow increase in voltage results in a very high increase in current. This phenomenon has been discussed in Chapter 4 [10–13].

18.3.3 Forces Acting on Bearing

The electrical interaction between two charges is expressed in terms of the force they exert on each other. The shaft and bearing, having equal charge Q on their surfaces and separated by the minimum film thickness h_o at the location of eccentricity ratio ε, the maximum force exerted on the bearing is expressed as [14]:

$$F = \frac{Q^2}{(4\pi\xi h_o^2)} \tag{18.4}$$

The minimum film thickness is given as [15]:

$$h_o = C_d \frac{(1-\varepsilon)}{2} \tag{18.5}$$

Using Eqns. (18.1) and (18.3), the net force acting on the bearing has been determined by Eqns. (18.4) and (18.5), and is given as:

$$F = \frac{5.1E^2L^2\xi\left\{\tan^{-1} 0.41\left[\frac{(1+\varepsilon)}{(1-\varepsilon)}\right]^{1/2}\right\}^2}{(1-\varepsilon)^3(1+\varepsilon)\psi^2 C_d^2} \tag{18.6}$$

The bearing attitude angle as per the short bearing theory is given as [15]:

$$\tan\phi = \frac{\left(1-\varepsilon^2\right)^{1/2}}{\varepsilon} \tag{18.7}$$

which gives:

$$\sin\phi = (1-\varepsilon^2)^{1/2} \quad \text{and} \quad Cos\,\phi = \varepsilon \tag{18.8}$$

From the Fig. 18.1, it is evident that the bearing reaction components of the force F acting along X and Y directions on the bearing are as follows:

$$F_X = F\ Sin\,\phi \quad \text{and} \quad F_Y = F\ Cos\,\phi \tag{18.9}$$

These components are expressed using Eqns. (18.6), (18.7) and (18.8) and are given as:

$$F_X = \frac{5.1E^2L^2\xi\left\{\tan^{-1} 0.41\left[\frac{(1+\varepsilon)}{(1-\varepsilon)^{1/2}}\right]\right\}^2}{(1-\varepsilon)^{5/2}(1+\varepsilon)^{1/2}\psi^2 C_d^2} \tag{18.10}$$

and

$$F_Y = \frac{5.1E^2L^2\xi\varepsilon\left\{\tan^{-1} 0.41\left[\frac{(1+\varepsilon)}{(1-\varepsilon)^{1/2}}\right]\right\}^2}{(1-\varepsilon)^3(1+\varepsilon)\psi^2 C_d^2} \tag{18.11}$$

On differentiating F_X and F_Y with respect to ε, the following derivatives are obtained:

$$\frac{dF_X}{d\varepsilon} = \frac{5.1E^2L^2\xi\tan^{-1}0.41\left[\frac{(1+\varepsilon)}{(1-\varepsilon)}\right]^{1/2}\left\{(2+3\varepsilon)\tan^{-1}0.41\left[\frac{(1+\varepsilon)}{(1-\varepsilon)}\right]^{1/2} - \frac{0.82\varepsilon(1-\varepsilon^2)^{1/2}}{(1.17-0.83\varepsilon)}\right\}}{\psi^2C_d^2(1-\varepsilon^2)^{3/2}(1-\varepsilon)^2} \tag{18.12}$$

$$\frac{dF_Y}{d\varepsilon} = \frac{5.1E^2L^2\xi\tan^{-1}0.41\left[\frac{(1+\varepsilon)}{(1-\varepsilon)}\right]^{1/2}\left\{(3\varepsilon^2+3\varepsilon+1)\tan^{-1}0.41\left[\frac{(1+\varepsilon)}{(1-\varepsilon)}\right]^{1/2} - \frac{0.83\varepsilon^2(1-\varepsilon^2)^{1/2}}{(1.17-0.83\varepsilon)}\right\}}{\psi^2C_d^2(1-\varepsilon^2)^2(1-\varepsilon^2)} \tag{18.13}$$

Also, from Fig. 18.1, it is evident that:

$$X = eSin\phi = \left(\frac{C_d}{2}\right)\varepsilon Sin\phi \tag{18.14}$$

$$Y = eCos\phi = \left(\frac{C_d}{2}\right)\varepsilon Cos\phi \tag{18.15}$$

X and Y can be expressed using Eqn. (18.8) and given as:

$$X = \left(\frac{C_d}{2}\right)\varepsilon(1-\varepsilon^2) \text{ and } Y = \varepsilon^2\left(\frac{C_d}{2}\right) \tag{18.16}$$

On differentiating Eqn. (18.16) gives:

$$\frac{d\varepsilon}{dX} = \frac{2(1-\varepsilon^2)^{1/2}}{C_d(1-2\varepsilon^2)} \tag{18.17}$$

and

$$\frac{d\varepsilon}{dY} = (C_d\varepsilon)^{-1} \tag{18.18}$$

18.4 Bearing Stiffness Coefficients

Assuming that a bearing is running in its equilibrium position as the origin (O_b), Y being vertically downwards and X being horizontal to the right as shown in the Fig. 18.1. Assuming also that initially shaft center velocities in X and Y directions are set to zero. If the shaft is given a small displacement delta X in the X direction, delta Y in Y direction, then additional forces delta F_x and delta F_Y will act on the shaft center O_j (Fig. 18.1).

Then stiffness coefficients by electric force in different directions can be calculated as:

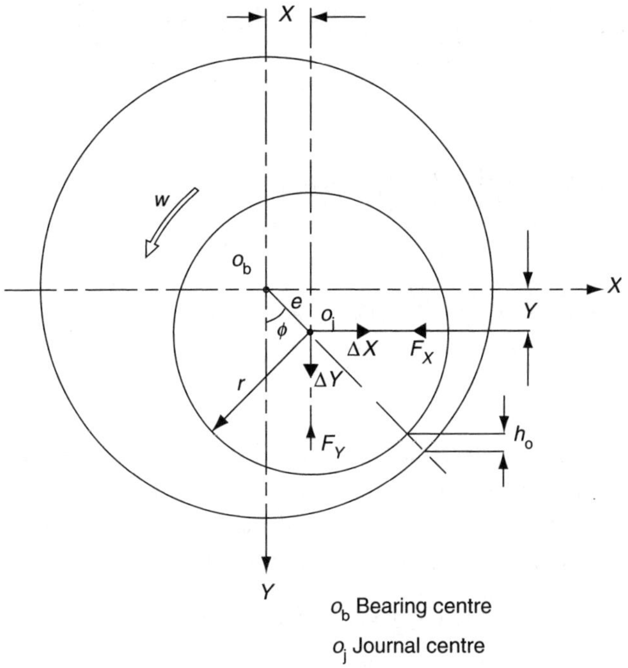

Fig. 18.1. Bearing coordinate system

$$(K_{XX})_e = \frac{dF_X}{dX} = \frac{dF_X}{d\varepsilon}\frac{d\varepsilon}{dX} \tag{18.19}$$

$$(K_{YY})_e = \frac{dF_Y}{dY} = \frac{dF_Y}{d\varepsilon}\frac{d\varepsilon}{dY} \tag{18.20}$$

$$(K_{XY})_e = \frac{dF_X}{dY} = \frac{dF_X}{d\varepsilon}\frac{d\varepsilon}{dY} \tag{18.21}$$

$$(K_{YX})_e = \frac{dF_Y}{dX} = \frac{dF_Y}{d\varepsilon}\frac{d\varepsilon}{dX} \tag{18.22}$$

By using Eqns. (18.16)–(18.18) and (18.19)–(18.22), the following equations for stiffness coefficients by electric force have been derived and given as:

$$(K_{XX})_e = \frac{10.2E^2L^2\xi\tan^{-1}0.41\left[\frac{(1+\varepsilon)}{(1-\varepsilon)}\right]^{1/2}\left\{(2+3\varepsilon)\tan^{-1}0.41\left[\frac{(1+\varepsilon)}{(1-\varepsilon)}\right]^{1/2} - \frac{0.82\varepsilon(1-\varepsilon^2)^{1/2}}{(1.17-0.83\varepsilon)}\right\}}{\psi^2C_d^3(1-2\varepsilon^2)(1-\varepsilon^2)(1-\varepsilon)^2} \tag{18.23}$$

$$(K_{YY})_e = \frac{5.1E^2L^2\xi \tan^{-1} 0.41 \left[\frac{(1+\varepsilon)}{(1-\varepsilon)}\right]^{1/2} \left\{ (3\varepsilon^2 + 2\varepsilon^2 + 1) \tan^{-1} 0.41 \left[\frac{(1+\varepsilon)}{(1-\varepsilon)}\right]^{1/2} - \frac{0.82\varepsilon^2(1-\varepsilon^2)^{1/2}}{(1.17 - 0.83\varepsilon)} \right\}}{\psi^2 C_d^3 \varepsilon(1-\varepsilon^2)^2(1-\varepsilon)} \tag{18.24}$$

$$(K_{XY})_e = \frac{5.1E^2L^2\xi \tan^{-1} 0.41 \left[\frac{(1+\varepsilon)}{(1-\varepsilon)}\right]^{1/2} \left\{ (2 + 3\varepsilon) \tan^{-1} 0.41 \left[\frac{(1+\varepsilon)}{(1-\varepsilon)}\right]^{1/2} - \frac{0.82\varepsilon(1-\varepsilon^2)^{1/2}}{(1.17 - 0.83\varepsilon)} \right\}}{\psi^2 C_d^3 \varepsilon(1-\varepsilon^2)^{3/2}(1-\varepsilon)^2} \tag{18.25}$$

$$(K_{YX})_e = \frac{10.2E^2L^2\xi \tan^{-1} 0.41 \left[\frac{(1+\varepsilon)}{(1-\varepsilon)}\right]^{1/2} \left\{ (3\varepsilon^2 + 2\varepsilon^2 + 1) \tan^{-1} 0.41 \left[\frac{(1+\varepsilon)}{(1-\varepsilon)}\right]^{1/2} - \frac{0.82\varepsilon^2(1-\varepsilon^2)^{1/2}}{(1.17 - 0.83\varepsilon)} \right\}}{\psi^2 C_d^3 (1-\varepsilon^2)^{3/2}(1-2\varepsilon^2)(1-\varepsilon)^2} \tag{18.26}$$

This may be noted that actual mechanical stiffness coefficients will be at right angle to that of the stiffness coefficients derived by electric forces. This is because of the field strength, which follows the standard Fleming's rule [8, 14]. Thus the actual stiffness coefficients in different directions are given as:

$$K_{XX} = (K_{YY})_e,\ K_{YY} = (K_{XX})_e$$

$$K_{XY} = -(K_{YX})_e,\ K_{YX} = (K_{XY})_e \tag{18.27}$$

18.4.1 Effective Stiffness

The load-carrying capacity of fluid film bearings is derived from the hydrodynamic pressures produced in the lubricant film by the shearing action in the oil film by the rotating journal in a bearing. At any given speed, the film reaction force is a function of the journal center position relative to the bearing center (i.e. bearing eccentricity) and, in addition, is a function of journal center velocity (Fig. 18.1). As the journal whirl orbit increases, its center no longer coincides with the static equilibrium position but moves gradually closer to the bearing center.

It should furthermore be recognized that the solution of the Reynolds equation normally is based on several simplifications, such as the assumption of constant viscosity and certain hypothesis for the mechanism of film rupture. Furthermore, the actual operating bearing temperature, and hence, the effective viscosity cannot be predicted accurately in actual practice. Hence, dynamic coefficients of a bearing cannot be determined very accurately. That is why the approximate representation of a bearing by effective stiffness and damping characteristics is satisfactory for most practical purposes.

For most practical purposes, the critical journal mass M_{crt} is determined as the upper limit for stability, and given as [16].

$$M_{\mathrm{crt}} = \frac{K_{\mathrm{cb}}}{V^2} \tag{18.28}$$

and

$$K_{\mathrm{cb}} = \frac{(K_{XX}B_{YY} + K_{YY}B_{XX} - K_{XY}B_{YX} - K_{YX}B_{XY})}{(B_{XX} + B_{YY})} \tag{18.29}$$

Also, instability whirl frequency is found as:

$$V^2 = \frac{[(K_{XX} - K_{\mathrm{cb}})(K_{YY} - K_{\mathrm{cb}}) - K_{XY}K_{YX}]}{(B_{XX}B_{YY} - B_{XY}B_{YX})} \tag{18.30}$$

By the electrical analogy, the effective stiffness K_{cbe} has been evaluated in Chapter 17, and is given as [6]:

$$K_{\mathrm{cbe}} = \frac{1}{C_{\mathrm{cb}}} \tag{18.31}$$

and also

$$B_{\mathrm{cb}} = R_{\mathrm{cb}} \tag{18.32}$$

When stiffness (K_{XX}, K_{YY}, K_{XY}, K_{YX}) and damping (B_{XX}, B_{YY}, B_{XY}, B_{YX}) coefficients are known, then Eqns. (18.29) and (18.30) enable determination of V and M_{crt}: first the product $M_{\mathrm{crt}}V^2$ from Eqns. (18.28) and (18.29), then V from the Eqn. (18.30) and finally M_{crt} from the Eqn. (18.28).

A cylindrical bearing tends to have frequency ratio V/w, of approximately 0.4–0.5 over most of the speed range as determined by the data published by Lund and Thomson [16], and Nica [17]. This can be used for the assessment of the M_{crt} by using Eqn. (18.28) under the given operating speed frequency on determining the effective stiffness. The effective stiffness can be determined much faster by the electrical analogy using Eqn. (18.31) as compared to that of conventional methodology by using Eqns. (18.29) and (18.30), which is evaluated after determining the four stiffness and four damping coefficients separately.

18.5 Theoretical Analysis for Damping Coefficients

Similar to the stiffness coefficients, theoretical analysis for bearing damping coefficients is carried out using the following procedure based on the electrical analogy.

18.5.1 Shaft Voltage

Shaft voltage can be expressed in terms of bearing current and resistance, and can be given as:

$$E = I_{\mathrm{b}} R_{\mathrm{cb}} \tag{18.33}$$

18.5.2 Forces on Bearings

Since the oil film resistance of a bearing lubricated with high-resistivity lubricant is much higher, and current does not flow through a bearing until the threshold voltage is reached [13], hence, on substituting shaft voltage by Eqn. (18.33) and bearing resistance by Eqn. (18.2) in Eqns. (18.10) and (18.11), the components of bearing forces are expressed as:

$$F_X = \frac{I_{\mathrm{b}}^2 \rho^2 \xi (1-\varepsilon^2)^{1/2}}{\pi C_{\mathrm{d}}^2 (1-\varepsilon)^2} \tag{18.34}$$

and

$$F_Y = \frac{I_{\mathrm{b}}^2 \rho^2 \xi \varepsilon}{\pi C_{\mathrm{d}}^2 (1-\varepsilon)^2} \tag{18.35}$$

On differentiating F_X and F_Y with respect to ε, the following derivatives are obtained:

$$\frac{\mathrm{d}F_X}{\mathrm{d}\varepsilon} = \frac{\xi \rho^2 I_{\mathrm{b}}^2 (2-\varepsilon^2-\varepsilon)}{\pi C_{\mathrm{d}}^2 (1-\varepsilon)^3 (1-\varepsilon^2)^{1/2}} \tag{18.36}$$

and

$$\frac{\mathrm{d}F_Y}{\mathrm{d}\varepsilon} = \frac{\xi \rho^2 I_{\mathrm{b}}^2 (1+\varepsilon)}{\pi C_{\mathrm{d}}^2 (1-\varepsilon)^3} \tag{18.37}$$

18.5.3 Finite Bearing Analysis

The finite bearing analysis has been used for deriving the following equations using numerical methods developed by Hays of General Motors [15]:

$$\frac{\mu L d (d/C_{\mathrm{d}})^2 (d\varepsilon/dt)}{2W} = a(1-\varepsilon)^b \tag{18.38}$$

which gives

$$\frac{S(d\varepsilon/dt)}{2n} = a(1-\varepsilon)^b \tag{18.39}$$

or

$$dt = \left[\frac{S}{2na(1-\varepsilon)^b}\right] d\varepsilon \tag{18.40}$$

where a and b are constants, and can be related to L/d ratio quite accurately for 1.33 $< L/d <$ 0.20 by the empirical formulae [15], and are given as:

$$a = 0.6\left(\frac{d}{L}\right) - 0.4 \quad \text{and} \quad b = 1.5 + 0.3\left(\frac{d}{L}\right)^{1/2} \tag{18.41}$$

18.6 Mathematical Model

Assume that a bearing is running in its equilibrium position as the origin (O_b), Y being vertically downwards and X being horizontal to the right as shown in the Fig. 18.1. Also assuming that initially shaft center velocities dX/dt and dY/dt are set to zero. If the shaft is given a small displacement delta X in the X direction, delta Y in Y direction, then additional forces delta F_X and delta F_Y will act on the shaft center O_j (Fig. 18.1).

Then damping coefficients by electric force in different directions can be calculated as:

$$(B_{XX})_e = \frac{dF_X}{(dX/dt)} = \frac{dF_X}{d\varepsilon}\frac{d\varepsilon}{dX}dt \tag{18.42}$$

$$(B_{YY})_e = \frac{dF_Y}{(dY/dt)} = \frac{dF_Y}{d\varepsilon}\frac{d\varepsilon}{dY}dt \tag{18.43}$$

$$(B_{XY})_e = \frac{dF_X}{(dY/dt)} = \frac{dF_X}{d\varepsilon}\frac{d\varepsilon}{dY}dt \tag{18.44}$$

$$(B_{YX})_e = \frac{dF_Y}{(dX/dt)} = \frac{dF_Y}{d\varepsilon}\frac{d\varepsilon}{dX}dt \tag{18.45}$$

By using Eqns. (18.36), (18.37), (18.17), (18.18) and (18.38)–(18.40), the following equations for damping coefficients derived by electric forces have been obtained:

$$(B_{XX})_e = \left[\frac{I_b^2\rho^2\xi S}{\pi C_d^3 na}\right] \int \left[\frac{(2-\varepsilon^2-\varepsilon)}{(1-\varepsilon)^{3+b}(1-2\varepsilon^2)}\right] d\varepsilon \tag{18.46}$$

$$(B_{YY})_e = \left[\frac{I_b^2 \rho^2 \xi S}{2\pi C_d^3 n a}\right] \int \left[\frac{(1+\varepsilon)}{\varepsilon(1-\varepsilon)^{3+b}}\right] d\varepsilon \tag{18.47}$$

$$(B_{XY})_e = \left[\frac{I_b^2 \rho^2 \xi S}{2\pi C_d^3 n a}\right] \int \left[\frac{(2-\varepsilon^2-\varepsilon)}{\varepsilon(1-\varepsilon)^{3+b}(1-\varepsilon^2)^{1/2}}\right] d\varepsilon \tag{18.48}$$

$$(B_{YX})_e = \left[\frac{I_b^2 \rho^2 \xi S}{\pi C_d^3 n a}\right] \int \left[\frac{(1+\varepsilon)(1-\varepsilon^2)^{1/2}}{(1-\varepsilon)^{3+b}(1-2\varepsilon^2)}\right] d\varepsilon \tag{18.49}$$

On expansion of bracketed terms contained (since $\varepsilon < 1$) in Eqns. (18.46)–(18.49), and subsequently integrating, the damping coefficients have been worked out. Using Eqns. (18.2) and (18.33), damping coefficients are given as:

$$(B_{XX})_e = P\left\{\frac{2\varepsilon+\varepsilon^2(2b+5)}{2} + \frac{\varepsilon^3(b^2+6b+12)}{3} + \frac{\varepsilon^4\left[\frac{(3b+7)+(3+b)(4+b)(2b+7)}{6}\right]}{4} + \ldots\right\} \tag{18.50}$$

where

$$P = \left[\frac{16 S\xi E^2 \delta^2 d^2}{n a C_d^3 \psi^2 (1-\varepsilon^2)}\right]\left\{\tan^{-1} 0.41\left[\frac{(1+\varepsilon)}{(1-\varepsilon)}\right]^{1/2}\right\}^2 \tag{18.51}$$

$$\begin{aligned}(B_{YY})_e = 0.5P\Bigg[&\log\varepsilon + \varepsilon(b+4) + \frac{\varepsilon^2(3+b)(6+b)}{4} + \frac{\varepsilon^3(3+b)(4+b)(8+b)}{18} \\ &+ \frac{\varepsilon^4(3+b)(4+b)(5+b)(10+b)}{96} \\ &+ \frac{\varepsilon^5(3+b)(4+b)(5+b)(6+b)(12+b)}{600} + \ldots\Bigg]\end{aligned} \tag{18.52}$$

$$\begin{aligned}(B_{XY})_e = 0.5P\Bigg\{&2\log\varepsilon + \varepsilon(2b+5) + \frac{\varepsilon^2(3+b)^2}{2} + \frac{\varepsilon^3\left[\frac{(3+b)(4+b)(2b+7)}{6-0.5}\right]}{3} \\ &+ \frac{\varepsilon^4\left[\frac{(3+b)(4+b)^2(5+b)}{12} - \frac{(3+b)}{2} - 0.5\right]}{4} \\ &+ \frac{\varepsilon^5\left[\frac{(3+b)(4+b)(5+b)(6+b)(2b+9)}{120} - \frac{(3+b)(3b+7)}{4} - \frac{3}{8}\right]}{5} + \ldots\Bigg\}\end{aligned} \tag{18.53}$$

$$(B_{YX})_e = P\left\{\frac{\varepsilon+\varepsilon^2(4+b)}{2}+\frac{\varepsilon^3[(3+b)(6+b)+3]}{6}\right.$$
$$+\frac{\varepsilon^4[9(5+b)+(3+b)(4+b)(8+b)]}{24}$$
$$\left.+\frac{\varepsilon^5[69+(3+b)(4+b)(5+b)(10+b)+18(3+b)(6+b)]}{120}+\ldots\right\} \quad (18.54)$$

It may be noted that actual mechanical damping coefficients will be at right angle to the damping coefficients derived by electric forces. This is because of the field strength, which follows the standard Fleming's rule [8, 14]. Thus the actual damping coefficients in different directions are given as:

$$B_{XX} = (B_{YY})_e,\ B_{YY} = (B_{XX})_e \quad \text{and} \quad B_{XY} = (B_{YX})_e,\ B_{YX} = (B_{XY})_e \quad (18.55)$$

18.7 Determination of Damping and Stiffness Coefficients

The damping coefficients B_{XX}, B_{YY},B_{XY} and B_{YX} have been determined using Eqns. (18.50)–(18.55) for the various values of Sommerfeld number, attitude angle, *L*/*d*, eccentricity and clearance ratios. Dimensionless damping coefficients given in Table 18.1, have been evaluated using radial clearance, angular speed of rotation and bearing load for *L*/*d*=0.5. The change in shaft voltage with eccentricity ratio generally measured in rotating machines is taken from 6.5 to 1.5 V for data deduction. The overall shaft voltage of lightly loaded bearing for each revolution is taken varying from 6.5 to 2.5 V as against overall 4.5 to 1.5 V for the bearing working with higher eccentricity ratios.

The following equations have been used on simplyfing the Eqns. (18.41), (18.50)–(18.55), for determining the damping coefficients for *L*/*d* = 0.5:

$$B_{XX} = 0.5P\left[\log\varepsilon + 5.92\varepsilon + 9.74\varepsilon^2 + 16.05\varepsilon^3 + 25.04\varepsilon^4 + 37.03\varepsilon^5 + 52.47\varepsilon^6 + \ldots\right] \quad (18.56)$$

$$B_{YY} = P\left[2\varepsilon + 4.42\varepsilon^2 + 9.07\varepsilon^3 + 16.35\varepsilon^4 + 27.45\varepsilon^5 + 43\varepsilon^6 + 65.22\varepsilon^7 + \ldots\right] \quad (18.57)$$

$$B_{XY} = P\left[\varepsilon + 2.96\varepsilon^2 + 6.99\varepsilon^3 + 14.63\varepsilon^4 + 26.44\varepsilon^5 + 45.78\varepsilon^6 + 47.34\varepsilon^7 + \ldots\right] \quad (18.58)$$

$$B_{YX} = 0.5P\left[2\log\varepsilon + 8.84\varepsilon + 12.10\varepsilon^2 + 17.37\varepsilon^3 + 24.12\varepsilon^4 + 30.95\varepsilon^5 + 44\varepsilon^6 + \ldots\right] \quad (18.59)$$

The values of damping coefficients determined by the developed approach (Table 18.1) have been compared with the values reported by Lund J. W. and Thomson K. K. [16] under identical conditions of operation (also given in Table 18.1).

The change in value of permittivity/dielectric constant with temperature has been considered for evaluation of the damping and stiffness coefficients of cylindrical

Table 18.1. Values of dimensionless damping coefficients of cylindrical bearing of $L/d = 0.5$, determined by the developed electrical approach, under various operating parameters

S	ε	ϕ°	$\overline{B_{XX}}$	$\overline{B_{XY}}$	$\overline{B_{YX}}$	$\overline{B_{YY}}$
3.937	0.114	83.45	− 0.34	− 1.01	− 2.1	1.83
		(77.32)	(8.58)	(1.93)	(1.93)	(18.44)
2.030	0.207	78.05	1.15	0.88	1.15	1.35
		(68.75)	(4.80)	(2.15)	(2.15)	(11.18)
0.917	0.372	68.16	1.49	1.15	1.55	1.49
		(57.45)	(3.23)	(2.06)	(2.06)	(7.70)
0.379	0.570	55.24	1.29	1.21	1.21	1.81
		(45.43)	(1.89)	(1.87)	(1.87)	(6.76)
0.244	0.655	49.08	1.59	1.29	1.13	2.95
		(40.25)	(1.54)	(1.82)	(1.82)	(6.87)
0.116	0.772	43.78	2.07	1.53	1.44	3.78
		(32.65)	(1.15)	(1.79)	(1.79)	(7.59)

$L/d = 0.5$, $L = 0.15\,\text{m}$, $d = 0.30\,\text{m}$, $\xi = (1.7 - 2.5) \times 8.854 \times 10^{-12} (\phi \text{m}^{-1})$, $\rho = 10^9\,\text{ohm m}$, $\psi = C_d/d = 0.001$, $w = 157\,\text{rad s}^{-1}$.

Note: Values in brackets correspond to the conventional approach, as determined by Lund J. W. and Thomson K. K. (1978)

bearings by using the variation given by Clark, M. Frank [18]. Also, change in oil viscosity with temperature has been considered for the analysis.

The stiffness coefficients K_{XX}, K_{YY}, K_{XY} and K_{YX} have been determined using Eqns. (18.23)–(18.27) for the various values of Sommerfeld number, attitude angle, L/d, eccentricity and clearance ratios. Dimensionless stiffness coefficients given in Table 18.2 have been evaluated using radial clearance and bearing load. The values of stiffness coefficients determined by the developed approach (Table 18.2) has been compared with that of the values reported by Lund J. W. and Thomson K. K. [16] under identical conditions of operation (Table 18.3).

The effective stiffness has been determined by using Eqns. (18.31) and (18.1) for cylindrical bearing using electrical analogy for different values of Sommerfeld number, eccentricity ratio, attitude angle. The data is shown in Table 18.3. Similarly, effective stiffness has been evaluated using conventional approach by using Eqn. (18.29) under the identical conditions. The various values for stiffness and damping coefficients for *XX*, *YY*, *XY* and *YX* directions are taken from the published data by Lund, J. W. and Thomson K. K. [16] for the bearing with C_d/d ratio as 0.001 and L/d as 0.5.

Table 18.2. Values of dimensionless stiffness coefficients of cylindrical bearing of $L/d = 0.5$ by developed electrical approach under various operating parameters.

S	ε	$\phi°$	$\overline{K_{XX}}$	$\overline{K_{XY}}$	$\overline{K_{YX}}$	$\overline{K_{YY}}$
3.937	0.114	83.45	0.42	−0.03	0.70	0.17
2.030	0.207	78.05	0.25	−0.08	0.39	0.14
0.917	0.372	68.16	0.22	−0.17	0.28	0.22
0.379	0.570	55.24	0.24	−0.92	0.20	0.56
0.244	0.655	49.08	0.40	−0.60	0.32	0.76
0.116	0.772	43.78	0.76	3.40	1.80	2.08

18.8 Results and Discussion

In this section damping and stiffness coefficients and effective stiffness have been discussed.

18.8.1 Damping Coefficients

The damping coefficients B_{XX}, B_{YY}, B_{XY} and B_{YX} as derived by electrical analogy, depend on shaft voltage, *L*/*d*, Sommerfeld number, rotating speed frequency, clearance and eccentricity ratios, and permeability of the used lubricant [Eqns. (18.50)–(18.55)].

Comparison of dimensionless values of damping coefficients of the cylindrical bearing, as derived by the developed and the conventional approach, shows similar pattern under identical values of Sommerfeld number, eccentricity ratio and attitude angle, as shown in Table 18.1. The difference in the value of damping coefficient in *YY* direction may be attributed to the circumferential positional difference of location and profile of the oil supply groove to the bearing.

Dimensionless coefficients in *XX* and *YY* directions vary between −0.34 and 2.07, and 1.83 and 3.78, respectively, as the eccentricity ratio changes between 0.114 and 0.772. Similarly, under the same variation of eccentricity ratios, values of dimensionless coefficients in *XY* and *YX* directions match closely with each other (Table 18.1). The difference in these values in *XY* and *YX* directions determined by developed methodology, with respect to that determined by the conventional approach, is because of the mutual electrical interaction of complex nature in *XY* and *YX* directions. Furthermore, variation of shaft voltage in each revolution and its effect on eccentricity ratio has not been considered precisely for analysis. This plays a significant role for the evaluation of the values of damping coefficients. In general,

Table 18.3. Values of stiffness coefficients of cylindrical bearing under various operating parameters

S*	ε*	ϕ°*	K_{XX}*	K_{XY}*	K_{YX}*	K_{YY}*	B_{XX}*	B_{XY}* = B_{YX}*	B_{YY}*	$K^{*}_{cb} \times 10^9$ ($N\,m^{-1}$)	$K_{cbe} \times 10^9$ ($N\,m^{-1}$)
3.937	0.114	77.32	1.89	−4.20	9.27	1.57	8.58	1.93	18.44	0.07	0.09
2.030	0.207	68.75	1.93	−2.5	5.57	1.65	4.80	2.15	11.18	0.11	0.11
0.917	0.372	57.45	1.85	−1.30	4.01	2.12	3.23	2.06	7.70	0.14	0.11
0.379	0.570	45.43	1.68	−0.43	3.64	3.33	1.89	1.87	6.76	0.17	0.12
0.244	0.655	40.25	1.64	−0.13	3.74	4.25	1.54	1.82	6.87	0.17	0.12
0.116	0.772	32.65	1.60	0.30	4.17	6.36	1.15	1.79	7.59	0.16	0.11

$L/d = 0.5$, $L = 0.15\,m$, $d = 0.30\,m$, $\xi = (1.7 - 2.5) \times 8.854 \times 10^{-12}$ ($\phi\,m^{-1}$), $\rho = 10^9$ ohm m, $\psi = C_d/d = 0.001$.

*parameters as per Lund J. W. and Thomson K. K. (1978).

difference between values of damping coefficients determined by the developed and existing method, may be attributed to the difference in the attitude angles and short bearing theory used for the analysis apart from the data of the shaft voltages used.

The values of damping coefficients of the cylindrical bearing have been determined by electrical analogy approach without complex analysis. This leads to establishing the stability regime of the bearings much faster under different operating parameters, as compared to the conventional approach.

18.8.2 Stiffness Coefficients

The stiffness coefficients K_{XX}, K_{YY}, K_{XY} and K_{YX} as derived by electrical analogy depend on shaft voltage, bearing length, clearance and eccentricity ratios, and permeability of the used lubricant [Eqns. (18.23)–(18.27)].

Comparison of dimensionless values of stiffness coefficients of the cylindrical bearing having $L/d = 0.5$, $C_{\mathrm{d}}/d = 0.001$ derived by the developed and the conventional approach indicate matching with acceptable proximity under different values of Sommerfeld number, eccentricity ratio and attitude angle as shown in Tables 18.2 and 18.3.

Dimensionless coefficients $\overline{K_{XX}}$, $\overline{K_{YY}}$ vary from 0.42 to 0.76 and 0.17 to 2.08, respectively, as per the developed methodology as against 1.89–1.60 and 1.57–6.36 respectively, as determined by conventional approach by J. W. Lund as the eccentricity ratio changes between 0.114 and 0.772. Similarly, under the same variation of eccentricity ratios, values of $\overline{K_{XY}}$ and $\overline{K_{YX}}$ coefficients change from −0.03 to 3.4 and 0.70 to 1.80, respectively, by the developed approach as against −4.20 to 0.30, and 9.27 to 4.17, respectively, by the conventional approach (Tables 18.2, 18.3). The difference in $\overline{K_{XY}}$ and $\overline{K_{YX}}$ values determined by developed methodology with that of the conventional approach is because of the mutual electrical interaction of complex nature in *XY* and *YX* directions. Furthermore, variation of shaft voltage in each revolution and its effect on eccentricity ratio and shaft voltage has not been considered precisely for analysis. This plays a significant role for the values of stiffness coefficients. In general, difference between values of stiffness coefficients determined by the developed and existing method may be attributed to the difference in the attitude angles (Tables 18.2 and 18.3) and short bearing theory used for the analysis apart from the values at shaft voltages used.

18.8.3 Effective Stiffness

As per the electrical analogy, it is established that the effective stiffness of a cylindrical bearing depends on bearing capacitance [Eqn. (18.31)] which is a function of operating parameters, bearing design parameters and permittivity of lubricant as shown in Eqn. (18.1). As per the conventional approach it has been established that the effective stiffness depends on the values of stiffness and damping coefficients in *XX*, *XY*, *YY* and *YX* directions as shown in the Eqn. (18.29). The values of effective stiffness under different operating parameters for the cylindrical bearing of *L*/*d* ratio as 0.5 are found to match closely using electrical analogy and conventional approaches. As shown in Table 18.3 effective stiffness as per electrical analogy vary between 0.07×10^9 and $0.17 \times 10^9\,\mathrm{N\,m^{-1}}$ as against between 0.09×10^9 and $0.12 \times 10^9\,\mathrm{N\,m^{-1}}$ by the conventional approach under identical operating conditions.

The value of stiffness coefficients and effective stiffness of the cylindrical bearing are determined by electrical analogy approach much faster without complex analysis. Also, effective stiffness leads to determination of critical journal mass by using Eqn. (18.28) to assess the stability regime for the bearings under different operating parameters.

18.9 Conclusions

Based on the above analysis, the following conclusions are drawn [19–22]:

(1) The stiffness and damping coefficients in different directions of a cylindrical journal bearing depends on eccentricity ratio, design parameters and relative permittivity of lubricant apart from the shaft voltage.

(2) The effective stiffness of cylindrical bearing depends on its capacitance which in turn is a function of bearing operating and design parameters, and permittivity of the lubricant.

(3) Dimensionless stiffness and damping coefficients, and effective stiffness of cylindrical bearings under various operating conditions determined by electrical analogy approach, are found to match with the values determined by the conventional approach, within acceptable proximity.

(4) It is easy and less time consuming to determine stiffness and damping coefficients and effective stiffness by developed electrical analogy approach as compared to the existing conventional approaches.

(5) The stability regime through critical journal mass can be established much faster using electrical analogy approach as against that of the conventional approach.

The technique developed here, and further development and extension of the methodology of electrical analogy, has the potential to deal precisely with the various associated tribological analysis for determination of stiffness and coefficients more reliably, much faster and may lead to close matching with the existing conventional techniques.

REFERENCES

1. Glienicke, J., Han, D. C. and Leonhard, M. (1980). Practical determination and use of bearing dynamic coefficients. *Tribol. Int.*, **13**, 297–308.
2. Knight, J. D. and Barrett, L. E. (1983). An approximate solution technique for multi-lobe journal bearings including thermal effects with comparison to experiments. *ASME Trans.*, **26**(4), 501–508.
3. Shang, L. and Dien I. K. (1989). A matrix method for computing the stiffness and damping coefficients of multi-arc journal bearings. *Tribol. Trans.*, **32**(3), 396–404.
4. Pujara, K. Kewal (1972). *Vibrations for Engineers*, Dhanpat Rai Sons, Delhi.
5. Prashad, H. (1991). Theoretical evaluation of capacitance, capacitive reactance, resistance and their effects on performance of hydrodynamic journal bearings. ASME, *J. Tribol.*, **113**, 762–767.
6. Prashad, H. (1996). Evaluation of dynamic coefficients of a two-lobe journal bearings using an electrical analogy approach. ASME, *J. Tribol.*, **118**, 657–662.
7. Prashad, H. (1997). Assessment of dynamic coefficients of three-lobe journal bearings through evaluation of electrical parameters – A new approach. *BHEL J.*, **18(**1), 40–48.
8. Prashad, H. (1996). Magnetic flux density distribution on the track surface of rolling-element bearings – An experimental and theoretical investigation. *Tribol. Trans.*, **39(**2), 386–391.
9. Bradford, M. (1988). *Report on Shaft Voltage in Electrical Machines*, ERA Technology Limited, UK.
10. Prashad, H. (1990). Theoretical analysis of the effects of instantaneous charge leakage on roller tracks of roller bearings lubricated with high-resistivity lubricants. ASME, *J. Tribol.*, **112(**1), 37–43.
11. Prashad, H. (1991). Theoretical evaluation of reduction in life of hydrodynamic journal bearings operating under the influence of different levels of shaft voltages. *Tribol. Trans.*, **34**(4), 623–627.
12. Prashad, H. (1988). The effects of viscosity and clearance on the performance of hydrodynamic journal bearings. *ASLE Trans.*, **31**(1), 113–119.
13. Prashad, H. (1987). Effects of operating parameters on the threshold voltage and impedance response of non-insulated rolling-element bearings under the action of electric current. *Wear*, **117**, 223–240.
14. Sydney, S. G. (1960). *Electricity and Magnetism*, Green and Co., London.
15. Cameron, A. (1987). *Basic Lubrication Theory*, Wiley Eastern Limited, New Delhi.
16. Lund, J. W. and Thomson, K. K. (1978). A calculation method and data for the dynamic coefficients of oil-lubricated journal bearings, in fluid film bearings and rotor bearing system design and optimization. ASME, New York, 1–28.
17. Nica, Al. (1985). *Sliding Bearings*, Allerton Press Inc., New York.
18. Clark, M. Frank (1962). *Insulating Materials for Design and Engineering Practice*, John Wiley and Sons Inc., New York.

19. Prashad, H. (2003). Evaluation of stiffness coefficients of cylindrical journal bearings by a non-conventional approach. *BHEL J.*, **24**(1), 34–45.
20. Prashad, H. (2002). Evaluation of damping coefficients of cylindrical journal bearings by electrical analogy – A non-conventional approach. *IE(I) J.-MC*, **83**, 72–77.
21. Prashad, H. (2001). Evaluation of damping coefficients of cylindrical journal bearings by electrical analogy – A non-conventional approach. *BHEL J.*, **22**(1), 51–60.
22. Prashad, H. (2000). Evaluation of stiffness coefficients of cylindrical journal bearings by electrical analogy – A non-conventional approach. *IE(I) J.-MC*, **81**, 55–61.

Nomenclature

a, b	bearing constants
B_{XX}, B_{YY}, B_{XY}, B_{YX}	dynamic coefficients in XX, YY, XY and YX directions, respectively, $\mathrm{N\,s\,m^{-1}}$
$\overline{B_{XX}}$, $\overline{B_{YY}}$, $\overline{B_{XY}}$, $\overline{B_{YX}}$	$B_{XX}C_r w/W$, $B_{YY}C_r w/W$, $B_{XY}C_r w/W$, $B_{YX}C_r w/W$, (dimensionless damping coefficients)
B_{cb}	effective damping of bearing, $\mathrm{N\,s\,m^{-1}}$
C_d	diametral clearance
C_r	radial clearance
C_{cb}	effective capacitance of cylindrical bearing ϕ ($\mathrm{A\,s\,V^{-1}}$)
d	shaft diameter, m
e	journal center eccentricity
E	shaft voltage, V
F	force, N
F_X, F_Y	bearing reaction force components in X and Y directions, respectively, N
h_o	minimum film thickness, m
I_b	bearing current
K_{cb}, K_{cbe}	effective stiffness of cylindrical bearing by conventional and electrical approach, respectively, $\mathrm{N\,m^{-1}}$
K_{XX}, K_{YY}, K_{XY}, K_{YX}	stiffness coefficients in XX, YY, XY and YX directions, respectively, $\mathrm{N\,m^{-1}}$
$(K_{XX})_e$, $(K_{YY})_e$, $(K_{XY})_e$, $(K_{YX})_e$	stiffness coefficients by electric force in XX, YY, XY and YX directions, respectively
$\overline{K_{XX}}$, $\overline{K_{YY}}$, $\overline{K_{XY}}$, $\overline{K_{YX}}$	$K_{XX}C_r/W$, $K_{YY}\,C_r/W$, $K_{XY}\cdot C_r/W$, $K_{YX}\,C_r/W$, (dimensionless stiffness coefficients)
L	bearing length, m
M_{crt}	critical journal mass
n	rotational speed, rps

P	bearing operational constant
Q	charge, Coulomb (A s)
r	radius of journal, m
R_{cb}	effective resistance of cylindrical bearing, ohm
t	time, s
S	Sommerfeld number $[(\mu ndL/W)\,(d/C_d)^2]$
V	whirl frequency, rad s^{-1}
w	$2\pi n$, angular speed of rotation, rad s^{-1}
W	static load, N
X, Y	journal center coordinates, m
$\delta = L/d$	length to diameter ratio
ε	eccentricity ratio (e/C_r)
ϕ	attitude angle of cylindrical bearing
ρ	resistivity of lubricant (ohm m)
ξ	permittivity/dielectric constant of lubricant $[\xi_o \xi_r\,(\phi\,m^{-1})]$
ξ_o	permittivity of free space $[8.854 \times 10^{-12}\,(\phi\,m^{-1})]$
ξ_r	relative permittivity of lubricant
$\psi = C_d/d$	diametral clearance ratio

Chapter 19

ELECTRICAL PARAMETERS AND LIFE ESTIMATION OF PIVOTED PAD THRUST BEARINGS UNDER THE INFLUENCE OF DIFFERENT LEVELS OF SHAFT VOLTAGES

19.1 Introduction

For the reliable performance of a thrust bearing, an adequate amount of oil must be provided to the interacting surfaces between the thrust collar/runner and pads under relative motion, which forms variable oil film thickness from leading to trailing edge of the pads [1].

The pivoted pad of a thrust bearing supported by a pivot is free to assume any angle with the rotating collar. The resultant force on the pad acts through a pivot point. However, if the pivot point coincides with the center of pressure of a fixed-pad system, there is no essential difference in both the systems [2]. For the reversible direction of operation, the pivot position is at the mid point of the pad. It is an experimental fact that the pad assumes an angle of inclination even with location at the center due to geometrical imperfections and slight convexity in the pad surface [1]. Also, thermal edge effects, causing a decrease in oil viscosity, change the angle of tilt, like offset pivoted pads during reversible operation [2]. In addition, the angle of tilt causes the variation in the ratio of oil film thickness at leading to trailing edge. This determines the load-carrying capacity of the bearing depending on sliding velocity, pad dimensions and oil viscosity, and establishes the bearing behavior.

It is evident that in thrust bearings, the variation in oil film between pads and thrust collar forms a capacitor of varying capacitance from the leading to trailing edge. This depends on permittivity of the lubricant, pad width, angle of tilt and the ratio of oil film thickness at leading to trailing edge. Besides this, variable oil film thickness offers variation in resistance along the pad profile depending on the resistivity of the oil. Thus, the variable oil film thickness forms the resistor–capacitor (RC) circuit of variable resistance and capacitance between the thrust collar and pads and offers impedance to a current flow in a hydrodynamic thrust bearing. By analyzing the RC circuit, the behavior of a thrust bearing can be predicted. However, similar studies have been carried out for roller and hydrodynamic journal bearings [3, 4], but no such studies have been carried out either for fixed or for pivoted pad thrust bearings.

This chapter brings out the theoretical studies undertaken to determine the capacitance and active resistance between the interacting surfaces of pivoted pads and thrust collar under different conditions of operation. Also, the capacitive reactance is analyzed for the effect of permittivity and the role of capacitance in the behavior of a pad. The effect of the number of pads on equivalent capacitance and resistance of a thrust bearing has also been analyzed. And the effect of different levels of angle of tilt under various operating conditions affecting the ratio of oil film thickness at leading to trailing edge has been computed to analyze how the variation in capacitance, resistance and capacitive reactance of a pivoted pad affect the behavior of thrust bearings.

This chapter also brings out the theoretical studies on the effects of instantaneous charge leakage between the thrust collar and a pad lubricated with high-resistivity oil under the influence of different levels of shaft voltages. The objective is to determine instantaneous rise in temperature of high 'points' between the thrust collar and a tilting pad due to instantaneous charge leakage in the zone of load-carrying oil film in each shaft rotation. The rise in temperature, thus achieved, is used to determine the contact stresses and to establish the reduction in thrust bearing life under the influence of different levels of shaft voltages before the initiation of craters on the liner surfaces as compared to the bearings operating without the influence of shaft voltages.

19.2 Theoretical Determination of Capacitance Between Pivoted Pad and Thrust Collar/Runner of a Bearing

The pivoted pad of a thrust bearing separated by oil film thickness h_2 at the leading and h_1 at the trailing edge from the rotating thrust collar forms a parallel plate capacitor of variable thickness in the presence of a lubricant of dielectric constant ξ. ΔC_p is the change in pad capacitance between two rectangular plates of width dX located in between the distances $X + \mathrm{d}X$ and X from the trailing edge

(Fig. 19.1) along the X-axis i.e. along pad length L, of a pad of width B, along Y-axis, and separated by a film thickness varying between $h + \Delta h$ and h (Fig. 19.1). This is given as:

$$C_{\mathrm{p}} = 2\xi \frac{[(X + dX)B - XB]}{(2h + \Delta h)} \tag{19.1}$$

The change in film thickness (Δh) is very small in comparison to dX, and hence is neglected.

The film thickness will vary from the trailing to leading edge (h_1 to h_2) for an offset pivoted pad along the X-axis (Fig. 19.1). This variation for the pivoted pad is given as [2]:

$$h = mX^2 \tag{19.2}$$

where m is a pad constant.

On differentiating Eqn. (19.2) and substituting in Eqn. (19.1), the change in capacitance can be expressed as:

$$\Delta C_{\mathrm{p}} = \frac{\xi B h^{-3/2} \mathrm{d}h}{2m^{1/2}} \tag{19.3}$$

The maximum change in capacitance depends on the film thickness between the trailing and leading edges, between h_1 and h_2. So on integrating between h_1 and h_2, the capacitance of a pad is given as:

$$C_{\mathrm{p}} = \frac{\xi B\,(h_2^{1/2} - h_1^{1/2})}{(m h_1 h_2)^{1/2}} \tag{19.4}$$

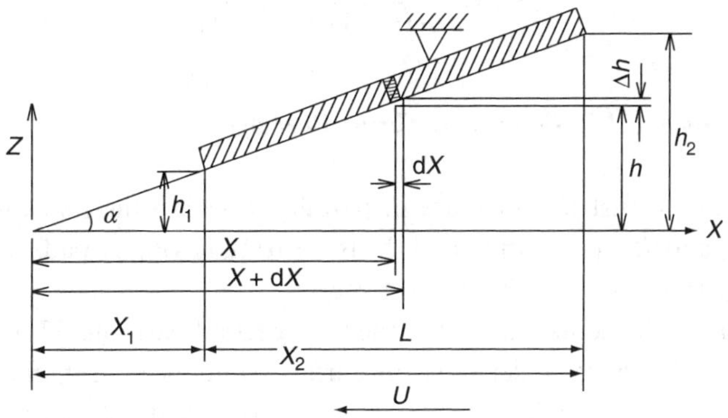

Fig. 19.1. Pad configuration in thrust bearing under rotation

On substituting the values of m, h_2/h_1, and h_2 from Appendix A, the pad capacitance is expressed as:

$$C_p = \frac{\xi B(H+1)(H^{1/2}-1)}{\alpha H} \quad (19.5)$$

19.3 Theoretical Determination of Active Resistance Between Thrust Collar/Runner and a Pivoted Pad

The change in conductance^{-1}/resistance $\Delta F_p^{-1}/\Delta R_p$ between two rectangular plates of width dX located in between the distances $X + dX$ and X from the trailing edge along the X-axis (along the length L) and width B (along Y-axis) of a pivoted thrust pad, and separated by a film thickness varying between $(h + \Delta h)$ and h (Fig. 19.1), is given as [5]:

$$\Delta F_p = \frac{\mathrm{d}XB}{\rho h} \quad (19.6)$$

The maximum change in conductance^{-1}/resistance depends on the film thickness between trailing and leading edges (h_1 and h_2). So on integrating between h_1 and h_2 and using Eqn. (19.2), the pad conductance is determined as:

$$F_p = \frac{1}{R_p} = \left(\frac{B}{2\rho m^{1/2}}\right)\int_{h_1}^{h_2} h^{-3/2}\mathrm{d}h \quad (19.7)$$

On integrating and substituting the values of m, h_2/h_1 and h_2 from Appendix A, R_p is expressed as

$$R_p = \frac{\rho\alpha H}{[B(H+1)(H^{1/2}-1)]} \quad (19.8)$$

19.4 Capacitance and Resistance of Thrust Bearings

The pads of thrust bearings are in parallel to each other, so capacitance and resistance (C_b and R_b) of a tilting pad thrust bearing having N pads, are determined as N times C_p and $1/N$ times R_p, respectively.

In operation, the actual contact in the zone of load-carrying oil film between the thrust collar and the pads depends on surface roughness, irregular oil film and comprises primarily the high 'points' contact. Also, vibrations of the oil film under different conditions of operation govern area of contact. In case of a stationary thrust collar, the actual area of contact is relatively large and consists mainly of quasi-metallic surfaces of the thrust collar and bearing pads. However, in rotation the actual high 'points' contacts between the thrust collar and pads of a bearing is much smaller. Under stationary conditions, resistance between thrust collar and pads is measured as 0.4–0.5 ohms, which enables to pass high current as compared to the operating conditions. Instantaneously, resistance under operating conditions (R_b) can be taken as equivalent to the high 'points' contact resistance under static condition (R_s). This is because for a fraction of a second (10^{-3} s – duration to pass zone of load-carrying oil film – as shown in Table 19.1) at the high 'points' the thrust collar and a pad is treated as stationary during rotation of the collar. This can also be explained because of temporary lack of slip at the asperities due to elastic deformation between the thrust collar and pads.

19.5 Determination of Capacitive Reactance and Impedance of Hydrodynamic Thrust Bearing, and Between Thrust Collar and an Individual Pad

The capacitive reactance between interacting surfaces of a thrust collar and a pad is given as [5]:

$$X_{cp} = \frac{1}{wC_p} \tag{19.9}$$

And the impedance of equivalent (RC) circuit of a pad is determined as:

$$Z_p = R_p - jX_{cp} \tag{19.10}$$

Similarly, capacitive reactance and impedance of a thrust bearing is given as:

$$X_{cb} = \frac{1}{wC_b}, \quad \text{and} \quad Z_b = R_b - jX_{cb} \tag{19.11}$$

The ratio of capacitive reactance to active resistance of a thrust bearing, and an individual pad and thrust collar is expressed as [Eqns. (19.9), (19.11), (19.8) and (19.5)]:

$$\frac{X_{cb}}{R_b} = \frac{X_{cp}}{R_p} = \frac{1}{w\xi\rho} \tag{19.12}$$

19.6 Life Estimation of Pivoted Pad Thrust Bearings

The effect of instantaneous charge leakage on the temperature rise at the contact between the thrust collar and a pivoted (tilting) pad of a hydrodynamic thrust bearing lubricated with high-resistivity lubricant operating under the influence of different levels of shaft voltages has been used to estimate the life of a pivoted pad thrust bearing. Charge leakage between high 'points' of the thrust collar and a pad liner during momentary contact in the zone of load-carrying oil film is used to establish the heat generated and instantaneous temperature rise in each shaft rotation. On the basis of temperature rise, an analysis is carried out for the rise of contact stresses by the instantaneous charge leakage, which leads to initiation of craters by the arcing effects resulting in subsequent damage of pad liner. The consequential reduction in thrust bearing life, in comparison to the bearing operating without the influence of shaft voltage, is then established. Subsequently, the safe limit of shaft voltage is assessed for a reliable operation of a thrust bearing operating under the influence of shaft voltage.

19.6.1 Determination of Width of Contact Between the Thrust Collar and a Tilting Pad

The high 'points' contact between the thrust collar and a tilting pad in the zone of minimum film thickness under the action of load p, form a rectangular surface of deformation on a pad liner having dimensions equal to width of contact (W_p) and pad width (B). The minimum width of contact causing pitting on the liner is expressed as:

$$W_p = \frac{p}{B\sigma_{sp}} \tag{19.13}$$

19.6.2 Duration of Contact Between a Pad and the Thrust Collar

Duration of surface contact between the thrust collar and a pad in crossing the zone of load-carrying oil film is determined as W_p/V_n, and given as:

$$t_p = \frac{p}{\pi B\sigma_{sp} d_m f_s} \tag{19.14}$$

19.6.3 Stored Charge on Bearing Surfaces

The equivalent capacitance of a bearing lubricated with high-resistivity oil (10^{14} ohm cm) opposes any change in the existing voltage and causes an electric charge to be stored. The instantaneous stored charge depends on capacitance and voltage across oil film, and is given as:

$$Q_b = VC_b = VC_pN = Q_pN \tag{19.15}$$

A change in charge results in current flow in the capacitor. It is this current, which opposes a voltage change across the capacitor and is determined as:

$$I_b = C_b \times \frac{dV}{dt} = \frac{NC_p dV}{dt} \tag{19.16}$$

where dV/dt is the rate of voltage change in volt per second.

19.6.4 Voltage Current Characteristics of Electric Arc

The dependence of voltage drop with the arc current is known as voltage-current characteristics. Voltage at which the arc discharge commences is called the ignition or storage voltage of the arc or the bearing threshold voltage [6]. It has been established that the voltage drop with the arc decreases as the arc current increases, which indicates that the gap resistance drops quicker than the current rises. The resistance of the arc gap and voltage drop with the arc for a short duration may be treated as constant and time invariant quantities. Furthermore, there is a great difference between the variation of the arc voltage and voltage gradient in the immediate neighborhood of the electrodes and with the rest of the arc. Also, voltage drops sharply within the cathode and anode region.

The above characteristics of the electric arc is used with the reasonable accuracy to analyze the phenomenon occurring in a hydrodynamical thrust bearing using lubricating oil of higher resistivity (10^{14} ohm cm.) and operating under the influence of different levels of shaft voltages. The make and break of the high 'points' contact between the thrust collar and pads in a load-carrying oil film during a contact duration of 10^{-3} s (Table 19.1) is taken as time invariant and voltage drop is treated as constant. Therefore during operation for an instant the resistance at the high 'points' contact is taken equivalent to that of bearing resistance under static conditions. The effect of shaft voltage on the Babbitt liner is taken up to the depth (H_d) of 0.001 mm as per the details available in the reference [7].

19.6.5 Lubricant Characteristics and Bearing Behavior

When a bearing using high-resistivity (ρ) lubricant (10^{14} ohm cm) having viscosity 43–48 cSt at 40°C and viscosity index 95 (min) with relative permittivity (ξ_r) of 2.2 operates under electric fields, the current does not begin to flow until the first threshold voltage is reached. At the first threshold voltage (the increase in voltage from zero to a specific maximum value with negligible current flow), the increase in current flow is nearly instantaneous. Beyond the first threshold voltage, even a further small increase in voltage results in a very high increase in current similar to journal and rolling-element bearings [6].

19.6.6 Heat Generated and Instantaneous Temperature Rise of Bearing Liner by Instantaneous Charge Leakage in the Zone of Load-Carrying Oil Film Between the Thrust Collar and Pads of a Thrust Bearing

When a bearing lubricated with high-resistivity oil is operated under the influence of shaft voltage, current does not flow through the bearing and charge gets accumulated on the surface of thrust collar until the shaft voltage is below the threshold voltage [6]. However, slow passage of charge takes place through the bearing when the high 'points' on the collar and the pads in the zone of load-carrying oil film come close or when the conducting particles bridge the oil film. These conducting paths are broken when high points become separated by greater thickness of oil film. Under such conditions arcing results due to instantaneous charge leakage under different operating conditions by vibration effects due to various reasons. This gives rise to instantaneous minimum resistance between the thrust collar and pads equivalent to that of thrust bearing resistance under static conditions (R_s).

The instantaneous charge leakage between the thrust collar and a pad generates an instant heat. The electrical energy q due to charge leakage Q for a duration t_p at the first threshold voltage V through a contact resistance R_s/N between the thrust collar and pads having capacitance C_p is given as:

$$q = \frac{I_p^2 \times R_s \times t_p}{N} \tag{19.17}$$

or in terms of charge and capacitance electrical energy can be expressed as [using Eqn. (19.15)]:

$$q = \frac{Q_p^2 \times t_p \times N}{C_p^2 \times R_s} \tag{19.18}$$

This energy influences the zone of load-carrying oil film (of width of contact W_p, pad width B and depth H_d), and increases instantaneous temperature of the mass $(W_p B H_d \rho_1)$ of the pad temperature after a certain number of cycles. Hence, by using Eqn. (19.18) and neglecting loss of heat by oil flow and other means, the following is evident

$$W_p \times B \times H_d \times T_{ip} \times \rho_1 \times c = \frac{Q_p^2 \times t_p \times N}{C_p^2 \times R_s} \tag{19.19}$$

So, the instantaneous temperature rise due to instantaneous charge leakage between the thrust collar and a pad on the high 'points' in the zone of load-carrying oil film during each contact is determined as [using Eqns. (19.13) and (19.14)]:

$$T_{\mathrm{ip}} = \frac{Q_{\mathrm{p}}^2 \times N}{\pi \times C_{\mathrm{p}}^2 \times f_{\mathrm{s}} \times R_{\mathrm{s}} \times B \times H_{\mathrm{d}} \times d_{\mathrm{m}} \times \rho_1 \times c} \tag{19.20}$$

19.6.7 Instantaneous Temperature Rise Due to Charge Leakage in the Zone of Load-Carrying Oil Film Between Thrust Collar and Pad in each Shaft Rotation

The number of times the thrust collar passes over the W_{p} in each shaft rotation is $\pi d_{\mathrm{m}}/W_{\mathrm{p}}$. Assuming that 0.2 per cent of the thrust collar has high 'points' which come in contact with the high 'points' of W_{p} – the width of contact on a pad liner in the zone of load-carrying oil film, which gives the instantaneous temperature rise at any single position on a pad liner in each shaft rotation as T_{ip} times the $\pi 10^3 d_{\mathrm{m}}/2W_{\mathrm{p}}$. This is simplified using Eqn. (19.20) and given as:

$$T_{\mathrm{p}} = \frac{10^3 \times Q_{\mathrm{p}}^2 \times N}{2 \times C_{\mathrm{p}}^2 \times f_{\mathrm{s}} \times R_{\mathrm{s}} \times W_{\mathrm{p}} \times B \times H_{\mathrm{p}} \times \rho_1 \times c} \tag{19.21}$$

For a bearing having N pads, the net charge (Q_{b}) and capacitance of a bearing (C_{b}) will be N times Q_{p} and C_{p} respectively, since the pads are in parallel with respect to the collar. Also, net pad width, and width of high 'points' contact, and H_{d} will be N times B, W_{p} and H_{d}, respectively. Substituting values of BW_{p} and H_{d} along with the values of Q_{b} and C_{b}, the instantaneous rise of thrust bearing temperature in each shaft rotation is given as [using Eqn. (19.15)]:

$$T_{\mathrm{b}} = \frac{10^3 \times Q_{\mathrm{p}}^2}{2C_{\mathrm{p}}^2 \times R_{\mathrm{s}} \times N^2 \times B \times W_{\mathrm{p}} \times f_{\mathrm{s}} \times H_{\mathrm{d}} \times \rho_1 \times c} \tag{19.22}$$

This instantaneous temperature rise would be identical between various high 'points' contact of the thrust collar and liner of thrust bearing pads in each rotation of the shaft. This could cause initiation of craters on the liner after a certain number of shaft revolutions.

19.6.8 Thermal Stresses Due to Thermal Transients on Bearing Liner

During operation of a thrust bearing, firstly from the initial conditions a transient thermal gradient due to charge leakage occurs. If at an instant, pad liner temperature is known, then transient thermal gradients after various time intervals can be determined. In addition, the thermal stresses can be evaluated for any instant of time. The overall temperature rise of bearing liner due to charge leakage for a short interval (10^{-3} s) approaches zero. However, instantaneous rise in contact stresses on the high 'points' of the pad liner are determined as [8]:

$$\sigma_{\rm ip} = \frac{-\Gamma \times E \times T_{\rm ip}}{(1 - \mu)} \tag{19.23}$$

and initial contact stresses without the influence of shaft voltage and charge leakage are given as:

$$\sigma_{\rm o} = \frac{-\Gamma \times E \times T_{\rm o}}{(1 - \mu)} \tag{19.24}$$

$\sigma_{\rm ip}$ is compressive during increase in temperature. The Babbitt liner temperature at the high 'points' contact with the thrust collar tends to grow instantaneously, but is restricted by adjacent material at a lower stabilized temperature. However, compressive stresses at the pad liner change to gradual tensile stresses at the bonded surface of liner with the bearing shell.

The operating temperature of a thrust bearing affects the contact stresses. But the instantaneous temperature rise of liner due to charge leakage in each shaft rotation may set residual contact stresses which are determined as:

$$\sigma_{\rm b} = \frac{-\Gamma \times E \times (T_{\rm b} + T_{\rm o})}{(1 - \mu)} \tag{19.25}$$

19.6.9 Bearing Damage Resulting From Shaft Voltages due to Instantaneous Charge Leakage

Shaft voltages and resulting bearing currents lead to damages caused by arcing across bearing pads and thrust collar. The main consequence of arcing in presence of high-resistivity oil is the wear of Babbitt liner and rapid aging of the lubricating oil. It is observed that oil films in the hydrodynamic thrust bearings cease to be insulators and behave like conductors when subjected to high voltages or fast changing voltages [9]. The voltage at which this happens is similar to threshold voltage phenomenon occurring in rolling-element bearings [6]. Also, this depends on oil film thickness, rate of change of voltage and is accompanied by an arc.

The current flow is intermittently broken as the high 'points' are separated by the motion of the collar, and this interruption takes place in a microsecond or less which causes high induced voltage and accumulation of charges across the oil film due to self inductance. Charge accumulation depends on bearing capacitance and may cause sustained arcing at the high 'points' contact between the thrust collar and the bearing pads resulting in metal fusing and damage of liner surface by craters formation. However, the collar is less sensitive to arcing effects due to its higher melting point.

19.6.10 Determination of Decrease in Life Span of Pad Liners due to the Initiation of Craters Under Various Levels of Shaft Voltages

Under ideal conditions of operation the life of the pad liners is very high because of hydrodynamical pressure developed in the squeeze film, which avoids direct metal-to-metal contact between the thrust collar and pad liners. However, during starts and stops there occurs shear force on the liners, which affects the fatigue life of pads. Also, oxidation, deterioration and aging of the circulating oil affect the bearing performance.

Due to charge leakage on the high 'points', the contact stresses increase in due course on the pad liners. The ratio of these stresses as compared to the original stresses without the effect of shaft voltages is determined as [by Eqns. (19.24) and (19.25)]:

$$\frac{\sigma_b}{\sigma_o} = 1 + \frac{T_b}{T_o} \tag{19.26}$$

The ratio of σ_b/σ_o gives the indication of decrease in the life of pad liners under the effect of shaft voltages as compared to the original life. And ratio of percentage increase in stresses as compared to the original stresses [$(\sigma_b - \sigma_o)/\sigma_o$ per cent] gives the percentage reduction in life of the pad liners on the high 'points' before the initiation of craters.

19.7 Data Deduction From Theoretical Analysis

Values of C_p, R_p, X_{cp}, and Z_p for the pivoted pad of the thrust bearing with different values of H varying from 2 to 5, and angle of tilt (α) varying from 1° to 3° have been computed using Eqns. (19.5), (19.8)–(19.10). Figures. 19.2–19.4 show the variation of capacitance, resistance and capacitive reactance of the pivoted pad at various angles of pad tilt and ratios of oil film thickness at leading to trailing edge of the thrust bearing having a pad width of 200 mm. Also, σ_b is determined for different values of shaft voltages varying from 0.25 to 2 V to determine the ratio of σ_b/σ_o of a bearing operating under stabilized regime and having parameters given in Table 19.1. The shaft voltages are taken less than first threshold voltage to avoid the effect of chemical decomposition of the oil. The various

Table 19.1. Instantaneous temperature rise, contact stresses, contact duration and reduction in bearing liner life due to charge leakage between the thrust collar and pads under the influence of different levels of shaft voltages and charge accumulation

Analytical values of temperature rise at different levels of charge accumulation and shaft voltages								
	(1)	(2)	(3)	(4)	(5)	(6)	(7)	(8)
V (V)	0.2	0.5	0.75	1.0	1.25	1.5	1.75	2
$Q_b \times 10^{-10}$ (C)	3.3	6.6	9.9	13	16.5	19.8	23	26
T_b° (C)	2.7	10.8	24.2	43	67.2	97	132	172
σ_b/σ_o	0.95	0.82	0.67	0.54	0.43	0.34	0.27	0.22
% reduction in bearing life	5	18	33	46	57	66	73	78
Ratio of temperature rise as against 0.25 V	1	4	9	16	25	36	49	64
Ratio of reduction in life as against 0.25 V	1	3.6	6.6	9.2	11.4	13.2	14.6	15.6

Bearing dimensions, operating parameters and analytical values $d_m = 300$ mm, $B = 200$ mm, $n = 1500$ rpm, $N = 8$, $f_s = 25$, $p = 12 \times 10^{14}$ N, $\rho_1 = 10^{14}$ ohm cm, $H_d = 0.001$ mm, $R_s = 0.5$ ohm, $\xi = 2.5 \times 8.85419 \times 10^{-12}$ (ϕ m^{-1}), $\xi_r = 2.5$, $T_o = 50\,^{\circ}$C, $\alpha = 1^{\circ}$, $H = 2.2$, $C_p = 16.3 \times 10^{-11}$ (ϕ), $C_b = 13.04 \times 10^{-10}$ (ϕ), $W_p \times B = 8.68 \times 10^{-3}$ m^2, $t_p = 1.8 \times 10^{-3}$ s.

parameters of a thrust bearing having eight pads of width 200 mm each are given in Table 19.1 [10]. The specific pressure (σ_{sp}) causing pitting on the tin-based Babbitt material is taken as $14 \times 10^6\,\mathrm{N\,m^{-2}}$ [11]. And by the field data analysis of various bearings operating under normal conditions, the stabilized bearing temperature is taken as 50°C.

19.8 Results and Discussion on Electrical Parameters and Bearing Life

19.8.1 Capacitance

The capacitance of the pivoted pad of a thrust bearing depends on the ratio of oil film thickness at the leading to trailing edge (H), angle of pad tilt (α), bearing width (B) and permittivity of lubricant (ξ), and is independent of pad length [Eqn. (19.5)]. Capacitance of a bearing is the product of the number of pads and the capacitance of an individual pad.

The capacitance of a pivoted pad decreases with an increase in the angle of tilt at the constant ratio of film thickness from the leading to trailing edge. However, the pad capacitance increases with increase in H at a fixed value of pad tilt. The relative increase of capacitance at the lower angle of tilt ($< 1°$) is much higher as compared to higher tilt ($> 2.5°$) as H varies from 2 to 5. This indicates that the capacitance of a centrally pivoted pad is much higher as compared to the capacitance of equivalent off set pivoted pad, (due to the lower value of pad tilt of a centrally pivoted pad as compared to the off set pivoted pad). The capacitance of a pad asymptotically decreases as the angle of tilt increases beyond 1.5° (Fig. 19.2).

For a pad of width 200 mm, the capacitance at $H = 2.3$ decreases from 331.0 pF at $\alpha = 0.5°$ to 55.2 pF at $\alpha = 3°$. On the contrary, at the constant angle of tilt ($\alpha = 0.5°$), the capacitance increases from 277.2 pF at $H = 2$ to 662 pF at $H = 5$. However, the relative variation of capacitance at $\alpha = 0.5°$ is 384.8 pF as against 64.1 pF at $\alpha = 3°$ as H varies from 2 to 5. In general, capacitance of a pad reduces asymptotically as the pad tilt increases from 0.5° to 1.5° and to 3°. The change is from 434.24 pF to 217.12 pF and 72.38 pF at $H = 3$ as shown in Fig. 19.2.

19.8.2 Resistance

The resistance of a pivoted pad depends on H, ρ, α, B, and is independent of length of the pad. The resistance of a thrust bearing decreases by number of pads multiplied by the resistance of an individual pad. The pad resistance increases with pad tilt at the constant value of film thickness from leading to trailing edge (H). However, resistance decreases with increase in H of the fixed value of angle of tilt. The relative change in resistance at a lower angle of pad tilt ($< 1°$) is

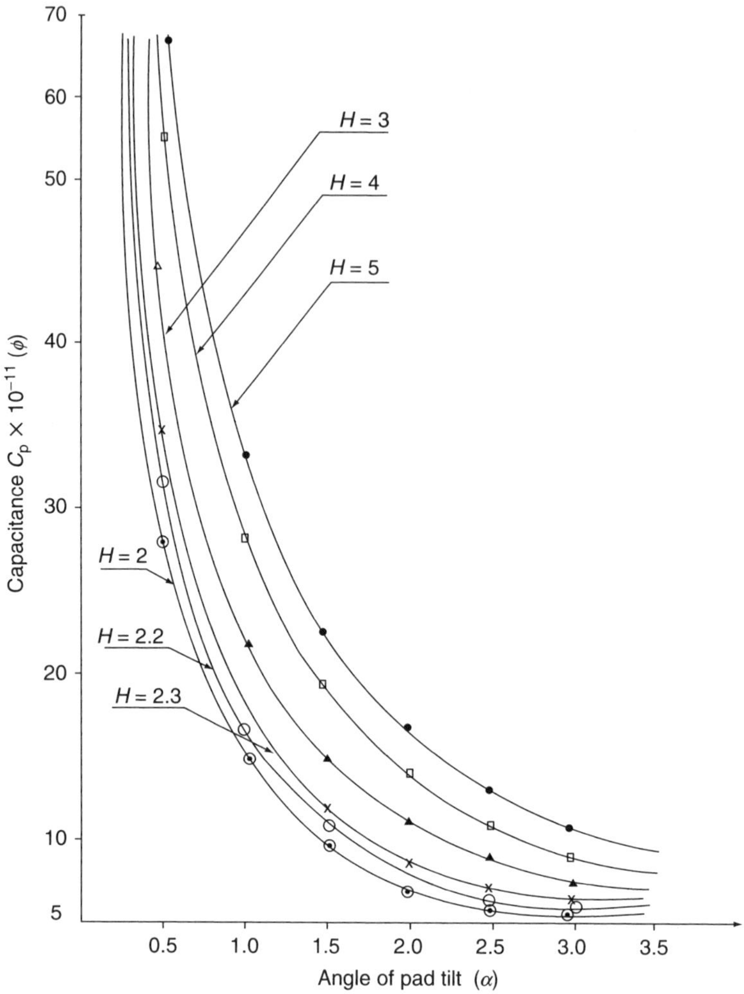

Fig. 19.2. Variation of capacitance (C_p) of pivoted pad with angle of tilt (α) at different ratios of film thickness at leading to trailing edges (H) of a thrust bearing of pad width 200 mm

much less as compared to higher angle of tilt (> 2.5°) as the H changes from 2 to 5 (Fig. 19.3).

For the pad width of 200 mm, the active resistance at $H = 2.3$ increases from 5.9×10^{10} ohm at $\alpha = 0.5°$ to 33.6×10^{10} ohm at $\alpha = 3°$. On the contrary, at constant $\alpha = 1.5°$, the resistance decreases from 21.2×10^{10} ohm at $H = 2$ to 8.75×10^{10} ohm at $H = 5$. The relative variation of pad resistance at $\alpha = 0.5°$ is

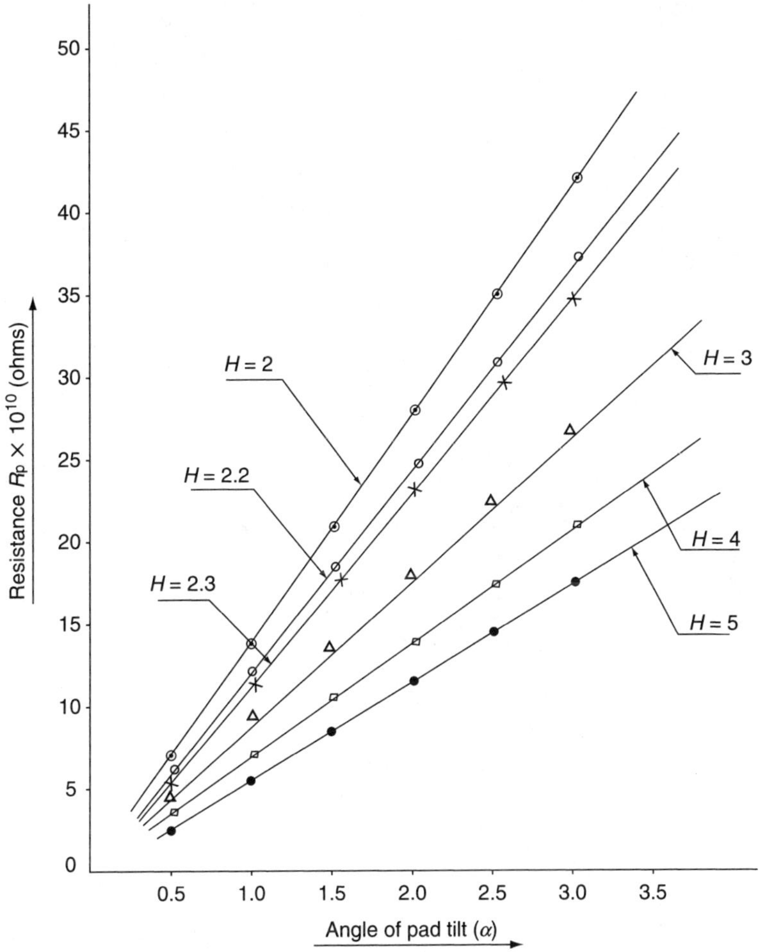

Fig. 19.3. Variation of resistance (R_p) of pivoted pad with angle of tilt (α) at different ratios of film thickness at leading to trailing edges (H) of a thrust bearing of pad width 200 mm

4.1×10^{10} ohm as compared to 24.6×10^{10} ohm at $\alpha = 3^\circ$ as the H varies from 2 to 5. In general, pad resistance increases at a uniform rate as α increases from 0.5° to 1.5° and to 3°. R_p increases from 4.45×10^{10} to 13.35×10^{10} and to 26.7×10^{10} ohms at $H = 3$ as shown in Fig. 10.3. Since α is much less for a centrally pivoted pad [2], so its resistance is much lower as compared to an equivalent off set pivoted pad (Fig. 19.3).

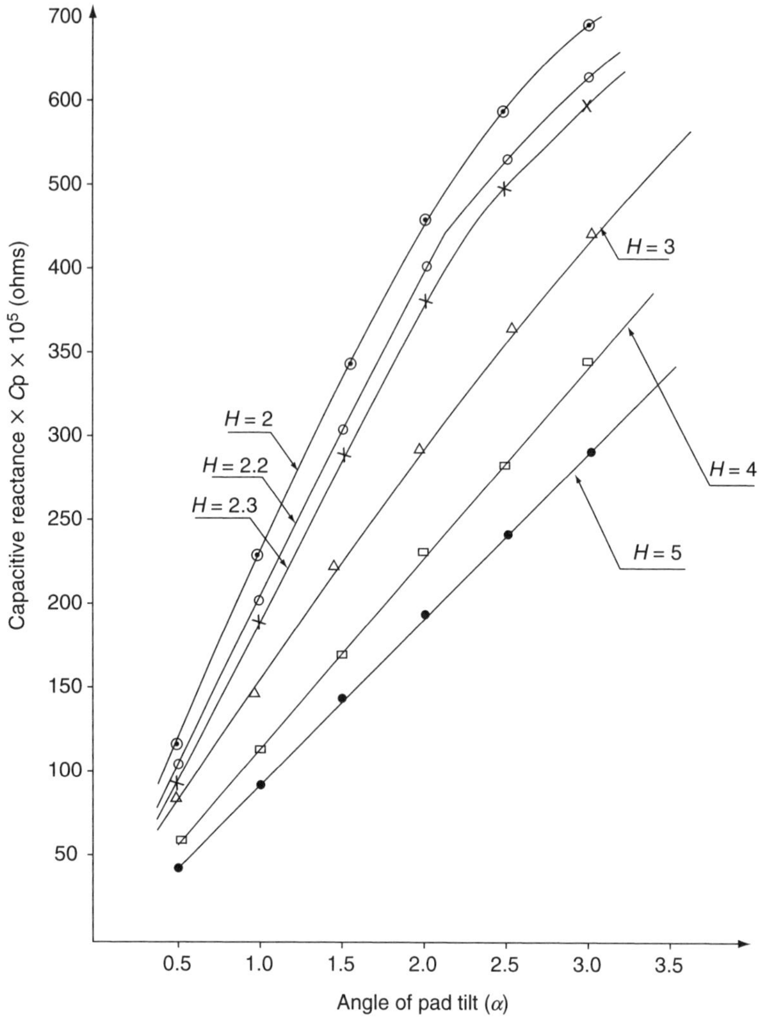

Fig. 19.4. Variation of capacitance reactance (X_{cp}) of pivoted pad with angle of tilt (α) at different ratios of film thickness at leading to trailing edges (H) of a thrust bearing of pad width 200 mm

19.8.3 Capacitive Reactance

The capacitive reactance of a pivoted pad (X_{cp}) reduces with increase in capacitance and frequency of applied voltage. Also, capacitive reactance of a bearing (X_{cb}) reduces with increase in number of pads times the capacitive reactance of an individual pad.

In general, X_{cp} like R_p increases with an increase in pad tilt at the constant ratio of film thickness at the leading to trailing edge (H). However, X_{cp} decreases with an increase in H at a constant value of pad tilt (Fig. 19.4). Also, the relative change in X_{cp} at a lower value of (<1°) is much less as compared to a higher value of pad tilt (>2.5°) as H changes from 2 to 5 as shown in Fig. 19.4. For a pad of width 200 mm, X_{cp} at $H = 2.3$ increases from 96.21×10^5 ohm at $\alpha = 0.5°$ to 577.36×10^5 ohm at $\alpha = 3°$ for $f = 50$ Hz. On the contrary at a constant value of $\alpha = 1.5°$, X_{cp} decreases from 344.71×10^5 ohm at $H = 2$ to 144.43×10^5 ohm at $H = 5$. But, the relative variation of X_{cp} at $\alpha = 0.5°$ is 66.74×10^5 as compared to 401.62×10^5 ohm at $\alpha = 3°$ as the H varies from 2 to 5. In general, X_{cp} of a pad increases with a uniform rate as α increases from 0.5° to 1.5° and to 3°. The change is from 73.34×10^5 ohm to 220.01×10^5 ohm and to 440×10^5 ohm at $H = 3$ as shown in Fig. 19.4. This indicates that the capacitive reactance of a centrally pivoted pad is lower than that of an equivalent off set pivoted pad.

19.8.4 Impedance

Resistance of a pivoted pad of the thrust bearing using oil of resistivity 10^{14} ohm cm and relative permittivity of 2.2 is approximately 10^4 times higher than its capacitive reactance (Figs 19.3 and 19.4). That is why the pad impedance (Z_p) is governed by the pad resistance (R_p), and contribution of capacitive reactance to the impedance is insignificant as determined by Eqn. (19.10). Similarly, impedance of bearing (Z_b) is governed by its resistance (R_b) as determined in Eqn. (19.11).

19.8.5 Variation of Capacitance, Resistance and Capacitive Reactance of the Pads for Optimum Performance of a Thrust Bearing

In general, the load-carrying capacity of a pad depends on pad dimensions, the operating speed of bearing and the lubricant viscosity [1]. These parameters govern oil film thickness at leading to trailing edge and the angle of tilt of the pad [1, 2]. The variation of H and angle of tilt of pad influence C_p, R_p and X_{cp}, which have been discussed in the previous sections. For lower coefficient of friction and higher load-carrying capacity the optimum variation of these parameters should be taken against $H = 2.3$ and $\alpha = 1°$ to 3° [1, 2]. The variation of capacitance, resistance and capacitive reactance of a pivoted pad of 200 mm width would be within 165.5 to 55.16 pF, 11.8×10^{10} to 33.6×10^{10} ohm and 192.43×10^5 to 577.36×10^5 ohm, respectively, for optimum performance characteristics (Figs. 19.2–19.4).

19.8.6 Ratio of Capacitive Reactance to Active Resistance

The ratio of capacitive reactance to active resistance of a pad and bearing depends on w, ρ, ξ and [Eqn. (19.12)]. It is independent of operating parameters and values of H and α. The ratio of capacitive reactance to active resistance is constant and is less than unity (1.64×10^{-4}) up to the first threshold voltage for a bearing using oil having resistivity 10^{14} ohm cm and relative permittivity of 2.2. This shows that the bearing accumulates the electric charges up to the first threshold voltage [3, 4].

19.8.7 Temperature Rise in the Zone of Load-Carrying Oil Film Between Thrust Collar and a Pad

The temperature rise in the zone of load-carrying oil film between the collar and a pad depends on the pad width, width of contact on pad in load-carrying oil film and the number of pads in a bearing. Temperature rise of a thrust bearing under the influence of different levels of shaft voltages varies inversely with the square of the number of pads and decreases with increase in frequency of rotation, but increases with the square of ratio of charge accumulation to the bearing capacitance.

For a thrust bearing having eight pads, each of width 200 mm and capacitance of $16.3 \times 10^{-11}(\phi)$ operating at 1500 rpm, the temperature rise at the high 'points' on the bearing surfaces increases as the charge accumulation/shaft voltage increases. As the shaft voltage increases from 0.25 to 2 V, temperature rise at the high 'points' in each shaft rotation (T_b) increases from 2.69 to 172 °C (Table 19.1). Up to shaft voltage of 0.5 V, T_b is 10.8 °C, which increases steeply to 4, 6.25, 9.0, 12.24 and 16 times as the shaft voltage rises to 1, 1.25, 1.5, 1.75 and 2 V respectively (Table 19.1). As compared to 0.25 V shaft voltage, temperature of thrust bearing increases to 4 to 64 times as the shaft voltage increases from 0.5 to 2 V (Table 19.1). However, below 0.25 V, temperature rise as well as rate of rise are much less as compared to that of above 0.25 V. The similar trend is found for various sizes of bearings under different conditions of operation.

19.8.8 Ratio of Thermal Stresses and Percentage Reduction in Bearing Life

Instantaneous thermal stresses due to thermal transients on the bearing surfaces depend on instantaneous rise in temperature [Eqn. (19.23)], which increases the contact stresses in due course [Eqn. (19.25)]. The ratio of original stresses without the effect of instantaneous charge leakage/shaft voltage to that of the bearings operating under different levels of shaft voltages (σ_o/σ_b) decreases from 0.95 to 0.22 as the shaft voltage increases from 0.25 to 2 V. Consequently, $(\sigma_b - \sigma_o)/\sigma_o$

per cent i.e. minimum reduction in bearing life before appearance of craters on liner increases from 5 to 78 per cent. However, up to 0.25 V shaft voltages, reduction in bearing liner life is only 5 per cent. Exceeding which liner life reduces very fast. The bearing life reduces from 3.6 to 15.6 times at the shaft voltage of 0.5 and 2 V, respectively as compared to that at 0.25 V shaft voltages. Below shaft voltage of 0.25 V, reduction in bearing life is very much less than at above 0.25 V (Table 19.1). It is therefore recommended that shaft voltage should not be allowed to exceed 0.25 V for safe and reliable operation of a hydrodynamic thrust bearing. This matches closely with the operational experience with the thrust bearings.

It may be noted that under the influence of lower shaft voltage the life expectancy of a thrust bearing is much higher before the craters are initiated on the liner. This is because of lesser rise in contact temperature on the high 'points'. In addition, decrease in charge accumulation and increase in bearing capacitance also result in lesser rise in contact temperature in the load-carrying zone of a bearing.

19.8.9 Comparison of Safe Shaft Voltage Between Journal and Thrust Bearings

Under the influence of same shaft voltage, failure of thrust bearing occurs much earlier than in case of the journal bearing. For the safe and reliable operation of a journal bearing, the safe reliable shaft voltage is 0.5 V as discussed in Chapter 15 [9] as against 0.25 V for thrust bearing (Table 19.1). This is attributed to the fact that, although, the duration of asperity high 'points' contacts of a pad with the thrust collar (t_p) is lower (1.8×10^{-3} s), but it occurs simultaneously in all the pads and increases to the product of the number of pad times the t_p. This makes it approximately three times higher than that of high 'points' contact in a journal bearing in each shaft rotation (5×10^{-3} s). Hence, in reduction in life of a thrust bearing as compared to that of a journal bearing [9].

19.9 Conclusions

From the various analysis and investigations discussed in this chapter, the following conclusions are drawn [10, 12]:

(1) The capacitance, resistance and capacitive reactance of a pivoted pad of a thrust bearing depend on the ratio of film thickness from the leading to trailing edge, angle of tilt and pad width and do not depend on pad length.

(2) The capacitance of the pad decreases with increase in angle of pad tilt versus the resistance and capacitive reactance, which increase, at the constant ratio of film thickness from the leading to trailing edge.

(3) The relative increase of capacitance versus the decrease in resistance and capacitive reactance, with an increase in film thickness from the leading to trailing edge at the constant angle of pad tilt, is much higher for lower angle of tilt as compared to a higher angle of tilt.

(4) The capacitance of a bearing increases with the number of pads times the capacitance of an individual pad, as compared to resistance and capacitive reactance, which decrease with the number of pads.

(5) For a centrally pivoted pad, capacitance is higher versus the resistance and capacitive reactance, which are lower as compared to an equivalent off set pivoted pad.

(6) For lower coefficient of friction and higher load-carrying capacity, the range of variation of capacitance, resistance and capacitive reactance of a pivoted pad of 200 mm width is 165.5–55.16 pF, $11.8 \times 10^{10} - 33.6 \times 10^{10}$ ohm and $192.43 \times 10^{5} - 577.36 \times 10^{5}$ ohm, respectively.

(7) The resistance of a tilted pad using oil of resistivity 10^{14} ohm cm and relative permittivity of 2.2 is approximately 10^4 times higher than its capacitive reactance. Also, impedance of a pad is mainly governed by its resistance rather than by its capacitive reactance.

(8) The ratio of capacitive reactance to active resistance depends on permittivity and resistivity of the lubricant, and this ratio is less than unity for the capacitive response of a pad and a thrust bearing.

(9) The width of contact between a pad and the collar in the load-carrying oil film depend on the liner properties, load on pad and the specific load causing pitting on liner surface.

(10) The instantaneous temperature rise of a pad in each shaft rotation due to charge leakage in the zone of load-carrying oil film depends on the pad width, width of contact on pad in load-carrying oil film, the number of pads in a bearing and frequency of rotation of the collar.

(11) Temperature rise under the influence of shaft voltage decreases with increase in frequency of rotation and varies inversely with the square of number of pads. However, it increases with square of the ratio of charge accumulation to the bearing capacitance.

(12) Instantaneous thermal stresses due to thermal transients on the bearing liner depend on instantaneous temperature rise due to charge leakage.

(13) Temperature rise of liner below 0.25 V shaft voltages is much less and does not have significant effect on the bearing life. Reduction in bearing life is 5 per cent at 0.25 V as against 78 per cent at 2 V shaft voltage.

(14) Reduction in the bearing life becomes increasingly significant as the shaft voltage goes up. At the shaft voltage of 2 V, the percentage reduction in bearing life is 15.6 times as much as that at 0.25 V.

(15) For safe, reliable operation and adequate life of a thrust bearing, shaft voltage of 0.25 V must not be exceeded. However, for a journal bearing shaft voltage up to 0.5 V is safe.

The analysis given in this chapter has a potential to predict the reduction in bearing liner life before the craters are initiated on its surface and influence of shaft voltage, capacitance and operating parameters on the bearing performance. And, also has a potential for optimum selection of pad parameters for safe and reliable operation of a thrust bearing.

REFERENCES

1. Gross, A., William (1975). *Fluid Film Lubrication*, A Wiley-Interscience Publication, John Wiley and Sons Inc., New York, USA.
2. Moore, F., Desmond (1975). *Principles and Applications of Tribology*, Pergamon Press, Oxford, New York, USA.
3. Prashad, H. (1988). Theoretical evaluation of impedance, capacitance, and charge accumulation on roller bearings operated under electrical fields. *Wear*, **125**, 223–239.
4. Prashad, H. (1991). Theoretical evaluation of capacitance, capacitive reactance, resistance and their effects on performance of hydrodynamic journal bearings. ASME, *J. Tribol.*, **113**, 762–767.
5. Lockhart, N. L. and Rice, O. E. (1976). *AC. Circuit Analysis*, Van Nostrand Reinhold, New York.
6. Prashad, H. (1987). Effects of operating parameters on the threshold voltages and impedance response of non-insulated bearings under the action of electrical currents. *Wear*, **117**, 223–240.
7. Rodstein, L. (1974). *Electrical Control Equipment*, Mir Publishers, Moscow.
8. Warnock, F. V. and Benham, P. P. (1965). *Mechanism of Solids and Strength of Materials*, Sir Isaac Pitman and Sons Ltd., London.
9. Prashad, H. (1991). Theoretical evaluation of reduction in life of hydrodynamic journal bearings operating under the influence of different levels of shaft voltages. *STLE*, **34**(4), 623–627.
10. Prashad, H. (1992). An approach to evaluate capacitance, capacitive reactance and resistance of pivoted pads of a thrust bearings. STLE, *Tribol. Trans.*, **35**(3), 435–440.
11. Kragelsky, I. V. (1981). Friction and Wear Lubrication, *Tribology Hand Book*, Mir Publishers, Moscow.
12. Prashad, H. (1992). Analysis of the effects of shaft voltages on life span of pivoted pad thrust bearings. *BHEL J.*, **13**(2), 1–12.

APPENDIX A
MATHEMATICAL DERIVATION OF CONSTANTS *m* AND h_2 IN THE FORM OF α, *L* AND *H*

Variation of film thickness (h) along X-axis is given as [Eqn. (19.2)]:

$$h = mX^2 \tag{A1}$$

From the Fig. 19.1, the following is evident

$$h_1 = mX_1^2 \tag{A2}$$

$$h_2 = mX_2^2 \tag{A3}$$

$$h_2 - h_1 = \alpha L \tag{A4}$$

$$\frac{h_2}{h_1} = H \tag{A5}$$

and

$$X_2 - X_1 = L \tag{A6}$$

From Eqns. (A1), (A2) and (A6), following is evident

$$h_2 - h_1 = mLX_2\left(1 + \frac{X_1}{X_2}\right) \tag{A7}$$

Using Eqns. (A2) and (A4), Eqn. (A7) is simplified as

$$\alpha = \left(\frac{m}{h_2}\right)^{1/2} (h_1 + h_2) \tag{A8}$$

where

$$\frac{X_1}{X_2} = \frac{h_1}{h_2} \text{ (Fig. 19.1)}$$

On simplifying Eqn. (A8), m is determined as:

$$m = \frac{\alpha H(H-1)}{L(H+1)^2} \tag{A9}$$

where h_2/h_1 is used as H and h_2, h_1 are simplified using Eqn. (A4), and given as:

$$h_1 = \frac{\alpha L}{(H+1)} \quad \text{and} \quad h_2 = \frac{\alpha LH}{(H-1)} \tag{A10}$$

APPENDIX B

PROPERTIES OF SAE 11 (Babbitt Material)

A. Chemical Composition

Tin	86%,	Antimony	6–7.5%
Lead	0.35%,	Copper	5–6.5%
Iron	0.08%,	Arsenic	0.10%
Bismuth	0.08%,	Zinc/Aluminum	Nil

B. Physical Properties

(1) Specific gravity (ρ) $= 7.46 \times 10^3 \text{Kg m}^{-3}$

(2) Specific heat (c)	$= 2.26 \times 10^2$ ws Kg^{-1} $°C^{-1}$
(3) Yield strength at 20 °C	$= 46.50 \times 10^6$ N m^{-2}
(4) Yield strength at 100 °C	$= 22.19 \times 10^6$ N m^{-2}
(5) Young's modulus of elasticity (E)	$= 40 \times 10^9$ N m^{-2}
(6) Poisson's Ratio (μ)	$= 0.36$

Nomenclature

B	pad width
c	specific heat of bearing liner
C_b	bearing capacitance
C_p	pad capacitance
d_m	thrust ring mean diameter
E	Young's modulus of elasticity
f	frequency of applied voltage (50 Hz)
f_s	shaft rotational frequency
F_b	bearing conductance (ohm^{-1})
F_p	pad conductance (ohm^{-1})
h	oil film thickness
h_o	minimum oil film thickness
h_1	oil film thickness at trailing edge
h_2	oil film thickness at leading edge
$H = h_2/h_1$	ratio of film thickness at leading to trailing edge
H_d	depth of craters on liner surface of a pad
I_b	bearing current
I_p	current through pad
L	pad length
m	pad constant, mm^{-1}
n	rpm
N	number of pads in a bearing
p	thrust load on each pad
P	bearing thrust load
q	electrical energy due to charge leakage between the thrust collar and a pad
Q_b	stored electric charge on bearing
Q_p	stored electric charge on a pad
R_b	bearing resistance in operating conditions
R_p	pad resistance in operating condition
R_s	bearing resistance in static conditions

t_p	duration of contact between thrust collar and a pad to cross zone of load-carrying oil film during operation
T_b	bearing temperature rise due to charge leakage in each shaft rotation
T_{ip}	instantaneous rise of pad temperature at high 'points' due to charge leakage during passage of thrust collar in the zone of load-carrying oil film
T_o	rise of bearing temperature without the effect of charge leakage
T_p	pad temperature rise at high 'points' due to charge leakage in each shaft rotation
V	shaft voltage
V_n	mean sliding speed of thrust collar (πdmf_s)
w	$2\pi f$
W_p	width of contact between the thrust collar and a pad in the zone of load-carrying oil film
X_{cb}	capacitive reactance of bearing
X_{cp}	capacitive reactance of a thrust pad
Z_b	bearing impedance
Z_p	pad impedance
Γ	cofficient of thermal expansion
α	angle of pad tilt
μ	Poisson's ratio
ρ	oil resistivity
ρ_1	density of bearing material
σ_{ip}	instantaneous rise in contact stresses due to charge leakage on a pad liner in the zone of load-carrying oil film
σ_o	contact stress on pad liner in the zone of load-carrying oil film without the effect of charge leakage
σ_p	contact stresses due to charge leakage on part of the pad liner in each shaft rotation
σ_{sp}	specific pressure causing pitting on the pad liner
ξ	permittivity/dielectric constant of oil ($\phi\,m^{-1}$)
ξ_r	relative permittivity of oil

Chapter 20

ANALYSIS OF PIVOTED PAD THRUST BEARINGS ON REPEATED STARTS AND STOPS OF A MACHINE OPERATING UNDER THE INFLUENCE OF SHAFT VOLTAGES AND THEORETICAL MODEL TO DETERMINE MINIMUM CYCLES FOR THE FORMATION OF CRATERS ON THE LINER SURFACE

20.1 Introduction

A number of surveys have indicated that the large numbers of bearing failures are on account of bearing current and related analysis [1–8]. Various authors have reported that the passage of current through hydrodynamic journal and thrust bearings in a zone of load-carrying oil film that causes craters to develop on the liner surface over time, and damages the bearing [9]. Also, it causes loss of load-carrying capacity of the bearing over time [10]. Theoretical analysis on the effects of instantaneous charge leakage on roller tracks of roller bearings and reduction in life of hydrodynamic journal as well as thrust bearings under the influence of electric current has been reported in Chapters 7, 8, 15 and 19 [11–13].

In this Chapter, a study is reported on the capacitive effect and life estimation of hydrodynamic pivoted pad thrust bearings on repeated starts and stops of a machine operating under the influence of shaft voltages to determine the increase in charge accumulation on pad liners of a bearing with time as soon as the machine is started and the gradual leakage of the accumulated charges on pad liners as the shaft voltage falls as soon as the power supply to the machine is switched off. Under these conditions the variation of shaft revolutions to accumulate charges and discharge of the accumulated charges on the liner surface of pads of a hydrodynamic pivoted pad thrust bearing at various levels of bearing to shaft voltage is analyzed. Also, the variation of safe limits of starts and stops with the ratio of bearing to shaft voltage is studied.

It has been established in Chapter 4 that at a certain threshold voltage depending on the resistivity of the oil, it creates an electrical breakdown in the oil film of a bearing causing instantaneous charge leakage leading to an arcing effect or allowing a passage of silent discharge through the oil film. Flow of current depends on shaft voltage and bearing impedance. In case of a journal bearing, impedance is a function of oil characteristics, resistivity, oil film thickness, eccentricity and clearance ratios, L/d ratio and capacitance of bearing. Capacitance itself is a function of bearing design and operating parameters [7]. It has also been established in Chapter 19 that the capacitance, resistance and capacitive reactance of a thrust pad depend on the ratio of film thickness at leading to trailing edge, angle of tilt, pad width, etc. [8]. The current flows within the shaft and bearings through the oil films [3]. The voltage over a bearing may be very low but local currents of very high order might flow depending on bearing impedance. Dirt, metallic particles, surface roughness and irregular oil film thickness reduce the impedance of the bearing circuit such that the small shaft voltages induce substantial bearing currents, which finally can damage the bearing by surface pitting, arcing effects or by deterioration, contamination and aging of the oil. In rotation, the oil film thickness of a tilting pad at certain location depends on pad constant, angle of tilt and variation of film thickness from leading to trailing edge [14]. Passage of current through a tilting pad thrust bearing in a zone of load-carrying oil film develops craters in due course and damages liner surface of a pad [15].

This Chapter, besides the effect of starts and stops, also brings out the theoretical model on the effects of instantaneous leakage of charge energy between the thrust collar and a pad lubricated with high-resistivity oil under the influence of different levels of shaft voltages so as to determine the minimum number of cycles before the formation of craters of different sizes on liner surface of the pads. The analysis can be used to assess the volume of craters formed on a pad in a known span of operation of a machine under the influence of the specified shaft voltage. The rise of temperature at the high ‘points’, which causes softening of the pad liner and reduction in its toughness, has been used to determine the minimum cycles before the formation of craters on a pad liner. Also, by the assessment of volume/dimensions and number of craters on the pad liner, shaft voltage in the known span of operation can be assessed.

20.2 Theoretical Analysis of Repeated Starts and Stops of a Machine

20.2.1 Theoretical Background

It is evident that in a hydrodynamic pivoted pad thrust bearing, the variation in oil film thickness between the pads and thrust collar forms capacitor of varying capacitance from leading to trailing edge. The capacitance of the capacitors, thus formed, depends on permittivity of lubricant, pad width, angle of tilt and ratio of film thickness at leading to trailing edge (H) [8], as discussed in Chapter 19. Besides this, variable oil film thickness offers resistance depending on the oil resistivity to the current flow along the pad profile. Thus, the variable oil film thickness forms the resistor–capacitor (RC) circuit of variable resistance and capacitance between thrust collar and pads and offers impedance to a current flow in the thrust bearings.

The pads of a thrust bearing are parallel to each other, so capacitance and resistance, C_b and R_b, of a tilting pad thrust bearing having N pads are determined as N times C_p and $1/N$ times R_p, respectively [8]. So the value of C_bR_b of the bearing is the same as that of C_pR_p of a pad of the bearing as analyzed in Chapter 19.

20.2.2 Time Required to Accumulate Charges on Bearing Liner After Start of a Machine

At the instant when the machine is started, the potential difference (V) across the thrust collar and liner surface of pads is zero. But this gradually increases and approaches the shaft voltage (E). While the shaft voltage increases, the charge (Q_b) on the liner surface of pads bearing lubricated with high-resistivity lubricant [$(\rho)10^{14}$ ohm cm] builds up. Till then V is changing, transient current is delivered from the shaft voltage.

To maintain continuity throughout the whole circuit, there is a rate of change of flux (dQ_b/dT) within the dielectric (lubricant), which is given as [16]

$$I_b = \frac{dQ_b}{dT} \tag{20.1}$$

And, the stored charges on the bearing pad surfaces is determined in Chapter 19 and is given as [8, 16]

$$Q_b = VC = VC_pM = Q_pM \tag{20.2}$$

And if the current is varying then [14]

$$I_b = C_b\frac{dV}{dT} = MC_p\frac{dV}{dT} \tag{20.3}$$

Also, it can be expressed as

$$I_b = \frac{E - V}{R_b} \tag{20.4}$$

So, from Eqns. (20.1)–(20.4), it is evident that

$$V = E - R_b C_b \frac{dV}{dT} \tag{20.5}$$

On integrating Eqn. (20.5) and applying the initial conditions as $T = 0$, $V = 0$, the solution of Eqn. (20.5) is obtained as

$$V = E(1 - e^{-T/C_b R_b}) \tag{20.6}$$

and so

$$Q_b = EC_b(1 - e^{-T/C_b R_b}) \tag{20.7}$$

The time taken (T_{cb}) to develop charge Q_b on liner surface of pads of a bearing having capacitance C_b and resistance R_b is expressed as [using Eqn. (20.6)])

$$T_{cb} = -C_b R_b \log_e(1 - a) \tag{20.8}$$

Similarly, the time taken to develop charge Q_p on a pad of the bearing is determined as

$$T_{cp} = -C_p R_p \log_e(1 - a) \tag{20.9}$$

From Eqns. (20.8) and (20.9), it is evident that $T_{cb} = T_{cp}$, since $C_b = NC_p$ and $R_b = R_p/N$ [8].

20.2.3 Current Passing Through Bearing

From Eqns. (20.2) and (20.4), shaft voltage is given as

$$E = I_b\ R_b + \frac{Q_b}{C_b} \tag{20.10}$$

Using Eqn. (20.1) and applying initial conditions as $T = 0$, $I = E/R_b$, the solution of Eqn. (20.10) is given as

$$I_b = \left(\frac{E}{R_b}\right) e^{-T/R_b C_b} \tag{20.11}$$

And potential drop across a bearing is given as

$$V = I_b\ R_b = Ee^{-T/R_b C_b} \tag{20.12}$$

Similarly, potential difference between thrust collar and a pad is given as

$$V = Ee^{-T/R_pC_p} \tag{20.13}$$

which is the same as potential difference between thrust collar and pads [Eqn. (20.12)] [8].

20.2.4 Time Required to Discharge the Accumulated Charges From Liner Surface of Pads of a Thrust Bearing During Stop

As soon as the power supply to a machine is switched off, magnetic flux disappears and the shaft voltage becomes zero instantaneously. The rate of discharge of bearing capacitor (formed during bearing operation) is determined by differentiating Eqn. (20.12), and given as

$$-\frac{dV}{dT} = \left(\frac{E}{C_b\ R_b}\right)e^{-T/C_bR_b} \tag{20.14}$$

At the instant, when machine is stopped ($T = 0$), the rate of change of bearing potential drop is determined as

$$\frac{dV}{dT} = -\frac{E}{C_b\ R_b} \tag{20.15}$$

In time constant (C_bR_b), the potential drop of bearing capacitor falls and is determined as [Eqn. (20.12)]

$$V = Ee^{-1} = 0.368E \tag{20.16}$$

The time required (T_{db}) to drop potential drop V to the value of 'a' times the shaft voltage (E) ($V = aE$) is given as [Eqn. (20.12)]

$$T_{db} = -R_bC_b \log_e a \tag{20.17}$$

Similarly, time required (T_{dp}) to drop potential drop to the value of 'a' times the shaft voltage (E) between thrust collar and a pad is given as

$$T_{dp} = -R_pC_p \log_e a \tag{20.18}$$

20.2.5 Duration/Time Taken by Thrust Collar to Cross the Zone of Load-Carrying Oil Film Between Thrust Collar and a Pad

The duration/time taken by the thrust collar to cross the zone of load-carrying oil film between thrust collar and a pad is determined as W_p/V_n, and analyzed in Chapter 19 and is given as [13]

$$t_{pn} = \frac{p}{\pi B F_s \sigma_{sp} d_m} \tag{20.19}$$

And the number of times, the thrust collar passes over W_p, the circumferential width of load-carrying oil film of the pad in each shaft rotation is determined as β_{pn} and is determined as [13]

$$\beta_{pn} = \frac{\pi \mathrm{d}_m}{W_p} \tag{20.20}$$

where

$$W_p = \frac{p}{B\sigma_{sp}} \text{ (reference[13]), and } p = \frac{P}{M} \tag{20.21}$$

20.2.6 Number of Shaft Rotation for the Charge Accumulation on Bearing Pad Liners After Start of a Machine

The number of times the thrust collar crosses the zone of load-carrying oil film of a pad (N_{pc}) for the charge accumulation Q_p after start of a machine can be determined as the ratio of time taken (T_{cp}) to accumulate charges to the time taken for the thrust collar to cross the zone of load-carrying oil film of a pad during operation (t_{pn}), and can be determined as

$$N_{pc} = \frac{T_{cp}}{t_{pn}} \tag{20.22}$$

By using Eqns. (20.9) and (20.19), the N_{pc} is expressed as

$$N_{pc} = -\frac{\pi B F_s \sigma_{sp} d_m C_p R_p \log_e (1 - a)}{p} \tag{20.23}$$

The number of shaft rotations (N_{pcn}) to accumulate the charges (Q_p) is determined as the ratio N_{pc} to β_{pn}, i.e. the number of times part of the circumference of thrust collar crosses the zone of load-carrying oil film of a pad to the number of such respective crosses of the thrust collar in each shaft rotation. Thus, N_{pcn} is expressed as follows using Eqns. (20.20)–(20.23):

$$N_{pcn} = -F_s C_p R_p \log_e (1 - a) \tag{20.24}$$

Similarly, the number of shaft revolutions to accumulate charges on liner surface of pads of a thrust bearing is determined as

$$N_{bcn} = -F_s C_b R_b \log_e (1 - a) \tag{20.25}$$

20.2.7 Number of Shaft Rotations for the Discharge of the Accumulated Charges After Stop of a Machine

The number of times part of the circumference of the thrust collar crosses the zone of load-carrying oil film of a pad to discharge of the accumulated charges from the bearing pad liner (N_{pd}) after the power supply to a machine is switched off, is determined similar to Eqn. (20.22) on using the time required for the discharge of the accumulated charges [Eqn (20.18)], and is given as

$$N_{\mathrm{pd}} = \frac{T_{\mathrm{dp}}}{t_{\mathrm{p}}} \tag{20.26}$$

Similar to the logic discussed as above for the charge accumulation, relations for N_{pd} and N_{pdn} (number of shaft rotations to discharge of the accumulated charges on the liner surface of a pad) are developed using Eqns. (20.18)–(20.21), and is given as

$$N_{\mathrm{pd}} = -\frac{\pi B F_{\mathrm{s}} \sigma_{\mathrm{sp}} d_{\mathrm{m}} R_{\mathrm{p}} C_{\mathrm{p}} \log_{\mathrm{e}} a}{p} \tag{20.27}$$

and

$$N_{\mathrm{pdn}} = -F_{\mathrm{s}} R_{\mathrm{p}} C_{\mathrm{p}} \log_{\mathrm{e}} a \tag{20.28}$$

Similarly, the number of shaft revolutions to discharge of the accumulated charges on the liner surface of the pads of a bearing is determined as

$$N_{\mathrm{bdn}} = -F_{\mathrm{s}} R_{\mathrm{b}} C_{\mathrm{b}} \log_{e} a \tag{20.29}$$

20.2.8 Determination of the Ratio of Number of Shaft Rotations for Charge Accumulation to Discharge of the Accumulated Charges

The ratio $N_{\mathrm{pc}}/N_{\mathrm{pd}}$ indicating the number of times, the part of the circumference of the thrust collar crosses the zone of load-carrying oil film for the charge accumulation to the discharge of the accumulated charges from the liner of a pad of the hydrodynamic tilting pad thrust bearing, and the ratio of the number of shaft revolutions ($N_{\mathrm{pcn}}/N_{\mathrm{pdn}}$) for the same, and the ratio of $N_{\mathrm{bcn}}/N_{\mathrm{bdn}}$ is determined using Eqns. (20.23)–(20.29), and is given as

$$\frac{N_{\mathrm{bcn}}}{N_{\mathrm{bdn}}} = \frac{N_{\mathrm{pc}}}{N_{\mathrm{bd}}} = \frac{N_{\mathrm{pcn}}}{N_{\mathrm{pdn}}} = \frac{\log_{\mathrm{e}}(1-a)}{\log_{\mathrm{e}} a} \tag{20.30}$$

20.2.9 Number of Starts and Stops Before the Initiation of Craters on the Liner Surface of Pads of a Tilting Pad Thrust Bearing

The number of starts and stops before the initiation of craters on the liner surface of pads of a thrust bearing (N_{ssb}) can be determined as the ratio of the net time required to initiate craters (C_{sb}/F_s) to the time lapse for charge and discharge of the accumulated charges (T_{cb} and T_{db}) in each start and stop of a machine to the number of cycles (C_{sp}) required before the machine comes to standstill condition after the power supply to the machine is off. Thus N_{ssb} can be determined as

$$N_{ssb} = -\frac{C_{sb}}{F_s(T_{cb} + T_{db})C_{sp}} \tag{20.31}$$

Similarly, N_{ssp}, the number of starts and stops before initiation of craters on liner surface of a pad of the thrust bearing is determined as

$$N_{ssp} = -\frac{C_{sb}}{F_s(T_{cp} + T_{dp})C_{sp}} \tag{20.32}$$

The number of cycles C_{sp}, before the machine comes to standstill condition, depends on the machine inertia, friction in bearings apart from the other factors. Various experimental investigations reveal that, in general, C_{sp} vary between $20F_s$ and $90F_s$. Using Eqns. (20.8), (20.9), (20.17) and (20.18) and relations $C_bR_b = C_pR_p$, the N_{ssb} and N_{ssp} are determined as

$$N_{ssb} = N_{ssp} = -\frac{C_{sb}}{F_sC_bR_b \log_e a(1-a)C_{sp}} \tag{20.33}$$

20.3 Theoretical Model to Determine Number of Cycles for the Appearance of Craters on the Track Surface of a Tilting Pad Thrust Bearing

20.3.1 Determination of Width of Contact Between the Thrust Collar and a Tilting Pad

Under the influence of load (p), the width of contact causing pitting on the pad liner is determined in Chapter 19, and expressed as [13]

$$W_p = \frac{p}{B\sigma_{sp}} \tag{20.34}$$

20.3.2 Duration of Contact Between the Pad and Thrust Collar

Duration of surface contact between the thrust collar and a pad to cross the zone of load-carrying/minimum oil film thickness is determined as W_p/V_n, and given in Chapter 19 as

$$t_p = \frac{p}{\pi B \sigma_{sp} d_m F_s} \tag{20.35}$$

20.3.3 Equivalent Bearing Capacitance and Stored Charges

The equivalent capacitance of a pivoted pad depends on angle of tilt, pad width, permittivity of oil and ratio of film thickness at leading to trailing edge. It is determined in Chapter 19 as [8]

$$C_p = \frac{\varepsilon B(H+1)(H^{0.5}-1)}{\theta H} \tag{20.36}$$

The equivalent capacitance of a bearing lubricated with high-resistivity oil (10^{14} ohm cm) opposes any change in the existing voltage and causes an electric charge to be stored. The instantaneous stored charge depends on bearing capacitance and voltage across oil film, and is determined in Eqn. (20.2).

20.3.4 Active Resistance, Capacitive Reactance and Impedance of the Thrust Bearings

The active resistance of load-carrying oil film of a hydrodynamic thrust bearing depends on oil resistivity, angle of tilt and ratio of film thickness at leading to trailing edge. This is determined in Chapter 19 as

$$R_p = \frac{\rho\theta H}{B(H+1)(H^{0.5}-1)} \tag{20.37}$$

And resistance (R_b) of a thrust bearing is determined as $1/M$ times the resistance of a single pad (R_p), since the pads are parallel to each other with respect to the thrust collar. Also, capacitive reactance of the interacting surfaces of the thrust bearing is determined as

$$X_b = \frac{1}{wC_b} = \frac{1}{wC_pM} = \frac{X_p}{M} \tag{20.38}$$

The equivalent circuit of a bearing consisting of a thrust collar and pads with the load-carrying oil film is considered as series resistor–capacitor (RC) circuit for further analysis.

In operation, the actual contact in the zone of load-carrying oil film between the thrust collar and a pad depends on surface roughness, irregular oil film and comprises primarily the high 'points' contact. Also, area of high 'points' contact is governed by vibrations of the oil film under different conditions of operation and is largely distributed around the pads width in the load-carrying zone.

20.3.5 Sliding Electric Contacts

The major differences between sliding contacts and most other contacts are the problems related to wear during extended period of operation. Good conduction between opposing contact members initiates essentially metal-to-metal contact, and such sliding characteristics produce surface changes like loss of surface finish, transfer of material from softer interacting metal surface, fretting corrosion, abrasion, adhesion, etc. at the contact interface. This has been discussed in Chapter 16 for hydrodynamic journal bearings in details and the same is valid for the thrust bearings.

20.3.6 Heat Generated and Instantaneous Temperature-Rise of Pad Liner by Instantaneous Leakage of Charge Energy in the Zone of Load-Carrying Oil Film Between the Collar and Pads

When a bearing lubricated with high-resistivity oil is operated under the influence of shaft voltage, current does not flow through the bearing and charge gets accumulated on the collar surface until the shaft voltage is below the threshold voltage. Also, an equal and opposite charge is induced on the liner surface due to self inductance, and equal and opposite to this charge is induced on the opposite surface of the pad liner and the outer peripheral surface of the thrust bearing, which is earthen for non-insulated bearing. The slow passage of charge takes place through the pad liners when the high 'points' on the collar and the liners in the zone of load-carrying oil film come close (due to the change in capacitance) or when the conducting particles bridge the oil film. These conducting paths are broken when high 'points' become separated by greater thickness of oil film. Under such conditions arcing results due to instantaneous leakage of charge energy.

The instantaneous charge leakage between the thrust collar and high 'points' of a pad generates an instantaneous heat. The discharge of a stored electrical energy q_p between collar and a pad of capacitance C_p due to charge leakage Q_p at the shaft/bearing voltage V is given as

$$q_{\mathrm{p}} = \frac{C_{\mathrm{p}} V^2}{2}$$

and

$$q_{\mathrm{b}} = \frac{C_{\mathrm{b}} V^2}{2} = M q_{\mathrm{p}} \tag{20.39}$$

The stored energy of charge (q_{p}) is dissipated to K high 'points' contact between thrust collar and a pad, and creates K craters in N_{p} cycles of rotation. The energy dissipated to K high 'points' in N_{p} cycles raise their temperature to the softening limits and leads to the formation of craters. Since in each contact the energy dissipation is q_{p}, so the net minimum energy dissipated (q_{n}) for the formation of K craters in N_{p} contacts is given as

$$q_{\mathrm{n}} = \frac{N_{\mathrm{p}} C_{\mathrm{p}} V^2}{2} \tag{20.40}$$

This dissipated energy influences K high 'points' (of radius r and depth H_{d}) in the zone of load-carrying oil film and increases temperature to T_{p} in N_{p} cycles of operation. Furthermore, each high 'point' of a bearing contacts the thrust collar by $(\pi d_{\mathrm{m}}/W_{\mathrm{p}})\Delta$ times in each shaft rotation. Hence, by using Eqn. (20.39) and neglecting loss of heat by oil flow and other means, the following is evident:

$$\frac{N_{\mathrm{p}} C_{\mathrm{p}} V^2}{2} = \frac{\pi r^2 H_{\mathrm{d}} K \rho c T_{\mathrm{p}} \pi d_{\mathrm{m}} \Delta}{W_{\mathrm{p}}} \tag{20.41}$$

The number of cycles to accumulate charges and discharge of the accumulated charges play an important role to determine the net cycles to create the craters on a liner. So considering the ratio of N_{bcn} to N_{bdn}, the minimum cycles for specified dimensions of craters is determined as $N_{\mathrm{bcn}}/N_{\mathrm{bdn}}$ times N_{p}. Furthermore, for a bearing having M pads, the net capacitance of the bearing (C_{b}) will be M times C_{p}. And the net volume of the craters and load on the thrust bearing will be M times V_{c} and p, respectively. Substituting these values in Eqn. (20.41), the minimum shaft revolutions for appearance of craters on liners of thrust bearing pads is given as [using Eqn. (20.33)]

$$N = \frac{2\pi V_{\mathrm{c}} \rho c d_{\mathrm{m}} T_{\mathrm{p}} B \sigma_{\mathrm{sp}}}{C_{\mathrm{p}} V^2 M p} \Delta \times \frac{N_{\mathrm{bcn}}}{N_{\mathrm{bdn}}} \tag{20.42}$$

In general, formation of craters occurs in the zone of minimum film thickness along the pad width. To determine minimum number of revolutions for initiation of craters on the liner, localized high 'point' temperature of 120°C – the softening temperature of the liner at which material toughness drastically reduces – is considered. Furthermore, the ratio of N_{bcn} to N_{bdn} depends on bearing to collar/shaft voltage and vary between 1 and 25 [17, 19]. For a thrust bearing, frequency of asperity contact between pads and collar is very high because of more number of pads. This leads to the frequent discharge before the full accumulation of charges take place. So, the ratio of N_{bcn}/N_{bdn} is taken as 2.5 considering the bearing to shaft voltage approximately as 0.65 [17, 18].

20.4 Data on Repeated Starts/Stops and on the Crater Formation

20.4.1 Repeated Starts and Stops of a Machine

The ratios N_{bcn}/N_{bdn}, N_{pcn}/N_{pdn} and N_{pc}/N_{pd} are determined for different values of bearing to shaft voltage ($a = V/E$) varying from 0.1 to 0.9 using Eqn. (20.30) and the variation of N_{bcn}/N_{bdn}, N_{pcn}/N_{pdn} and N_{pc}/N_{pd} as against bearing to shaft voltage (a), is shown in Fig. 20.1.

The number of starts and stops of a machine before the initiation of craters on the liner surface of pads is determined for $a = 0.1 - 0.9$ using Eqn. (20.33). Figure 20.2 indicates the variation of number of starts and stops (N_{ssb}) and (N_{ssp}) before initiation of craters on the pad liners at various levels of bearing to shaft voltage (V/E) of the hydrodynamic tilting pad thrust bearing of $d_m = 300$ mm having 8 pads

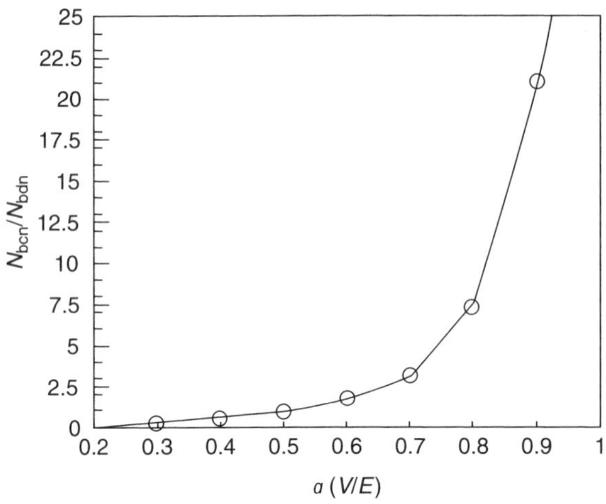

Fig. 20.1. Variation of the ratio of shaft revolutions to accumulate and discharge of accumulated charges (N_{bcn}/N_{bdn}) at various levels of bearing of shaft voltages a (V/E) on the liner surface of a tilting pad thrust bearing under the influence of shaft voltages

($M = 8$) at 1500 rpm ($F_s = 25\,s^{-1}$) having C_p and R_p as $(16.3 \times 10^{-11})\phi$, and 12×10^{10} ohm, respectively, at $\theta = 1$ and $H = 2.2$ as determined in Chapter 19 [8, 13]. C_{sp} for the machine has been established as approximately $20F_s$. And C_{sb} based on the temperature-rise and thermal stress calculations [13] before initiation of craters on liner surface of tilting pads of a thrust bearing is approximately worked out as 5×10^8 cycles.

20.4.2 Cycles for Formation of Craters Due to Leakage of Charge Energy

N, the minimum number of cycles for formation of craters due to leakage of charge energy, is determined using Eqn. (20.42) for different values of shaft voltages varying from 0.25 to 2 V of a bearing having parameters given in Table 20.1. The shaft voltages are taken less than first threshold voltage to avoid the effect of chemical decomposition of the oil. The various parameters of thrust bearing and properties of bearing liner given in Table 20.1 are taken from the reference [8], and shown in Chapter 19.

20.5 Results and Discussion

20.5.1 Duration and Number of Shaft Rotations Required to Accumulate Charges on Pad Liners and Discharge of the Accumulated Charges

The time (T_{cb}) required to accumulate charges (Q_b) on the liner of pads of a hydrodynamic thrust bearing depends on the bearing capacitance (C_b) and the resistance (R_b). Also, this is a function of the natural logarithm of the difference of shaft and bearing voltage to the shaft voltage [Eqn. (20.8)]. Similarly, the time (T_{db}) required to discharge of the accumulated charges depends on C_b, R_b and is a function of natural logarithm of the ratio of bearing to the shaft voltage [Eqn. (20.17)]. Also, the time (T_{cp} and T_{dp}) required to accumulate charges and discharge of the accumulated charges (Q_p) on a pad is the same as that of T_{cb} and T_{db} on all the pads because pads of a thrust bearing are parallel to each other and $C_b = MC_p$ and $R_b = R_p/M$ [Eqns. (20.8), (20.9), (20.17) and (20.18)].

The number of times a thrust collar crosses the zone of load-carrying oil film of a pad to accumulate and discharge of the accumulated charges (N_{pc} and N_{pd}) depends on the shaft rotational frequency, the pad width, the load per unit area (p), and the natural logarithm of the shaft to bearing voltage, besides C_p, R_p and the properties of liner material [Eqns. (20.23) and (20.27)]. However, the shaft revolutions to accumulate charges and discharge of the accumulated charges (N_{pcn} and N_{pdn}) on a pad depends on C_p, R_p, a and F_s [Eqns. (20.24) and (20.28)]. N_{bcn} and N_{bdn} depend on C_b, R_b, a and F_s [Eqns. (20.25) and (20.29)]. And the ratio of shaft revolutions to

Table 20.1. Minimum cycles for the formation of craters of specified dimensions on the liner surface of pads of a hydrodynamic thrust bearing due to leakage of charge energy between the collar and pads of the thrust bearing under the influence of different levels of shaft voltages

Analytical values of minimum cycles at different levels of charge accumulation and shaft voltages								
	(1)	(2)	(3)	(4)	(5)	(6)	(7)	(8)
V (V)	0.25	0.5	0.75	1.0	1.25	1.5	1.75	2.0
$Q_p \times 10^{-11}$(C)	4.1	8.2	12.2	16.3	20.4	24.5	28.5	32.6
$N \times 10^9$	6.72	1.68	0.74	0.42	0.27	0.19	0.14	0.11
Years	8.64	2.16	0.96	0.54	0.35	0.24	0.18	0.14
Ratio of increase in reduction in life as against 0.5 V	–	–	2.25	4	6.2	9	12	15.5
Ratio of increase in reduction in life as against 0.25 V	–	4	9	16	25	36	48	62

Bearing dimensions, operating parameters and analytical values: $d_m = 300$ mm, $B = 200$ mm, $M = 8$, $n = 1500$ rpm, $F_s = 25$, $\sigma_{sp} = 14 \times 10^6$ N m^{-2}, $p = 12 \times 10^4$ N, $\rho = 10^{14}$ ohm cm, $H_d = 0.50$ mm, $r = 0.25$ mm, $V_c = 1.9 \times 10^{-9}$ m^3, $\xi = 2.5 \times 8.85419 \times 10^{-12}$ ϕ m^{-1}, $\xi_r = 2.5$, $C_p = (16.3 \times 10^{-11})\phi$ [Eqn. (20.36)] [8], $t_p = 10^{-3}$ s, $\theta = 1°$, $H = 2.2$, $K = 20$, $\rho_1 = 7.46$ kg m^{-3}, $c = 2.26 \times 10^2$ W s kg^{-1} °C^{-1}, $\Delta = 0.012\,(0.01 - 0.02)$, $N_{bcn}/N_{bdn} = 2.5$.

accumulate and discharge of the accumulated charges on the liner surface of a single pad/and all the pads is independent of various operating parameters and depends only on the ratio of difference of natural logarithm of shaft and bearing voltage to the voltage across bearing, to the ratio of the natural logarithm of shaft and the bearing voltage [Eqn. (20.30)].

The ratio of shaft rotations to accumulate and discharge of the accumulated charges on liner surface of pads (N_{bcn}/N_{bdn}) increases with *V*/*E*. As *V*/*E* increases from 0.1 to 0.9, N_{bcn}/N_{bdn} and N_{pcn}/N_{pdn} increase from 0.05 to 20.91 (Fig.20.1). It is evident that for the successive higher value of *a*, the ratio of N_{bcn}/N_{bdn} and N_{pcn}/N_{pdn} is higher than for the immediate lower value of *V*/*E* (Fig. 20.1).

20.5.2 Number of Start and Stop Cycles Before the Initiation of Craters on the Bearing Pad Liners Due to Discharge of the Accumulated Charges

The number of starts and stops before the initiation of craters on the liner surface of pads (N_{ssb}) of a hydrodynamic thrust bearing depends on the number of cycles required to initiate craters, shaft rotational frequency, capacitance and resistance of the bearing, the number of cycles before machine comes to standstill condition after the power supply to the machine is put off (C_{sp}) and natural logarithm of *a* (*V*/*E*) [Eqn. (20.32)].

The number of starts and stops to initiate craters on the liner surface of pads (N_{ssb}) decreases as the ratio of bearing to shaft voltage (*a*) increases. For the tilting pad thrust bearing of 300 mm mean diameter, number of starts and stops to initiate craters (N_{ssb} and N_{ssp}) on the liner surface of a pad/all the pads decreases from 1471 to 849 as the bearing to shaft voltage (*a*) increases from 0.5 to 0.9 (Fig. 20.2).

20.5.3 Number of Cycles Before the Formation of Craters of Various Sizes at Different Shaft Voltages on the Liner Surface

The minimum number of cycles for formation of craters of specified sizes/net volume of craters on the liner surface of pads of a thrust bearing depends on the pad width, load on the pad and the properties of liner material. The net cycles for formation of craters on pads reduce with increase in number of pads. This is because of the net increase in width of contact between collar and pads. Also, *N*, minimum number of cycles for formation of craters varies inversely to the square of shaft voltage and capacitance of a bearing [Eqn. (20.42)]. Bearing capacitance in turn depends on the angle of tilt, ratio of film thickness at leading to trailing edge, pad width and the oil permittivity [Eqn. (20.36)].

For 300 mm thrust ring mean diameter bearing having 8 pads of width 200 mm each and capacitance of a pad as 16.3×10^{-11} F operating at 1500 rpm, the *N* for the same volume of craters decreases as shaft voltage increases from 0.25 to 2 V. At 0.5 V

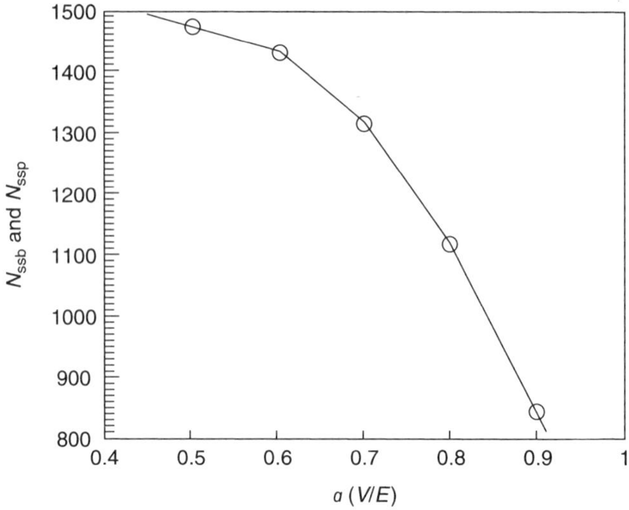

Fig. 20.2. Variation of number of stars and stops (N_{ssb}/N_{ssp}) of a machine before formation of craters on liner surface of the hydrodynamic tilting pad thrust bearing operating under the influence of electric currents at various levels of bearing to shaft voltage a (V/E)

shaft voltage, N is more than 1.68×10^9 cycles, which decreases steeply to 0.11×10^9 cycles as voltage increases to 2 V (Table 20.1). However, at 0.25 V (below 0.5 V) N is more than 6.72×10^9 cycles. Similar trend is found for various sizes of bearings under different conditions of operation.

20.5.4 Effect of Shaft Voltage on Minimum Duration of Operation and Reduction in Bearing Life

The minimum life span and operating cycles of operation of a bearing decreases as the shaft voltage increases. At 1500 rpm, the minimum years of operation of the thrust bearing for formation of craters of different sizes of net volume 1.96×10^{-9} m on a pad decreases from 2.16 to 0.14 years as the shaft voltage increases from 0.5 to 2 V. However, at 0.25 V shaft voltage, the life of pads of the thrust bearing is more than 8.64 years.

In general, as compared to 0.5 V shaft voltage, reduction of pad life increases from 2.25 to 15.5 times as the shaft voltage increases from 0.75 to 2 V. Whereas at 0.25 V, the life of pads is significantly higher (Table 20.1). But it reduces 4–62 times as compared to 0.25 V, as the shaft voltage increases from 0.5 to 2 V. It is therefore recommended that shaft voltage should not be allowed to be more than 0.25 V for safe, reliable operation and adequate life of a hydrodynamic thrust bearing.

However, a journal bearing can safely work under the shaft voltage of 0.5 V, and has about 6 times higher life than the tilting pad thrust bearing [12], as brought out in Chapter 15 (Table 20.1). This matches closely with the recommendation of leading manufacturers of the rotating machines and field experiences.

It may be noted that under the influence of lower voltage, the life expectancy of a thrust bearing is much higher before the craters are initiated on the liner. This is because of lower rise in contact temperature on the high 'points'. In addition, the decrease of charge accumulation and the increase in bearing capacitance also result in the lower rise in contact temperature in the load-carrying zone of a bearing [Eqn. (20.41)]. This subsequently increases the number of cycles for appearance of craters on the pad liner surface of a bearing [Eqn. (20.42)].

The analysis can also be used to ascertain the shaft voltage after establishing the volume of craters and the duration of operation of a machine under the given operating conditions.

20.6 Conclusions

Based on the analysis carried out in this Chapter, the following conclusions are drawn [18, 19]:

1. The time required to accumulate and discharge of the accumulated charges on the liner surface of pads of a thrust bearing depends on the bearing capacitance and resistance, and is a function of the natural logarithm of the ratio of bearing to shaft voltage.
2. The time/number of shaft revolutions required to accumulate and discharge of the accumulated charges on the liner surface of a single pad is the same as that of all the pads.
3. The number of shaft rotations to accumulate and discharge of the accumulated charges on liner surface of pads depends on the shaft rotational frequency besides capacitance and resistance of a bearing, and the ratio of bearing to shaft voltage.
4. As the ratio of bearing to shaft voltage increases, the ratio of shaft rotation to accumulate and discharge of the accumulated charge on the pads increases.
5. With increase of bearing to shaft voltage, the number of starts and stops to initiate craters on the liner surface of pads of a hydrodynamic tilting pad thrust bearing decreases.
6. The minimum number of cycles for the formation of different sizes of craters of specified volume on the pads liner of a thrust bearing depends

on the pad width, load on the pad, the liner properties, the number of pads and vary inversely to the square of shaft voltage, and the pad capacitance.

7. Above 0.25 V of shaft voltage, the minimum cycles for appearance of craters reduces rapidly. Reduction in bearing life increases to 62 times at 2 V shaft voltage as compared to 0.25 V for the identical volume of crater formation.
8. For the thrust bearing operating at 1500 rpm, the years of operation for appearance of craters of the specific volume reduces from 8.64 and 2.16 to 0.14 years as the shaft voltage increases from 0.25 V and 0.5 V to 2 V, respectively.
9. For the safe, reliable operation and adequate life of a thrust bearing, shaft voltage of 0.25 V must not be exceeded.

This analysis besides establishing the effect of capacitive response of the bearings and charge accumulation on the bearing surfaces, is useful for transient performance analysis under the effect of shaft voltages.

REFERENCES

1. Verma, S. P. and Gupta, V. (1980). Shaft potential in modern turbo-generators with particular reference to static excitation. Department of Electrical Engineering, University of Saskatchewan, Saskatoon, Canada S 7 N OWO.
2. Anderson, S. (1968). Passage of electric current through rolling bearings. *Ball Bearing J.*, **153**, 6–12.
3. Chu, P. S. Y. and Cameron, A. (1967). Flow of electric current through lubricated contacts. *ASLE Trans.*, **10**, 226–234.
4. Prashad, H. (1987). Effects of operating parameters on the threshold voltages and impedance response on non-insulated rolling-element bearings under the action of electrical currents. *Wear*, **117**, 223–240.
5. Prashad, H. (1991). Theoretical and experimental investigations on the pitch and width of corrugations on the surfaces of ball bearings operated. *Wear*, **143**, 1–14.
6. Prashad, H. (1988). Theoretical evaluation of impedance, capacitance and charge accumulation of roller bearings operated under electrical fields. *Wear*, **125**, 223–239.
7. Prashad, H. (1991). Theoretical evaluation of capacitance, capacitive reactance, resistance and their effects on performance of hydrodynamic journal bearings. ASME, *J. Tribol.*, **113**, 762–767.
8. Prashad, H. (1992). An approach to evaluate capacitance, capacitive reactance and resistance of pivoted pads of a thrust bearing. STLE, *Tribol. Trans.*, **35**(3) 435–440.
9. Jones-Conway, J. M. and Leopard, A. J. (1976). Plain bearing damage. Proceedings of the Fourth Turbomachinery Symposium, 55–63.
10. Verma, S. P., Girgis, P. S. and Fleming, R. J. (1975). The problems and failure caused by shaft potentials and bearing currents in turbogenerators. Method of Prevention Report,

Power System Research Group, Department of Electrical Engineering, University of Saskatchawavan.
11. Prashad, H. (1990). Theoretical analysis of the effects of instantaneous charge leakage of roller tracks of roller bearings lubricated with high-resistivity lubricants under the influence of electric current. ASME, *J. Tribol.*, **112**, 37–43.
12. Prashad, H. (1991). Theoretical evaluation of reduction in life of hydrodynamic journal bearings operating under the influence of different levels of shaft voltages. STLE, *Tribol. Trans.*, **34**(4), 623–627.
13. Prashad, H. (1992). Analysis of the effects of shaft voltages on the life span of pivoted pad thrust bearings. *BHEL J.*, **13**(2), 1–12.
14. Moore, F. D. (1975). *Principles and Application of Tribology*, Pergamon Press, Oxford, New York, USA.
15. Jones-Conway, J. M. and Leopard, A. J. (1976). Plain bearing damage. Fourth Turbomachinery Symposium, CG-409/76, 55–63.
16. Cotton, H. (1983). *Advanced Electrical Technology*, A.H. Wheeler and Company, Great Britain.
17. Prashad, H. and Rao, K. N. (1994). Analysis of capacitive effects and life estimation of journal bearings on repeated starts and stops of a machine operating under the influence of shaft voltages. *Tribol. Trans.*, **37**(5), 641–645.
18. Prashad, H. (1998). Analysis of pivoted pad thrust bearings on repeated starts and stops of a machine operating under the influence of shaft voltages. *IE(I) J.-MC*, **79**, 42–47.
19. Prashad, H. (1994). Theoretical model to determine minimum cycles before formation of crates due to leakage of charge energy on the liner surface of tilting pads of a thrust bearing. *IE(I) J.-MC*, **75**, 82–86.

Nomenclature

a	ratio of potential difference across bearing to shaft voltage ($a = V/E < 1$)
B	pad width
C	specific heat of bearing liner
C_b	bearing capacitance
C_p	pad capacitance
C_{sb}	number of cycles before initiation of craters on liner surface of tilting pads of thrust bearing
C_{sp}	number of cycles before the machine comes to standstill condition after the power supply is put off
d_m	mean diameter of thrust bearing
E	shaft voltage
f	frequency of applied voltage (50 Hz)
h_o	minimum oil film thickness
F_s	shaft rotational frequency

h_2 oil film thickness at leading edge
h_1 oil film thickness at trailing edge
H h_2/h_1
H_d depth of craters on liner
I_b current passing through bearing
I_p current through a pad
K number of craters on a pad
M number of pads in bearing
n rpm
N minimum shaft revolutions/cycles before formation of craters of specified sizes on liner surface of all the pads of thrust bearing
N_p minimum shaft revolutions/cycles before formation of craters of specified sizes on liner surface of a pad
N_{pc} number of times the part of circumference of thrust collar crosses the load-carrying oil film of a pad during operation to accumulate charges (Q_p) on the liner surface of a pad
N_{pcn} number of shaft rotations to accumulate charges (Q_p) on a pad
N_{bcn} number of shaft rotations to accumulate charges (Q_b) on liner surface of pads
N_{pd} number of times the part of circumference of thrust collar crosses the load-carrying oil film of a pad to discharge of the accumulated charges (Q_p)
N_{pdn} number of shaft revolutions to discharge of the accumulated charges (Q_p) from the liner surface of a pad
N_{bdn} number of shaft revolutions to discharge of the accumulated charge (Q_b) from liner surface of pads
N_{ssp} number of starts and stops before the formation of craters on liner surface of a pad
N_{ssb} number of starts and stops before the formation of craters on liner surface of pads
p thrust load on each pad
P bearing thrust load
q_n electrical energy dissipated to K points due to charge leakage between collar and a pad liner in N_p cycles (W s)
q_p stored electric energy between thrust collar and a pad (W s)
q_b stored electric energy between thrust collar and pads (W s)
Q_p electric charge accumulation on a pad liner
Q_b electric charge accumulation on liner surface of pads
r radius of crater
R_p resistance of a pad
R_b bearing resistance

t_p	duration/time taken by the thrust collar to cross the zone of load-carrying oil film of a pad during operation
T	time
T_{cb}	time required to accumulate charge (Q_b) on liner surface of pads
T_{cp}	time required to accumulate charge (Q_p) on a pad liner
T_{db}	time required to discharge of the accumulated charges from liner surface of pads
T_{dp}	time required to discharge of the accumulated charges from a pad liner
V	potential drop across bearing
V_n	mean sliding speed of thrust collar ($\pi d_m F_s$)
V_c	net volume of different size of craters on a pad ($\pi r^2 K H_d$)
w	$2\pi f$
W_p	width of contact with minimum film thickness between thrust collar and a pad in the zone of load-carrying oil film
X_b	capacitive reactance of bearing (ohm)
X_p	capacitive reactance of a thrust pad (ohm)
β_{pn}	number of times part of the circumference of thrust collar passes in the zone of load-carrying oil film of a pad in each shaft rotation
θ	angle of pad tilt
ρ	resistivity of lubricant
ρ_1	density of bearing material ($kg\,m^{-3}$)
σ_{sp}	specific pressure causing pitting on the pad liner ($N\,m^{-2}$)
ξ	permittivity/dielectric constant of oil ($\phi\,m^{-1}$)
ξ_r	relative permittivity of oil
Δ	ratio of asperity to the net contact surface of pads and collar in each shaft revolution

Chapter 21

ANALYSIS OF INDUCTIVE EFFECTS OF BEARINGS UNDER THE INFLUENCE OF SHAFT VOLTAGES

21.1 Introduction

Shaft voltages are developed along the rotating shaft and between shaft and ground in all kinds of rotating machinery. Depending on the source and the magnitude of shaft voltages, large current may be produced, which may cause serious damage and electro-erosion to the bearing surfaces (Chapter 3) [1–3]. Impedance, capacitance and charge accumulation on the surfaces of rolling-element and hydrodynamic (journal and thrust) bearings operating under the influence of shaft voltages have been evaluated (Chapters 6, 15 and 19) [4–6]. Theoretical analysis on the effects of instantaneous charge leakage on roller tracks of roller bearings and reduction in life of hydrodynamic journal bearings has been reported (Chapter 7, 15 and 16) [7, 8]. Also, capacitive effect of bearings using high-resistivity lubricant on repeated starts and stops of a machine under the influence of shaft voltages has been analyzed (Chapters 8, 16 and 20) [9].

In this chapter, a study is reported on the inductive effect of bearings under start/stop regime of a machine operating under the influence of shaft voltages in growth/increase and decay/decrease of the bearing current. Under these conditions, ratio of the time for growth of the bearing current to that for its decay under start/stop regime at various ratios of transient to steady-state value of bearing current has been analyzed, and the effect of self-inductance on the bearing surfaces studied.

21.2 Inductance of Bearing Curcuit

In general, the electric circuit of a hydrodynamic journal bearing consisting of shaft and bearing can be taken similar to a circuit consisting of two coaxial cylinders of diameters d_1 and d_2 and length equal to the bearing length or width. In a rotating system, magnetic field is confined to the space between the bearing and the shaft. As a circular path taken around the shaft and the bearing indicates equal and opposite current flow, the line integral of the magnetic field round the path is zero [10]. Furthermore, within the shaft, the magnetic field is zero, because a closed path does not enclose any current. It follows that the resultant field is confined to the space/film thickness between the shaft and the bearing (for journal and thrust bearings), to the film thickness between inner and outer races with rolling-elements, and to the free space between inner race and outer race of a rolling-element bearing.

At an external point, the magnetic fields due to the shaft and the bearing are equal and opposite. This indicates that the field due to current at the external point is the same as through the current flowing along the axis. From this, it follows that the inductance of coaxial cylinder/per unit length (in cm) is given as [10]

$$L = 2 \times 10^{-9} \log_e \left(\frac{d_2}{d_1} \right) \tag{21.1}$$

From this analogy, the inductance of a journal bearing having diametral clearance C_d, bearing length l and shaft diameter d, is expressed as

$$L_j = 2 \times 10^{-9} \times L \times \log_e [\{1 + (C_d/d)\}] \tag{21.2}$$

On similar lines, relation for inductance of a tilting pad thrust bearing and a rolling-element bearing can be determined.

21.3 Analysis of Inductive Effect

21.3.1 Theoretical Background

Magnetic flux develops in all kinds of electrical machines due to dissymmetry of the magnetic circuits, which closes in the circumference over the yoke and induces the voltage on the shaft. The shaft voltage depends on the rate of change of magnetic flux linked with the circuit. And this, in turn, leads to the change of current flowing through the bearing depending on the resistance, inductance and capacitance of the oil film of a bearing. The increase of bearing current indicates that the induced emf on the bearing surface is in opposite direction to the current itself [10]. Also, the charge accumulation on the bearing surfaces occurs for a bearing using high-resistivity lubricant (10^{14} ohm cm), and bearing circuit behaves predominantly like a resistive–capacitive (RC) circuit (Chapters 5, 7, 8, 17 and 19) [5, 7].

On the contrary, a bearing using low-resistivity lubricant (10^{7} ohm cm), behaves like a inductive–resistive (LR) circuit, and increase of the flow of current with time through the bearing depends on the bearing inductance and resistance as the shaft voltage changes. The inductance of a bearing depends on the oil film thickness/gap between the shaft and the bearing, and the length/width of the bearing, while resistance of a bearing is a function of oil resistivity, operating parameters and bearing dimensions (Chapters 4–6) [4–6].

21.3.2 Time Required to Growth/Increase of Current Flow in a Bearing

At the instant when a machine is started, the electromotive force (E) acts on the shaft. The resultant electromotive force on overcoming the resistance of bearing circuit (R_b), gives rise to transient varying current (I_b) in the bearing, which is determined by the relation given as [10]

$$E - R_b I_b = L\frac{dI_b}{dT} \tag{21.3}$$

This general equation of electromotive force for shaft-bearing circuit, having resistance and inductance (R_b and L), is solved on integrating and applying the initial conditions before starting the machine as $T = 0$, $E = 0$. The time (T_g) required to increase of bearing current is expressed as

$$T_g = -\frac{L}{R_b}\log_e\left(\frac{E - R_b I_b}{E}\right) \tag{21.4}$$

And the transient current through a bearing is given as

$$I_b = \frac{E}{R_b}(1 - e^{-R_b T_g/L}) \tag{21.5}$$

Since $E/R_b = I_{bs}$ is the final steady-state value of bearing current, so Eqn. (21.5) can be written as

$$I_b = I_{bs}(1 - e^{-R_b T_g/L}) \tag{21.6}$$

And the time required for the transient current I_b to approach A times the steady-state value of bearing current I_{bs}, is given as [from Eqn. (21.4)]

$$T_g = -\frac{L}{R_b}\log_e(1 - A) \tag{21.7}$$

where

$$\frac{I_b}{I_{bs}} = A(A < 1)$$

Equation (21.6) shows how the transient bearing current (I_b) grows and approaches the steady-state value (I_{bs}) in LR circuit.

For the steady-state value of bearing current, I_b approaches $I_{bs}(I_b = I_{bs})$, and Eqn. (21.6) is transformed as

$$e^{-(R_b/L)\cdot T_g} = 0 \tag{21.8}$$

which shows that

$$T_g = \infty \tag{21.9}$$

Also, the rate of growth of the current is determined by differentiating Eqn. (6) and is given as

$$\frac{dI_b}{dT_g} = \frac{R_b}{L}(I_{bs} - I_b) \tag{21.10}$$

And, for the time constant when $T = T_g = L/R_b$, the transient current through a bearing using Eqn. (21.6) is expressed as

$$I_b = I_{bs}(1 - e^{-1}) = 0.632\, I_{bs} \tag{21.11}$$

21.3.3 Time Required for Decay/Decrease of Current Flow in a Bearing

As soon as power supply to a machine is switched off, although the steady value of bearing current I_{bs} has been reached, the magnetic flux disappears and electromotive force/shaft voltage reduces instantaneously. And the transient bearing current (I_b) in a circuit is determined with the initial conditions of shaft voltage as zero.

Thus, Eqn. (21.3) is transformed using $E = 0$, and given as

$$L\frac{dI_b}{dT} + R_b I_b = 0 \tag{21.12}$$

At the instant, when power supply to a machine is switched off, i.e. at $T = 0$, and $I_b = I_{bs}$, transformation and integration of Eqn. (21.12) gives the time (T_d) required for decay/decrease of bearing current as

$$T_d = -\frac{L}{R_b}\log_e\left(\frac{I_b}{I_{bs}}\right) \tag{21.13}$$

and

$$I_b = I_{bs} e^{-R_b T_d / L} \tag{21.14}$$

or

$$A = \frac{I_b}{I_{bs}} = e^{-R_b T_d / L} \tag{21.15}$$

It is evident that the growth and decay of bearing current is complementary, because the sum of growing and decaying value of current in the identical time period (when $T_g = T_d$) is always I_{bs} [Eqns. (21.6) and (21.14)].

The rate of the decay of current is determined by differentiating Eqn. (21.14), and is given as

$$\frac{dI_b}{dT_d} = -\frac{R_b}{L} I_b \tag{21.16}$$

21.3.4 Ratio of Time Required for Growth of Current to that for Decay

The ratio of T_g to T_d is determined by Eqns. (21.4) and (21.13) and is expressed as

$$\frac{T_g}{T_d} = \frac{\log_e (1 - I_b/I_{bs})}{\log_e (I_b/I_{bs})} = \frac{\log_e (1 - A)}{\log_e A} \tag{21.17}$$

21.4 Data Deduction

The ratio of time duration for growth and decay of bearing current (T_g/T_d) is determined for different values of transient to steady-state current ($A = I_b/I_{bs}$) varying from 0.2 to 0.9 using Eqn. (21.17). This is shown in Fig. 21.1.

This data is significant to analyze because it has relevance to transient current at any instant and to steady-state value of current for all those operating conditions where on/off is frequent. For such transient analysis, time of the current growth and time of the current decay is very significant to establish the value of transient current at any instant of time. The data can also be used to establish the number of revolutions before the failure of bearing takes place under the inductive effect, as has been used for rolling-element and hydrodynamic bearings under the capacitive effect in the start/stop regime of a machine (Chapters 8 and 16) [9, 11].

21.5 Discussions

21.5.1 Duration for the Growth and Decay of Bearing Current

Duration for the growth of bearing current during the start regime of a machine depends on the bearing inductance (L) and bearing resistance (R_b). Also, this is a function of the ratio of the transient value to that of the steady-state value of natural logarithm of bearing current [Eqn. (21.7)]. Similarly, time required to decay of the bearing current during stop regime of a machine is a function of L, R_b, and natural logarithm of the ratio of the transient to steady-state value of bearing current [Eqn. (21.13)].

As seen from Eqn. (21.17), the ratio of duration of growth of the bearing current to that of decay (T_g/T_d) is independent of all the parameters. It depends only on the natural logarithm of the ratio of the transient bearing current to the steady-state bearing current.

It is evident that the ratio of duration to growth and decay of bearing current (T_g/T_d) increases in the ratio of transient to steady-state value (I_b/I_{bs}) of current. As the I_b/I_{bs} increases from 0.2 to 0.9, T_g/T_d increases from 0.14 to 20.91 (Fig. 21.1). It is evident that for the successive higher value of A (I_b/I_{bs}), the ratio T_g/T_d is higher than for the immediate lower value of I_b/I_{bs} (Fig. 21.1).

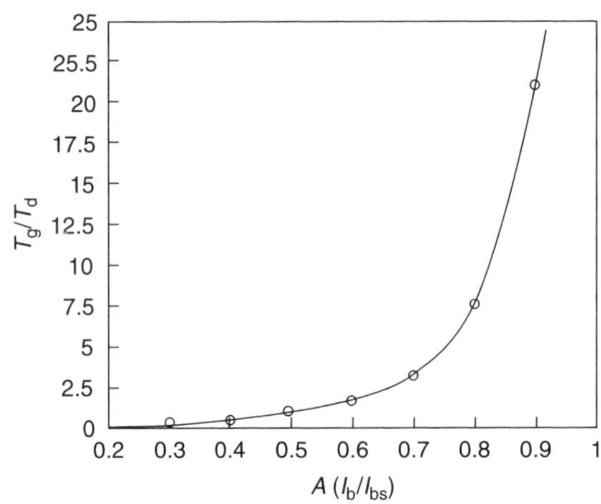

Fig. 21.1. Variation of duration of the ratio of growth and decay of bearing current (T_g/T_d) at various levels of transient to steady-state value of current A (I_b/I_{bs}) of a bearing operating under influence of shaft voltages

21.5.2 Rate of Growth and Decay of Bearing Current

The rate at which the transient bearing current approaches its final steady-state value depends on the ratio of bearing resistance to its inductance, and not independently on separate values of bearing resistance (R_b) and inductance (L) as is evident from Eqn. (21.10). Usually, inductance of a bearing (L) is a very small quantity as compared to bearing resistance (R_b), and depends basically on bearing length/width, and natural logarithm of ratio of the bearing diameter to shaft diameter [Eqn. (21.2)]. The rate of growth of bearing current (dI_b/dT_g) is very high since $R_b > L$) for a bearing using low/high resistivity lubricant ($10^7 - 10^{14}$ ohm cm) for inductance–resistance (LR) bearing circuit [Eqn. (21.10)]. This causes transient bearing current to grow instantaneously in a fraction of a second, but it takes a longer duration to reach the steady-state value. In fact, I_b approaches I_{bs} asymptotically, but never reaches it. This is evident from Eqn. (21.9), which shows T_g, the time required for growth/increase of bearing current, is infinite. The same is also evident from Fig. 21.1, which indicates that for I_b/I_{bs} to be unity, T_g/T_d is infinite. Moreover, bearing resistance (R_b) increases as the machine picks up the speed, because of the increase in oil film thickness of bearing with speed [4–6]. This further adds to increase the rate of increase of current with time during the start regime.

On the contrary, greater the ratio of bearing inductance to bearing resistance (L/R_b), the more slowly the bearing current decay as soon as the power supply to the machine is switched off [Eqn. (21.13)]. However, the rate of decay of bearing current depends on transient value of the current and the ratio of bearing resistance to inductance (R_b/L) [Eqn. (21.16)]. Under stop regime, as soon as power supply to the machine is switched off, the speed reduces, which in turn reduces oil film thickness [4–6]. Consequently, the resistance of the bearing reduces. This increases the L/R_b ratio and bearing current decreases slowly because of increase in T_d [Eqn. (21.13)]. However, for a bearing using high-resistivity lubricant (10^{14} ohm cm), where R_b will be higher as compared to the bearing using low-resistivity lubricant (10^7 ohm cm), the decay of current will be relatively fast [Eqn. (21.16)]. The electromotive force due to inductance is proportional to the rate of change of current [10], and this may cause a spark because of diminishing film thickness and create craters on the bearing surface due to the arcing effect under the stop regime of a machine, in bearing using high-resistivity lubricant apart from capacitive and vibration effects [5, 9]. However, for a bearing using low-resistivity lubricant, the chances of bearing failure due to sparking/arcing effect causing pitting on bearing surface in LR circuit will be remote. In the bearings of 200/500 MW turbo-generators, high-resistivity lubricant (10^{14} ohm cm) are used, because of which, in those bearings, the capacitive effect plays a more predominant role than the inductive effect. The same has been discussed in Chapter 16 [11].

21.6 Conclusions

Based on the above analysis and discussion, the following broad conclusions have been drawn [12]:

1. Time required for growth and decay of the transient bearing current depends on the bearing resistance and inductance, and is a function of natural logarithm of the transient value to the steady-state value of bearing current.
2. The ratio of duration of growth of the bearing current to that of its decay depends on the natural logarithm of the ratio of the transient bearing current to the steady-state bearing current.
3. As the ratio of transient value to the steady-state value of the bearing current increases from 0.2 to 0.9, the ratio of time required for growth of the bearing current to that of its decay increases from 0.14 to 20.91.
4. The rate of growth of transient bearing current to final steady-state value, and the rate of decay of the current, depends on the ratio of bearing resistance to its inductance and not on their separate values.
5. Inductance of a bearing depends on its length/width and clearance ratio. In general, bearing inductance is a very small quantity as compared to the bearing resistance.
6. During start-up regime, ratio of L/R_b is less as compared to the stop regime of the machine.
7. The rate of decay of the bearing current is slower than the rate of its growth.
8. The probability of damage by arcing effect in a bearing using low-resistivity lubricant is remote as against a bearing using high-resistivity lubricant.

This analysis besides establishing the inductive effect of bearing, is useful for transient performance analysis under the effect of shaft voltages.

REFERENCES

1. Verma, S. P., Girgis, R. S. and Fleming R. J. (1975). The problems and failures caused by shaft potentials and bearing-currents in turbogenerators. Chewon, Saskatoon, Sask, Canada.

2. Chu, P. S. Y. and Cameron, A. (1967). Flow of electric current through lubricated contacts. *ASLE Trans.*, **10**, 226–234.
3. Jones-Conway, J. M. and Leopard, A. J. (1976). Plain bearing damage. Proceedings of the Fourth Turbomachinery Symposium, 55–83.
4. Prashad, H. (1988). Theoretical evaluation of impedance, capacitance, and charge accumulation of roller bearings operated under electrical fields. *Wear*, **125**, 223–239.
5. Prashad, H. (1991). Theoretical evaluation of capacitance, capacitive reactance, resistance and their effects on performance of hydrodynamic journal bearings. Transactions of the ASME, *J. Tribol.*, **113**, 762–767.
6. Prashad, H. (1992). An approach to evaluate capacitance, capacitive reactance, and resistance of pivoted pads of a thrust bearing. STLE, *Tribol. Trans.*, **35(**3), 435–440.
7. Prashad, H. (1990). Theoretical analysis of the effects of instantaneous charge leakage of roller tracks of roller bearings lubricants under the influence of electrical currents. ASME, *J. Tribol.*, **112**, 37–43.
8. Prashad, H. (1991). Theoretical evaluation of reduction in life of hydrodynamic journal bearings operating under the influence of different levels of shaft voltages. STLE, *Tribol. Trans.*, **34(**4), 623–627.
9. Prashad, H. (1992). Theoretical analysis of capacitive effect of roller bearings on repeated starts and stops of a machine operating under the Influence of shaft voltages. ASME, *J. Tribol.*, **114**, 818–822.
10. Sydney, G. S. (1965) *Electricity and Magnetism*, Longmans, Green and Co., New York.
11. Prashad, H. and Rao, K. N. (1994). Analysis of capacitive effect and life estimation of hydrodynamic journal bearings on repeated starts and stops of machine operating under the influence of shaft voltages. STLE, *Tribol. Trans.*, **37**(3), 641–645.
12. Prashad, H. (1994). Analysis of inductive effects of bearings under the influence of shaft voltages. *BHEL J.*, **15**(1), 26–31.

Nomenclature

A	ratio of transient to steady value of bearing current ($I_b/I_{bs} < 1$)
C_d	bearing diametral clearance
d	shaft diameter
d_1, d_2	diameter of coaxial cylinders
E	shaft voltage
I_b	transient bearing current
I_{bs}	final steady bearing current
L	inductance of a bearing
l	bearing length
L_j	inductance of journal bearing

R_b	bearing resistance
T	time
T_g	time required for growth/increase of bearing current
T_d	time required for decay/decrease of bearing current

INDEX